MOLECULAR AND CELLULAR CONTROLS OF HEMATOPOIESIS

May 31, 89

ANNALS OF THE NEW YORK ACADEMY OF SCIENCES
Volume 554

MOLECULAR AND CELLULAR CONTROLS OF HEMATOPOIESIS

Edited by Donald Orlic

The New York Academy of Sciences
New York, New York
1989

♾ *The paper used in this publication meets the minimum requirements of American National Standard for Information Sciences—Permanence of Paper for Printed Library Materials, ANSI Z39.48-1984.*

Cover (paper edition): *A scanning electron micrograph depicting growth of a myeloid colony adjacent to the nylon mesh fibers (see Naughton & Naughton, pages 125–140).*

Library of Congress Cataloging-in-Publication Data

Molecular and cellular controls of hematopoiesis / edited by Donald Orlic.

p. cm. — (Annals of the New York Academy of Sciences, ISSN 0077-8923 ; v. 554)

Papers presented at a conference held Feb. 13–15, 1988, in Ajijic, Mexico, and sponsored by the Autonomous University of Gudadalajara and New York Medical College.

Bibliography: p.

Includes index.

ISBN 0-89766-506-6 (alk. paper). — ISBN 0-89766-507-4 (pbk. : alk. paper)

1. Hematopoiesis—Regulation—Congresses. 2. Growth promoting substances—Physiological effect—Congresses. 3. Genetic regulation—Congresses. I. Orlic, Donald. II. Universidad Autónoma de Guadalajara. III. New York Medical College. IV. New York Academy of Sciences. V. Series.

Q11.N5 vol. 554

[QP92]

500 s—dc20

[612.1′1]

B-B

Printed in the United States of America

ISBN 0-89766-506-6 (cloth)

ISBN 0-89766-507-4 (paper)

ISSN 0077-8923

ANNALS OF THE NEW YORK ACADEMY OF SCIENCES

Volume 554
May 1, 1989

MOLECULAR AND CELLULAR CONTROLS OF HEMATOPOIESIS[a]

Editor
DONALD ORLIC

Scientific Organizing Committee
H. BROXMEYER, G. CHAVEZ, A. GORDON,
F. MARTINEZ-SANDOVAL, D. ORLIC, AND R. SHADDUCK

CONTENTS

[a]The papers in this volume were presented at a conference entitled Molecular and Cellular Controls of Hematopoiesis, which was sponsored by the Autonomous University of Guadalajara and New York Medical College and held in Ajijic, Mexico, on February 13–15, 1988.

Part III. Effects of Viruses on Hematopoiesis

Part IV. Stromal Cell Regulation of Hematopoiesis

Part V. Colony-Stimulating Factors for Megakaryocytopoiesis and Granulopoiesis

Part VI. Red Cell Development and Senescence

Part VII. Growth Factors in Clinical Hematology

Financial assistance was received from:

- AMGEN
- CONNAUGHT LABORATORIES, LTD.
- COULTER ELECTRONICS CORPORATION
- E. I. DU PONT DE NEMOURS AND COMPANY
- KIRIN BREWERY COMPANY, LTD.
- MERCK SHARPE AND DOHME RESEARCH LABORATORIES
- PHARMACIA DELTEC, INC.
- PHARMACIA LKB NUCLEAR MICROTOMY, INC.
- SMITH KLINE AND FRENCH LABORATORIES
- TECHNICON INSTRUMENTS CORPORATION
- TOYOBO NEW YORK, INC.

Preface

This conference on hematopoietic controls has focused on growth factors and other regulator molecules. Some of the extraordinary progress of the past few years in our understanding of the structure and biological activities of regulators such as erythropoietin and granulocyte-macrophage colony-stimulating factor (GM-CSF) were presented at this binational conference by a widely acknowledged group of researchers from the United States and Mexico. Their achievements, made possible in large part by technological advances in molecular biology, in many respects represent stellar advances from the accomplishments of an entire era dating back to the 1950s.

Isolation of a large number of hematopoietic factors has led to testing for synergistic action. Clearly, the *in vitro* mixes now being used are more closely related to the microenvironmental milieu in which blood cells normally develop. The concept of "one cell–one molecule" as a model for studying the principal events of blood cell differentiation must be reevaluated as laboratory after laboratory demonstrates the interdependence that exists between the hemopoietins, interleukins, and interferons, to name only a few.

Finally, one of the most rewarding aspects of a conference such as this on regulatory factors in hematopoiesis is the recent progress seen in clinical applications. Several studies in this volume report some of the first efforts to treat anemic and leukemic patients with recombinant human erythropoietin and recombinant human GM-CSF. It may soon be possible to match our understanding of hematopoietic cell response to growth factors with progress in clinical treatment.

I wish to thank the members of the Scientific Organizing Committee for their work (with special thanks to Dr. G. Chavez), the institutions who generously gave financial assistance for their contributions, the speakers for their papers, the Autonomous University of Guadalajara for providing the beautiful conference center in Ajijic, Mexico, as our venue, and New York Medical College for its support.

Donald Orlic

Factors That Affect the Rate of Erythropoietin Production by Extrarenal Sites

WALTER FRIED

Department of Medicine
Rush Medical College
1753 West Congress Parkway
Chicago, Illinois 60612

In 1957 Jacobson *et al.*[1] observed that the plasma erythropoietin titer of nephrectomized rats does not rise in response to hypoxia, and concluded that the kidneys are the site of erythropoietin production. With the development of more sensitive bioassays for erythropoietin, it became apparent that anephric animals,[2–4] including humans,[5] produce a small but detectable amount of erythropoietin in response to intense hypoxia. Furthermore, anephric fetal and neonatal animals produce almost as much erythropoietin as do unoperated ones.[6,7] Accordingly, extrarenal sites (predominantly in the liver[7,8]) are the primary source of erythropoietin in fetal animals and retain the ability to produce small amounts of erythropoietin in mature animals.

In this paper, I will review some studies of factors that affect the rate of extrarenal erythropoietin production.

In 1964 Nathan *et al.*[9] reported their observations of erythropoiesis in anephric patients on chronic hemodialysis. They emphasized that the hematocrits of these patients fall postnephrectomy but then level off and become stabilized at a very subnormal level. After reaching this new steady state, the number of erythroid progenitors in the marrow appear to be near normal. The apparent explanation of this observation is that extrarenal sites are stimulated by anemia to produce enough erythropoietin to sustain a rate of erythropoiesis that is comparable to that in persons with intact kidneys, *i.e.*, in very anemic patients extrarenal sites produce as much erythropoietin as the kidneys produce in nonanemic persons. This, in turn, suggested the hypothesis that erythropoietin production by extrarenal sites is regulated primarily by the rate of oxygen supply: oxygen demand as is that by the kidneys, but extrarenal sites produce only a fraction of the amount of erythropoietin produced by the kidneys in response to comparable conditions (in rats this fraction has been estimated to be 1/10th). This hypothesis has been supported by experimental evidence.[10] Although the rate of erythropoietin production by both renal and extrarenal sites is determined primarily by the tissue's oxygen supply relative to its oxygen requirements, these sites differ in their response to some conditions. For example, protein deprivation causes erythropoietin production to decrease in unoperated rats,[11,12] whereas it has either no effect or increases that in anephric rats;[13,14] and androgens increase the rate of erythropoietin production in unoperated rats[15] but do not significantly change that of anephric ones.[16] Two types of experimental manipulation have been shown to result in profound increments in the plasma erythropoietin titers of hypoxic anephric rats. These are experimental interventions that cause liver damage[17–19] and the administration of renin or angiotensin.[20,21] Effects of three types of hepatic injury on extrarenal erythropoietin production were studied in rats and results of representative experiments are shown in FIGURE 1.

Groups of rats had either surgical removal of 70% of the liver; intragastric adminis-

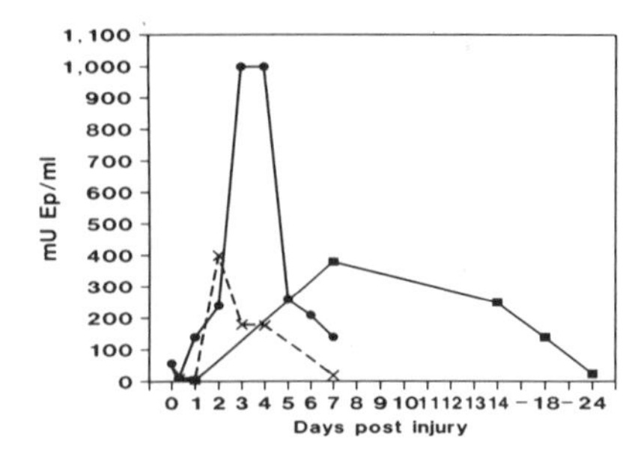

FIGURE 1. Effect of partial hepatectomy, CCl_4 and bile duct ligation on extrarenal erythropoietin production. Numbers on horizontal axis represent days postliver injury that rats were nephrectomized and then exposed to 0.425 atmosphere for 6 hours. Numbers on vertical axis represent mU erythropoietin/ml of pooled plasma obtained posthypoxia. Each data point represents mean of results from 6 assay mice. ●———● CCl_4-treated rats; x---x partially hepatectomized rats; ■———■ bile duct-ligated rats.

tration of a 20% solution of CCl_4 in sesame oil in a single dose of 0.5 ml/100 gm body weight; or ligation of the common bile duct. At various times afterwards the rats were nephrectomized and 1 hour postoperatively they were exposed to 0.425 atmosphere for 6 hours. They were then exsanguinated and the erythropoietin content of the plasma was assayed in posthypoxic polycythemic mice. The plasma erythropoietin titers of anephric rats, made hypoxic less than one day after partial hepatectomy or administration of CCl_4, remained below detectable levels (less than 30 mU/ml) posthypoxia. Hypoxia-induced plasma erythropoietin levels of partially hepatectomized rats rose to levels that were 6-fold higher than those of sham-operated rats two days postoperatively and then declined during the subsequent five days. Hypoxia-induced plasma erythropoietin levels of CCl_4-treated rats rose to levels that were 14-fold higher than those of sesame oil-treated rats three days posttherapy and rapidly fell five days posttherapy. Rats, made hypoxic seven days after ligation of their common bile duct, had plasma erythropoietin titers that were 5-fold higher than those of sham-operated rats. Fourteen days postoperatively, the plasma erythropoietin levels of bile duct-ligated rats were still almost 4-fold greater than those of sham-operated rats. However, 24 days postoperatively, at a time when the rats with ligated bile ducts are very ill, their plasma erythropoietin titers were barely detectable.

In the case of the CCl_4-treated and partially hepatectomized rats, increased erythropoietin production occurs during periods when there is active hepatocyte regeneration, and also Kupffer cell hyperplasia.[22,23] Bile duct ligation, on the other hand, is not prominently associated with either hepatocyte regeneration or with Kupffer cell hyperplasia. Instead, the most consistent histologic change is bile duct dilation and hyperplasia. The site of erythropoietin production in the liver has not yet been identified, nor has it yet been possible to detect erythropoietin in liver extracts.[14] Nevertheless, there is experimental evidence implicating the Kupffer cells as the source of production.[22,23] The studies in rats with ligated bile ducts suggest that biliary duct epithelium should be considered as a possible site of erythropoietin production, particularly since the renal tubular epithelium has been suggested as the probable site of renal erythropoietin production.[24]

Rats, infused with angiotensin, also demonstrate a marked increment in the production of erythropoietin from extrarenal sites in response to hypoxia.[21] Optimal effects on erythropoietin production have been observed when miniosmopumps, containing angiotensin in a concentration that is calculated to release 5 micrograms of angiotensin per hour, are implanted subcutaneously 18 to 20 hours prior to exposing anephric rats to hypoxia.[21] It is noteworthy that this same dose of angiotensin has little or no detectable effect on erythropoietin production by sham operated rats. Although infusion of angiotensin increases erythropoietin production to a level comparable to that in CCl_4 treated rats, the mechanism of action of these two agents differ. A consistent but poorly understood property of extrarenal erythropoietin production is that it falls to an undetectable level if the hypoxic stimulus is initiated ten or more hours after nephrectomy.[25] FIGURE 2 shows the results of experiments in which rats were exposed to 0.425 atmosphere for 6 hours, beginning either 1 or 18 hours after nephrectomy. One group received 0.5 ml/100 gm of a 20% solution of CCl_4 in sesame oil 72 hours prior to beginning hypoxia. Another group had mini-osmopumps, containing angiotensin at a concentration calculated to deliver 5 micrograms per hour, implanted subcutaneously 20 hours prior to initiating hypoxia. A third group (controls) received no treatment prior to nephrectomy and exposure to hypoxia. Posthypoxia, all rats were exsanguinated, and the plasmas of each group were pooled and assayed for erythropoietin. The plasma of the control group, made hypoxic 1 hour postnephrectomy, contained about 100 mU of erythropoietin per ml; whereas that of control rats, made hypoxic 18 hours postnephrectomy, contained no detectable erythropoietin activity. The CCl_4- and angiotensin-treated rats, made hypoxic one hour postnephrectomy, had 3- to 4-fold higher plasma erythropoietin titers than did comparable controls. However, when made hypoxic 18 hours postnephrectomy, the plasma erythropoietin titers of CCl_4-treated rats were only about 1/4 as great as those of rats made hypoxic 1 hour postnephrectomy; whereas angiotensin-treated rats made hypoxic 18 hours postnephrectomy had 2.5 times as much erythropoietin in their plasma as did those made hypoxic 1 hour postnephrectomy. Accordingly angiotensin, but not hepatic injury, prevented the decline in erythropoietin production (and even increased it) which occurs when the stimulus to erythropoietin production is initiated 18 hours postnephrectomy. One hypothesis to explain these findings is that angiotensin must be present to permit extrarenal erythropoietin production to occur. Following nephrectomy, the site of renin production is removed, and therefore angiotensin II can no longer be generated. Consequently, the angiotensin content of the liver gradually falls below the permissive level, and hepatic erythropoietin decreases to levels that are no longer detectable. If this is correct, then administration of captopril, which inhibits the enzyme responsible for conversion of angiotensin I to angiotensin II, should also cause a decrease in extrarenal erythropoietin production. To test this, captopril (provided by E. R. Squibb & Sons, Inc., Princeton, NJ) was dissolved in a 5% glucose solution to a concentration of 0 to 37.5 mgm/ml as indicated on the abscissa of FIGURE 3. One ml of this solution was then administered per intragastric tube twice daily for 3 days prior to nephrectomy and again at the time of the operation. The rats were then exposed to 0.425 atmosphere for 7 hours, following which their plasma was collected, pooled, and assayed for erythropoietin. This experiment was also performed using rats that received CCl_4 three days prior to nephrectomy. (This was done to increase the plasma erythropoietin level of the controls that received no captopril, and thereby make it easier to detect an inhibitory effect.) Results are shown in FIGURE 3. The plasma erythropoietin levels of captopril-treated rats were about 1/2 as high as those of comparable treated controls, but were still significantly elevated. These results are consistent with the hypothesis that angiotensin is required for optimal extrarenal erythropoietin production to occur. However, the erythro-

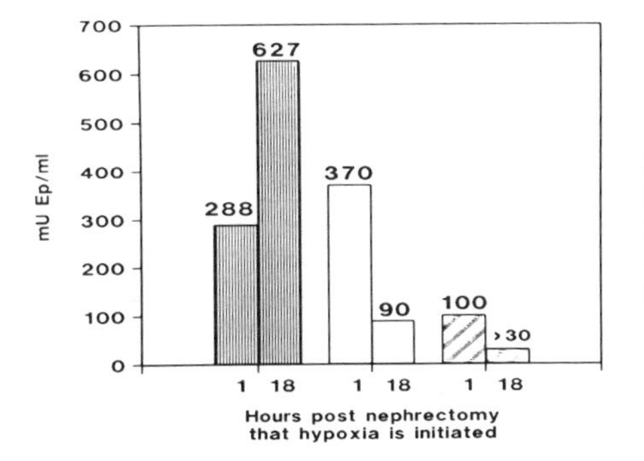

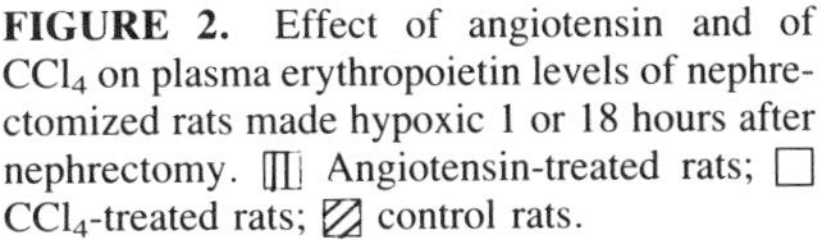
FIGURE 2. Effect of angiotensin and of CCl_4 on plasma erythropoietin levels of nephrectomized rats made hypoxic 1 or 18 hours after nephrectomy. ▥ Angiotensin-treated rats; □ CCl_4-treated rats; ▨ control rats.

poietin content of plasma from captopril-treated rats appeared to be higher than those of rats made hypoxic 18 hours postnephrectomy (data shown in FIG. 2). Therefore, either the captopril was not as effective as nephrectomy in reducing the hepatic angiotensin content, or angiotensin depletion is not the only postnephrectomy change that affects extrarenal erythropoietin production.

Angiotensin II has potent direct vasoconstrictor activity and also promotes the release of biologically active substances from endothelial cells. One of these substances is prostaglandin E_2 which has been reported to cause increased erythropoietin production by both renal[26] and extrarenal[27] sites. The following experiments were designed to ascertain whether the angiotensin-induced increase in extrarenal erythropoietin is mediated by an increase in PGE_2 secretion. It is important to reemphasize that the dose and mode of angiotensin administration used in these experiments, continuous perfusion at a rate of 5 micrograms per hour, has little or no effect on erythropoietin production in nonnephrectomized rats.

Indomethacin, a cyclooxygenase inhibitor is a potent inhibitor of prostaglandin synthesis and has been reported to decrease renal erythropoietin production.[28] In the studies, results of which are illustrated in FIGURE 4, indomethacin was dissolved in saline at

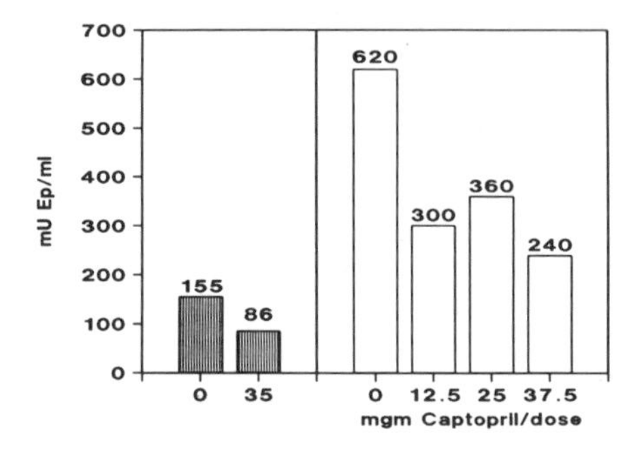

FIGURE 3. Effect of captopril on extrarenal erythropoietin production by CCl_4-treated and nontreated rats. Numbers over each bar indicate the mean mU erythropoietin/ml of plasma based on assay in 6 polycythemic mice. □ Rats that received CCl_4 three days prior to nephrectomy and exposure to hypoxia; ▥ rats that did not receive CCl_4.

concentrations of 0 to 5.0 mgm/ml. One ml of this solution was injected intraperitoneally 18 hours prior to nephrectomy and again at the time of nephrectomy. One hour later, rats were exposed to 0.425 atmosphere for 6 hours. Their plasma was then collected, pooled, and assayed for erythropoietin. None of the indomethacin-treated groups of nephrectomized rats had lower plasma erythropoietin titers than did the controls. Those that received doses of 3.3 or 5.0 mg of indomethacin had plasma erythropoietin titers that were 4 times higher than those of controls. The erythropoietin titers of the indomethacin-treated nonnephrectomized groups were not significantly lower than those of controls that received no indomethacin.

FIGURE 5 shows the effect of angiotensin, and of indomethacin alone and in combination on plasma erythropoietin titers of nephrectomized rats exposed to 0.425 atmosphere for 6 hours. Indomethacin (3 mg/ml) was injected intraperitoneally 18 hours prior to nephrectomy and again at the time of surgery; and angiotensin was perfused at a rate of 5 micrograms per hour via miniosmopumps implanted subcutaneously 18 hours prior to nephrectomy. The plasma erythropoietin titers of rats that received either indomethacin or angiotensin significantly exceeded those of untreated controls; and the plasma erythropoietin titers of rats that received both substances significantly exceeded those of rats that received only indomethacin or angiotensin.

Since indomethacin is a cyclooxygenase inhibitor, it should inhibit the production of $PGF_{2\alpha}$ as well as PGE_2. The biological effects of $PGF_{2\alpha}$ are often opposite to those

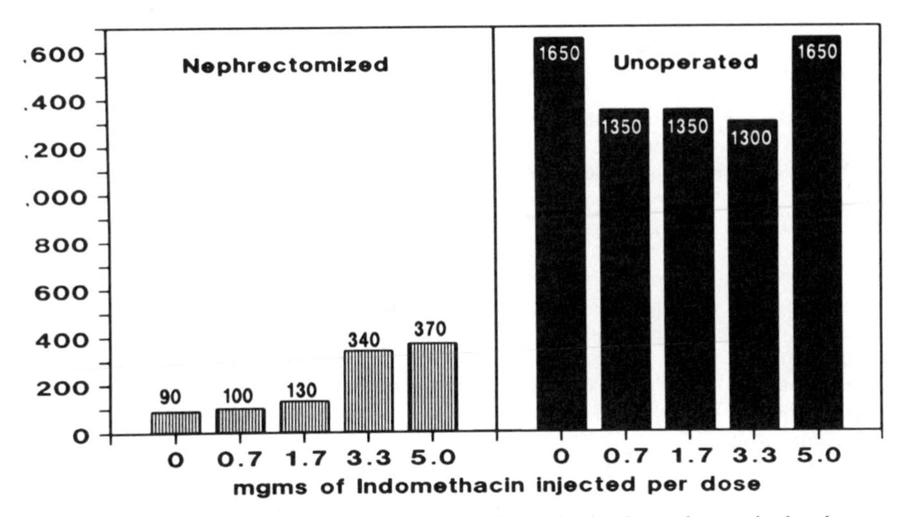

FIGURE 4. Effect of indomethacin on erythropoietin production by nephrectomized and unoperated rats. Numbers above or in the bars indicate mean mU erythropoietin/ml plasma based on assay in 6 polycythemic mice.

of PGE_2. There is some evidence that this is the case for their effect on extrarenal erythropoietin production.[27] Measurement of the PGE_2 and the $PGF_{2\alpha}$ content of livers from indomethacin-treated rats by specific radioimmunoassay demonstrated a more marked suppression of hepatic $PGF_{2\alpha}$ than of hepatic PGE_2 titers.[14] Accordingly, indomethacin may increase extrarenal erythropoietin production by it preferentially suppressing hepatic production of $PGF_{2\alpha}$, a potent inhibitor of extrarenal erythropoietin production.

FIGURE 6 shows the results of experiments designed to determine the effects of PGE_2, $PGF_{2\alpha}$, and indomethacin, alone and in combination, on extrarenal erythropoietin production. Indomethacin was administered as in the previous experiment (3 mg/ml intraperitoneally 18 hours prior to nephrectomy and again during the operative procedure). PGE_2 and $PGF_{2\alpha}$ were dissolved in saline to a concentration of 12.5 mgm/ml. Miniosmopumps were then loaded with this solution and transplanted subcutaneously 72 hours prior to nephrectomy. The perfusion rate for these pumps was calculated to be 12.5 micrograms per hour. One hour postnephrectomy, the rats were exposed to 0.425 atmosphere for 6 hours. Their plasmas were then pooled and assayed for erythropoietin. The plasma erythropoietin titers of rats treated with PGE_2 did not differ significantly from those of control rats, whereas those treated with $PGF_{2\alpha}$ were significantly lower than

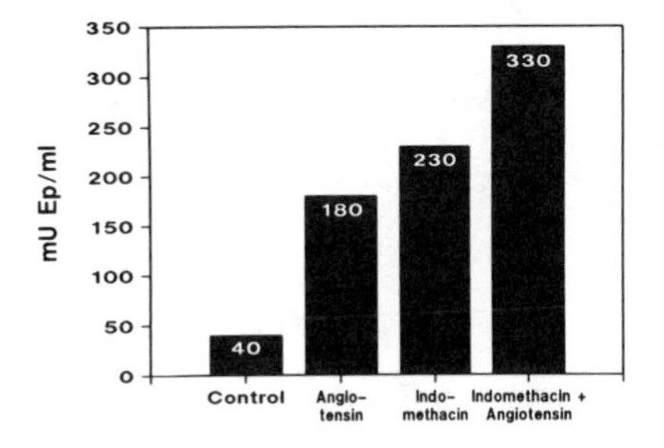

FIGURE 5. Effect of angiotensin and/or indomethacin on extrarenal erythropoietin production. Numbers in the bars represent mean mU erythropoietin/ml plasma based on assay in 6 polycythemic mice.

those of controls. Rats treated with indomethacin alone had significantly higher plasma erythropoietin titers than did untreated controls, whereas those treated with both indomethacin and $PGF_{2\alpha}$ had slightly lower plasma erythropoietin levels than did controls. These results support the hypothesis that indomethacin increases extrarenal erythropoietin production by decreasing the hepatic titer of $PGF_{2\alpha}$, a potent inhibitor of extrarenal erythropoietin production. They are incompatible with the hypothesis that angiotensin stimulates extrarenal erythropoietin production by increasing the secretion of PGE_2 by endothelial cells in the liver.

To determine the feasibility of attributing the effect of angiotensin on extrarenal erythropoietin production to its vasoactive properties, we compared the effects of angiotensin and of norepinephrine on extrarenal erythropoietin production, systemic blood pressure, and hepatic blood flow (TABLE 1). Angiotensin was administered via subcutaneously implanted miniosmopumps at a rate of 5 micrograms per hour. Norepinephrine was also administered by subcutaneously implanted miniosmopumps at a flow rate of 30 micrograms per hour. Miniosmopumps loaded with saline were implanted subcutaneously into control rats. All rats were nephrectomized 18 hours after implanting the pumps. Plasma erythropoietin titers were measured in rats that were exposed to 0.425 atmosphere for 6 hours beginning 1 hour postnephrectomy. Mean blood pressures were measured using a tail vein cuff. Hepatic blood flow was measured in nembutal-anesthetized rats using ^{99}Tc colloid as follows: the ^{99}Tc colloid was injected intravenously by tail vein. Two tenths of an ml of blood was then collected from the retroorbital sinus into calibrated capillary tubes at 0.5-minute intervals for two minutes, then every minute for the subsequent two minutes. Exactly 10 minutes after injection of the ^{99}Tc colloid, the rats were sacrificed. The livers were removed and weighed. An aliquot of the liver was removed, weighed and placed into a counting tube. The T ½ of the plasma clearance was determined. This was corrected for the fraction of the total ^{99}Tc colloid that was extracted by the liver, *i.e.*, at each interval tested the amount of remaining ^{99}Tc colloid was multiplied

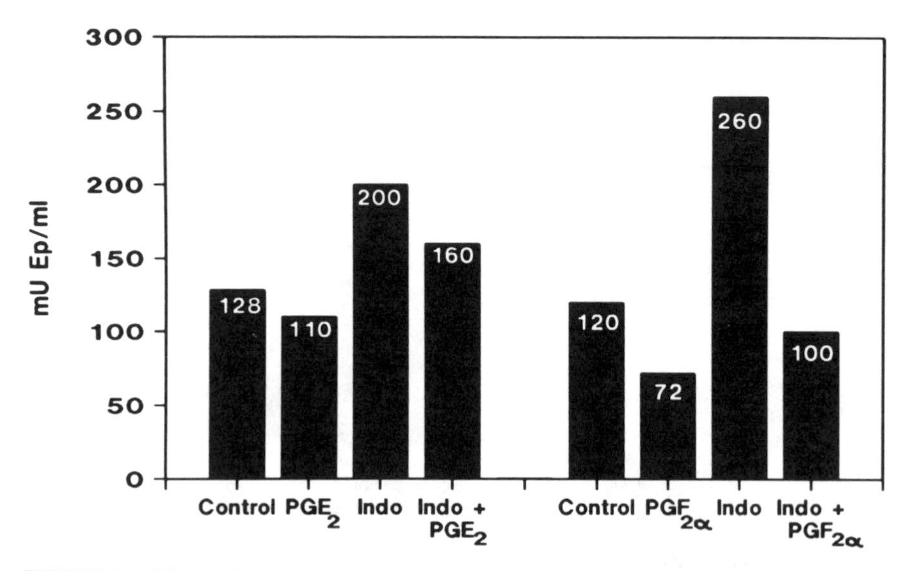

FIGURE 6. Effect of PGE_2, $PGF_{2\alpha}$ and indomethacin, individually and in combination on extrarenal erythropoietin production. Numbers in bars represent mean mU erythropoietin/ml plasma, based on assay in 6 polycythemic mice. Indo = indomethacin.

TABLE 1. Effect of Angiotensin and of Norepinephrine on Extrarenal Ep, Mean Blood Pressure, and Hepatic Blood Flow

	Plasma Ep[a] (mU/ml)	Mean Blood Pressure (Measured via Tail Cuff)	Clearance of ^{99}Tc Colloid from Plasma (T ½ in Minutes)	Amount of ^{99}Tc Colloid Extracted by Liver (as % of Injected Dose)
Control	50	119 (4)[b]	0.45 (3)	96.3 (5)
Angiotensin (5 micro grams per hour)	440	125 (5)	1.0 (4)	76.5 (7)
Norepinephrine (30 micrograms per hour)	71	154 (6)	0.51 (4)	86.3 (7)

[a]All rats were nephrectomized one hour before being made hypoxic (0.43 atmosphere) for 6 hours. Their plasma was then collected and pooled for Ep assay.
[b]Numbers in parentheses indicate the number of rats studied.

by the fraction of the injected dose that was extracted by the liver in 10 minutes. The hepatic extractable fraction was calculated by determining the total counts in the liver after 10 minutes and dividing by the number of counts injected. ^{99}Tc colloid under normal conditions is removed entirely by Kuppfer cells in the liver at a rate that is proportional to hepatic blood flow. Decreased total hepatic blood flow results in a decrease in the rate of clearance of ^{99}Tc colloid from the plasma and also in a decrease in the fraction that is extracted by the liver. The plasma erythropoietin titers of rats that received norepinephrine were higher than those of controls, but significantly lower than those of rats that received angiotensin. The mean blood pressure of norepinephrine-treated rats significantly exceeded that of angiotensin-treated ones, which was slightly higher than that of controls. However, the parameters of hepatic blood flow of angiotensin-treated rats were significantly lower than those of the other two groups, *i.e.*, the plasma clearance of ^{99}Tc colloid was decreased as was the total amount extracted by the liver. These results are compatible with the concept that angiotensin stimulates extrarenal erythropoietin production by reducing hepatic blood flow, although they do not provide definitive proof of it.

In summary, extrarenal sites located in the liver are capable of producing small but significant amounts of erythropoietin in intensely hypoxic anephric rats. Erythropoietin production by extrarenal sites increases several fold in magnitude following liver damage and also in response to angiotensin infusion. The results of studies reviewed here suggest that possibility that angiotensin's effect on extrarenal erythropoietin production is mediated by decreased hepatic blood flow. The data also unexpectedly show that indomethacin increases extrarenal erythropoietin production probably by suppressing $PGF_{2\alpha}$ levels in the liver. This suggests the hypothesis that extrarenal erythropoietin production is modulated by the amount and mixture of prostaglandins in the liver.

REFERENCES

1. Jacobson, L. O., E. Goldwasser, W. Fried & L. Plzak. 1957. Role of the kidney in erythropoiesis. Nature **179:** 622–634.
2. Gallagher, N. I., J. M. McCarthy & R. D. Lange. 1961. Erythropoietin production in uremic rabbits. J. Lab. Clin. Med. **57:** 281–289.
3. Erslev, A. J. 1958. Erythropoietin function in uremic rabbits. Arch. Intern. Med. **101:** 407–412.
4. Mirand, A. C. & T. C. Prentice. 1957. Presence of plasma erythropoietin in hypoxic rats with or without kidneys and/or spleen. Proc. Soc. Exp. Biol. Med. **96:** 49–51.

5. Naets, J. P. & M. Wittek. 1968. Presence of erythropoietin in the plasma of one anephric patient. Blood **31:** 249–251.
6. Carmena, A. O., D. Howard & F. Stohlman. 1968. Regulation of erythropoietin. XXII. Production in the newborn animal. Blood **32:** 376–382.
7. Zanjani, E. D., I. Foster, H. Burlington & L. R. Wasserman. 1977. Liver as the primary site of erythropoietin formation in the fetus. J. Lab. Clin. Med. **89:** 640–644.
8. Fried, W. 1972. The liver as a source of extrarenal erythropoietin production. Blood **40:** 671–677.
9. Nathan, D. G., E. Schupak & F. Stohlman. 1964. Erythropoiesis in anephric man. J. Clin. Invest. **43:** 2158–2164.
10. Fried, W., T. Kilbridge, S. Krantz, T. McDonald & R. D. Lange. 1969. Studies on extrarenal erythropoietin. J. Lab. Clin. Med. **73:** 244–248.
11. Reissman, K. R. 1964. Protein metabolism and erythropoiesis. II. Erythropoietin responsiveness and erythropoietin formation in protein deprived rats. Blood **23:** 146–153.
12. Anagnostou, A., S. Schade, M. Ashkenaz, J. Barone & W. Fried. 1977. Effect of protein deprivation on erythropoiesis. Blood **50:** 1093–1097.
13. Anagnostou, A., S. Schade & W. Fried. 1978. Effect of protein deprivation on extrarenal erythropoietin production. Blood **51:** 549–533.
14. Fried, W., J. Barone-Varelas & C. Morley. 1984. Factors that regulate extrarenal erythropoietin production. Blood Cells **10:** 287–304.
15. Fried, W. & C. W. Gurney. 1965. Erythropoietic effect of plasma from mice receiving testosterone. Nature **206:** 1160–1161.
16. Fried, W. & T. Kilbridge. 1969. Effects of testosterone and cobalt on erythropoietin production. J. Lab. Clin. Med. **74:** 623–629.
17. Naughton, B. A., S. M. Kaplan, M. Roy & A. S. Gordon. 1977. Hepatic regeneration and erythropoietin production in the rat. Science **196:** 301–302.
18. Anagnostou, A., S. Schade, J. Barone & W. Fried. 1977. Effect of partial hepatectomy on extrarenal erythropoietin production in rats. Blood **50:** 459–426.
19. Fried, W., J. Barone-Varelas, S. Schade & A. Anagnostou. 1979. Effect of carbon tetrachloride on extrarenal erythropoietin production. J. Lab. Clin. Med. **93:** 702–705.
20. Gould, A. B., S. Goodman, R. DeWolf, G. Onesti & C. Swartz. 1980. Interrelation of the renin system and erythropoietin in rats. J. Lab. Clin. Med. **96:** 523–534.
21. Naughton, B. A., D. J. Birmbach, P. Liu, G. A. Kolks, M. Z. Tung, S. J. Piliero & A. S. Gordon. 1979. Reticuloendothelial system hyperfunction and erythropoietin production in the regenerating liver. J. Surg. Oncol. **12:** 227–242.
22. Caro, J. & A. J. Erslev. 1984. Biologic and immunologic erythropoietin in extracts from hypoxic whole rat kidneys and in their glomerular and tubular fraction. J. Lab. Clin. Med. **102:** 922–931.
23. Schooley, J. C. & L. J. Mahlmann. 1972. Erythropoietin production in the anephric rat. I. Relationship between nephrectomy, time of hypoxic exposure, and erythropoietin production. Blood **39:** 31–38.
24. Schooley, J. C. & L. J. Mahlmann. 1971. Stimulation of erythropoiesis in plethoric mice by prostaglandins and its inhibition by antierythropoietin. Proc. Soc. Exp. Biol. Med. **138:** 523–524.
25. Naughton, B. A., G. K. Naughton, P. Liu, J. M. Arce, S. Piliero & A. S. Gordon. 1982. The effects of prostaglandins on extrarenal erythropoietin production. Proc. Soc. Biol. Med. **170:** 231–236.
26. Gross, D. M., W. Jubiz & J. W. Fisher. 1977. Released erythropoietin and prostaglandins E by the dog kidney during hypoxic hypoxia and the effects of indomethacin. Fed. Proc. **36:** 1052.

External Messengers and Erythropoietin Production[a]

JAMES W. FISHER AND MUNEHISA UENO

Department of Pharmacology
Tulane University School of Medicine
1430 Tulane Avenue
New Orleans, Louisiana 70112

INTRODUCTION

We have been interested for several years in the mechanism of hypoxic stimulation of kidney production of erythropoietin (Ep).[1,2] Our recently developed model is shown in FIGURE 1, in which we propose that hypoxia of the kidney induced by hypobaria, anemia, ischemia or molecular deprivation of oxygen with cobalt results in the release of several "external messenger" substances.[1,2] We use the word "external messenger" to indicate that these chemical agents, which are also called "autacoids," are released by a cell locally or at a remote site to stimulate Ep production directly in an adjacent cell in the kidney. These external messages activate a membrane receptor on the surface of the Ep-producing cell which initiates a cascade of events within the cell leading to the release of second messenger substances such as cyclic AMP, calcium and inositol phosphate.

The important chemical agents known to be released that trigger production of Ep are adenosine,[3,4] eicosanoids[3,4] (PGE_2, PGI_2 and its metabolite 6-keto PGE_1),[5] oxygen-derived free radicals such as hydrogen peroxide (H_2O_2), superoxide ion (O_2^-)[6,7] and beta-2-adrenergic agonists.[8–10] These agents are probably released during hypoxia depending upon the severity of the hypoxic stimulus. The eicosanoids and beta-2-adrenergic agonists probably play a role in a milder hypoxic stimulus; whereas adenosine and the oxygen free radicals, H_2O_2 and O_2^-, are probably involved in a more severe hypoxic stimulus to trigger Ep production. These external messenger substances could act alone or in concert to trigger membrane receptors which activate a stimulatory G protein in the membrane leading to the activation of adenylate cyclase, which causes the generation of cyclic AMP from ATP. Cyclic AMP causes a dissociation of the regulatory unit from the catalytic head of protein kinase A. Important phosphoproteins generated by kinase A in the kidney may lead both to increased biosynthesis of Ep at the level of transcription of messenger RNA in a renal cell as well as the actual release process of Ep from this cell.

For several years, we worked with eicosanoids[5] and beta-adrenergic agents[8–10] in studies of their role in the production of Ep in response to hypoxia. However, when mice were exposed to hypoxia and radioiron incorporation directly measured, there was an elevation in radioiron incorporation with increasing severity of the hypoxic stimulus, which was partially but not completely blocked by the cyclooxygenase inhibitors, indomethacin and meclofenamate.[5] Likewise, when we exposed rabbits to hypoxia for 18 hr, and studied the effects of the beta-adrenergic blocking drugs such as butoxamine, which selectively blocks the beta-2-adrenergic receptors, the effects of hypoxia on the increased plasma levels of erythropoietin were also only partially blocked.[9] Thus, both beta-2-adrenergic activation and eicosanoid stimulation play some role in hypoxia-induced Ep

[a]Supported by United States Public Health Service Grant DK13211.

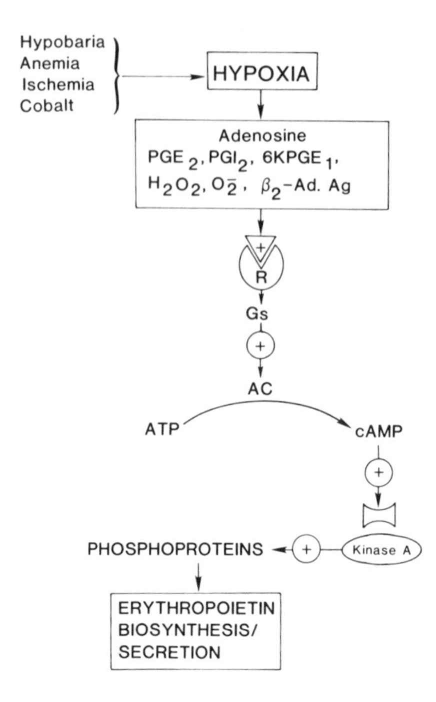

FIGURE 1. Schematic model for the role of external messengers and hypoxia in the regulation of kidney production of erythropoietin. Erythropoietin biosynthesis/secretion can be switched on by hypoxia through the release of several chemical agents that activate receptors in the cell membrane to increase stimulatory G proteins (Gs): prostaglandin E_2 (PGE_2), prostacyclin (PGI_2), 6-ketoprostaglandin E_1 ($6KPGE_1$), hydrogen peroxide (H_2O_2), superoxide (O_2^-), and β_2-adrenergic agonists (β_2-Ad. Ag). Gs activates adenylate cyclase (AC), which increases cyclic 3′,5′-adenosine monophosphate (cAMP). cAMP activates kinase A to phosphorylate proteins (phosphoproteins), which are important in the transcriptional and/or translational stages of Ep biosynthesis and/or secretion. (From Fisher.[2] Reprinted by permission from the *Annual Review of Pharmacology and Toxicology*.)

production, but are probably not the primary stimulus in controlling Ep production in response to hypoxia.

For the above reasons, we have more recently focused on adenosine, an important nucleoside which has been reported to be involved in many cardiovascular[11,12] and renal[13] mechanisms. Adenosine triphosphate (ATP) metabolism is initiated in hypoxic tissues causing the depletion of ATP through the limited availability of oxygen for requisite ATP production as outlined in FIGURE 2. The depletion of cellular ATP levels leads to an increased concentration of AMP. AMP is metabolized to adenosine, inosine and hypoxanthine. Hypoxanthine and xanthine serve as oxidizable purine substrates for xanthine dehydrogenase or xanthine oxidase.[7] Xanthine is metabolized to oxygen-free radicals in tissues which are first exposed to an oxygen deficit and are then reperfused with oxygenated blood.[7,14–16] We postulate that adenosine is produced in a moderate to severe hypoxic stimulus and the oxygen free radicals H_2O_2 and O_2^- ion are produced in a more severe hypoxic stimulus following reperfusion. Both adenosine and oxygen free radicals play a significant role in hypoxic stimulation of kidney production of erythropoietin *in vivo* as renal blood flow waxes and wanes due to the regional redistribution of flow during intermittent hypoxia.

METHODS

Exhypoxic Polycythemic Mouse Assay

Exhypoxic polycythemic mice were employed to determine the erythropoietic activity of adenosine and other test compounds. The details of the methods used in these studies have been described elsewhere.[16] Briefly, HAM/ICR (Charles River Breeding Laboratories, Inc., CD-1, Wilmington, MA) strain female mice were placed for 2 weeks in a hypobaric chamber. On posthypoxic days, 3, 4, 5, and 6, the hemisulfate salt of ADE or the ADE agonists, i.v. (Sigma Chemical Co., St. Louis, MO), theophylline, i.p. (Sigma), albuterol i.p. (Schering Co., Bloomfield, NJ), and dipyridamole, i.p. (Sigma) were administered as single doses in a 0.5 ml volume. The mice were exposed to reduced atmospheric pressure (0.42 atm) for 4 hr on the 6th posthypoxic day after their removal from the hypobaric chamber to increase Ep production. Human urinary Ep was administered subcutaneously in final doses of 200, 400 and 800 milliunits/mouse (total dose divided into two injections in mice on the 6th and 7th posthypoxic days) to construct an Ep standard curve. Each mouse was given 0.5 μCi of radioactive iron (^{59}Fe citrate, i.v.) on the 8th posthypoxic day. 48 hours after receiving ^{59}Fe (posthypoxic day 10), the mice were exsanguinated via cardiac puncture and the percentage of ^{59}Fe incorporation in red cells determined, assuming a 7.5% blood volume/body weight. Mice with hematocrits of less than 51% were excluded from the assay to avoid the possibility of *in vivo* spontaneous Ep production. Ep concentrations were determined from an Ep dose-response curve for each experiment.

Renal Carcinoma Cell Culture System

Renal carcinoma (RC) cells obtained from a lung metastasis in a patient with erythrocytosis were serially transplanted into BALB/c athymic nude mice. The carcinoma cells were prepared in a monolayer cell culture and maintained in Eagle's MEM supplemented with 10% fetal bovine serum (FBS), 0.1 mM nonessential amino acid, 1 mM sodium pyruvate, 100 U/ml penicillin G and 100 μg/ml streptomycin in a humidified atmosphere of 5% CO_2 and 95% air at 37°C and passaged every 2–3 weeks. After at least five passages of the subculture, the cells were harvested by trypsinization and dispersed into 4.5 cm^2 multiwells at a concentration of 2×10^5 cells/ml culture medium. The Ep activity

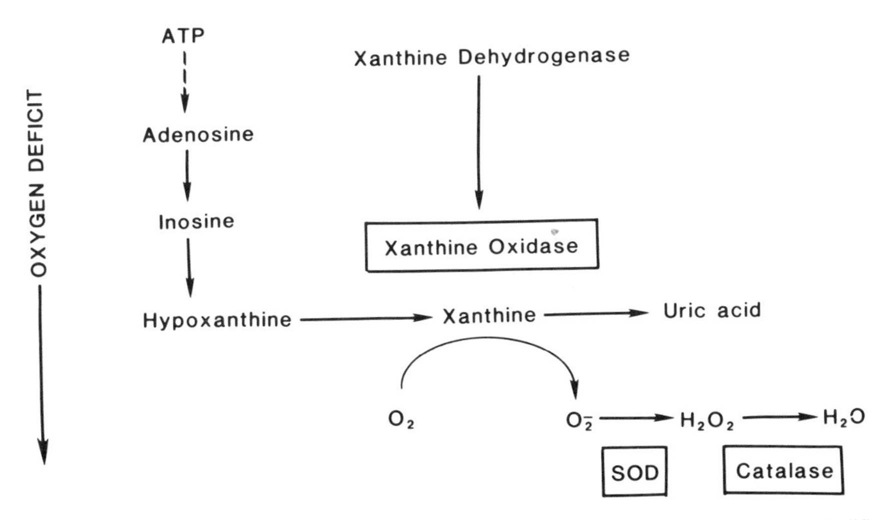

FIGURE 2. A proposed mechanism for oxygen free radical production during hypoxia. Modification of model of J. McCord.[7]

was determined with a sensitive radioimmunoassay as described previously. The spent culture medium was frozen at −80°C until used in the Ep assay. The cells reached confluent density in 6 days and saturation density in 18 days after seeding. The culture medium with 10% FBS contained undetectable levels of Ep. RC cells in early confluency, which was from the 6th to 9th day after seeding when Ep activity should be less than 50 mU/ml/24 hr, were used in the present studies. The details of our monolayer cell culture system have been described previously.[17] The cells appeared to be epithelial-like when examined under a phase contrast microscope; and electronmicroscopically the cells retained ultrastructural characteristics similar to the original tumor cells showing tight junctions, cell surface microvilli, a cytoplasm with diffusely scattered glycogen particles and sparse lipid droplets, and a nucleus with infolding showing the presence of heterochromatin.

^{51}Cr Release Assay

The renal carcinoma cells were preincubated for 90 min with 100 μCi $Na_2\,^{51}CrO_4$ (ICN Biomedicals, Inc.) at 37°C after trypsinization, and after washing twice with PBS the cells were plated at 2×10^4 cells/well in 96-well flatbottomed microtiter plates (Flow Laboratories, Inc.) and incubated with the test compounds for 24 hr at 37°C. Aliquots were taken from each well and the amount of ^{51}Cr released from the cells was determined using a gamma counter. Spontaneous release and maximum release were determined from cells incubated in complete culture medium and in 0.5% Triton X-100 (Sigma), respectively. Cytotoxicity was expressed as the percentage of ^{51}Cr release and was calculated with the following formula:

$$\% \text{ cytotoxicity} = \frac{(A - B)}{(C - B)} \times 100$$

where A is the test release, B is spontaneous release, and C is the maximum release. The 24-hr spontaneous release of incorporated ^{51}Cr is approximately 28 to 32% of the maximum release. More than 95% of the cells were viable after 24-hr incubation in control medium.

Chemical Agents

Xanthine, xanthine oxidase (Grade 1), glucose oxidase (Type V), thymol-free bovine catalase (17,000 U/mg) and bovine erythrocyte superoxide dismutase (3,000 U/mg protein) were all purchased from Sigma Chemical Co. and the hydrogen peroxide (30%) from Mallinckrodt, Inc.

Statistical Analyses

Tukey's multiple range test for the comparison of several treatments with a single control was used for the statistical analyses. A p value of less than 0.05 was considered to be statistically significant.

RESULTS

We used several criteria to prove that adenosine is involved in hypoxic stimulation of Ep production. The first criterion is whether adenosine synthesis and release from the kidney is seen with hypoxia, and furthermore whether this increase is correlated with Ep production. As indicated in TABLE 1, Miller *et al*[3] have shown that ischemic hypoxia induced in the dog kidney after renal artery occlusion caused a significant increase after 5 and 10 min in adenosine which is metabolized to inosine and hypoxanthine. It is interesting that very high levels of hypoxanthine (282 ng/mg per kidney), which is a precursor for oxygen free radicals such as H_2O_2 and O_2^- ion, are seen after 10 min renal artery occlusion in the dog. In preliminary studies, we found a significant increase in adenosine levels in the kidney in response to 4-hr hypoxia in the dog (0.42 atm) (to be published). For example, after exposure for 4 hr at 0.42 atm in a hypobaric chamber the levels of adenosine increased from a control level of 169 ng/mg to 211 ng/mg. Further studies are needed to confirm this preliminary experiment, but it appears that adenosine levels are significantly increased in the kidney following exposure to hypoxia, and this increase is correlated with a very early rise in Ep production. However, with further reperfusion of the kidney with oxygenated blood, perhaps the hypoxanthine which has

TABLE 1. Adenosine, Inosine, and Hypoxanthine in nmol/g Wet Weight ± SE in Normal Dog Kidneys After Renal Artery Occlusion

Dog	(N = 6)	5 min (N = 6)	10 min (N = 2)
Adenosine	7.6 ± 3.4	18.8 ± 5.0	25.9 ± 8.1 $p < 0.05$
Inosine	4.3 ± 1.9	20.5 ± 6.6	84.5 ± 28.7 $p < 0.005$
Hypoxanthine	11.0 ± 6.0	75.3 ± 21.6	281.8 ± 88.6 $p < 0.001$

Miller W.L. et al. Circ. Res. 43:390. 1978

been elevated by the hypoxic stimulus is metabolized by xanthine oxidase to the oxygen-free radicals H_2O_2 and O_2^- ion to further influence Ep production.

The next criterion which seems to us to be important to prove that adenosine is involved in hypoxia-induced Ep production is "mimicry" where exogenous adenosine is demonstrated to directly stimulate Ep production. Does adenosine itself have direct effects on Ep production? The protocol used for these studies involved placing mice in a hypobaric chamber (0.42 atm) for 14 days and on the 3rd–6th day, adenosine or the analogs were injected daily. On the 6th posthypoxic day the mice were exposed to hypoxia for 4 hr and compared with the response to hypoxia alone or the Ep standard, on radioiron incorporation. Note in FIGURE 3 that when adenosine hemisulfate was administered in dosages of 100, 400, 1600 nmol/kg per day for 4 days prior to and at the time of exposure to hypoxia that a significant increase in radioiron incorporation was seen with 400 and 1600 nmol/kg per day when compared with hypoxia controls. This indicates that *in vivo* the effects of hypoxia on Ep production is enhanced by adenosine.

The next criterion deals with specificity of the response to adenosine on erythropoietin production and involves the use of selective adenosine A_1 and A_2 agonists on erythropoietin production and/or secretion. For this work we used the selective adenosine receptor agonist CHA (N^6-cyclohexyladenosine), an analog of adenosine which is selective for A_1 receptors and decreases adenylate cyclase activity. CHA has a high affinity for the

A_1 receptor in a nanomolar concentration range. We also used the selective A_2 agonist NECA (5′-N-ethylcarboxamideadenosine), an analog of adenosine which is more selective for A_2 receptors, has a lower affinity in a micromolar range but stimulates adenylate cyclase activity. FIGURE 4 shows the results of dosages of adenosine, CHA and NECA ranging between 10 to 6400 nmol/kg/day i.v. Note that adenosine concentrations of 400–1600 nmol/kg per day produced a significant increase in radioiron incorporation in response to hypoxia. On the other hand, the A_2 selective agonist NECA which triggers adenylate cyclase produced a much more marked increase in radioiron incorporation at dosages of 10–200 nmol/kg; whereas CHA had no significant effect on radioiron incorporation. CHA would have been expected to suppress Ep production in response to hypoxia. In considering the reasons why CHA did not suppress Ep production in response to hypoxia in our posthypoxic polycythemic mouse model, we considered the possibility that cyclic AMP levels needed to be elevated higher in the kidney than that seen with hypoxia alone in order to assess the inhibitory response to the A_1 agonist CHA. Therefore, we combined hypoxia with a potent beta-2-adrenergic agonist albuterol, which is well known to stimulate adenylate cyclase to increase cyclic AMP levels. Note in FIGURE 5 that when we used both hypoxia and albuterol at dosages of 25 to 100 μg/kg per day i.p. a more significant increase in radioiron incorporation occurred than with hypoxia alone. On the other hand, the albuterol/hypoxia effect was significantly suppressed by 100

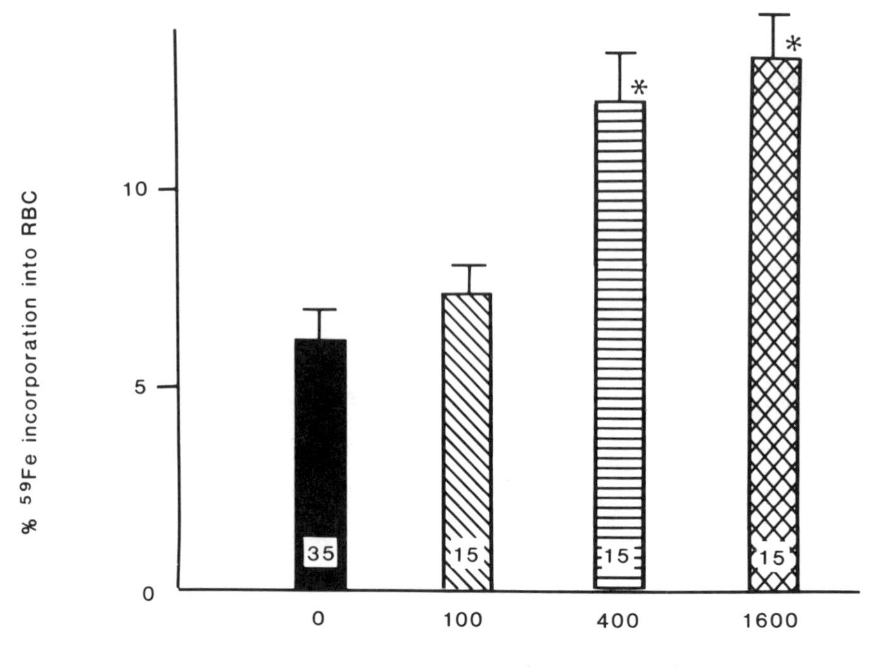

FIGURE 3. The effects of adenosine and hypoxia on radioiron incorporation in red cells of exhypoxic polycythemic mice. *Asterisk:* $p < 0.05$ when compared with hypoxia control. Numbers at the base of each bar represent number of mice. All groups were exposed to 4 hr hypoxia on the 6th day out of the hypobaric chamber. *Left hand column* represents 4 hr hypoxia control.

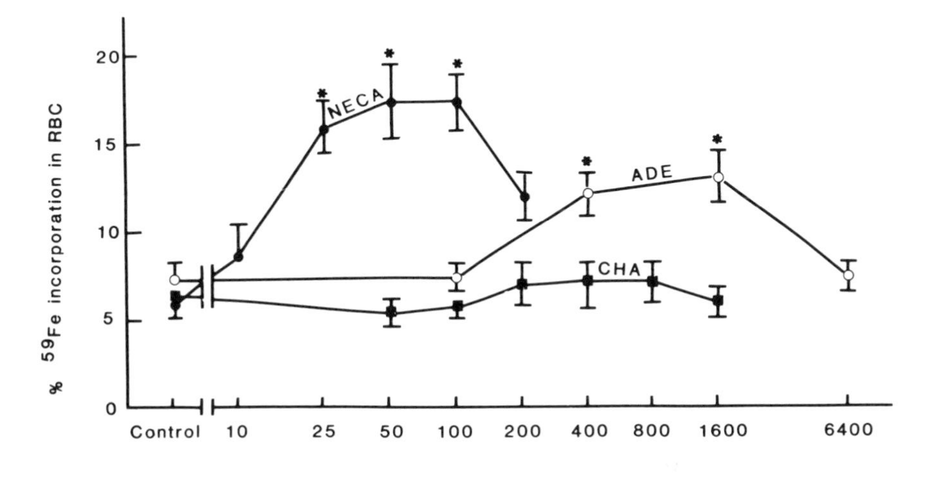

FIGURE 4. Effects of ADE (○) 100–6400 nmol/kg/day, i.v.), CHA (■) (50–1600 nmol/kg/day, i.v.), and NECA (●) 10–200 nmol/kg/day, i.v.) on radioiron incorporation in red cells in response to 4 hr hypoxia on the 6th posthypoxic day. The control value is the radioiron incorporation for untreated mice exposed to reduced atmospheric pressure (0.42 atm) for 4 hr on posthypoxic day 6. Each value represents the mean ± SEM of 11 to 34 mice. *Asterisk:* significantly different from control ($p < 0.05$). The mean hematocrit for the control nonhypoxic mice was 60.5 ± 0.9 (N = 35). The mean hematocrit values for all of the treatment groups ranging between 61.3 and 64.1 were not significantly different. (From Ueno *et al.*[4] Reprinted by permission from *Life Sciences.*)

nmol/kg per day of the A_1 selective agonist CHA. This means to us that *in vivo* in the kidney there may be predominantly A_2 receptors and a relatively low number of A_1 receptors indicating that adenosine's effect is more selective for the A_2 receptor in this system *in vivo*.

In order to study further this A_1 receptor agonist effect of CHA directly on Ep-producing cells in culture, we turned to our Ep-producing RC line. The technique used in these studies has been reported previously[17] and involves a monolayer cell culture in Eagle's MEM and bovine serum albumen. These cells are passaged every 2–3 weeks. The cells were plated in 4.5 cm^2 wells, CHA was added at early confluency, and the 24-hr spent culture medium was assayed with our sensitive radioimmunoassay for Ep.[19] When concentrations of CHA ranging between 10^{-7} and 10^{-3} M were used in this RC cell line, a significant decrease

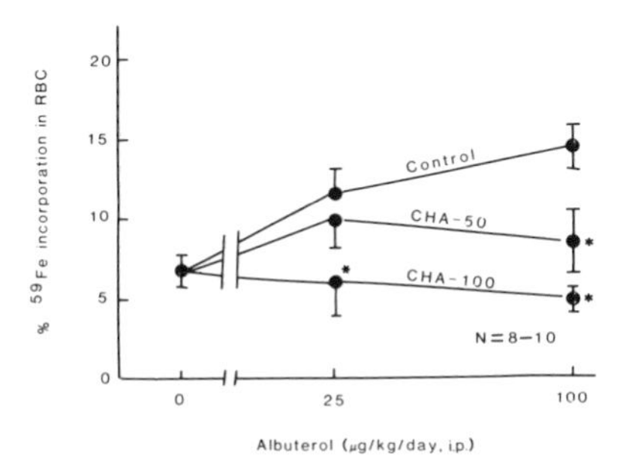

FIGURE 5. The inhibitory effects of CHA (50 and 100 nmol/kg/day, i.v.) on albuterol (25 and 100 μg/kg/day, i.p.) enhancement in radioiron incorporation in red cells in response to 4 hr hypoxia. Each value represents the mean ± SEM of 8 to 10 mice. The control value is the mean radioiron incorporation in mice exposed to reduced atmospheric pressure (0.42 atm) alone for 4 hr on posthypoxic day 6. There was no significant change in hematocrits of mice in each group. *Asterisk:* significantly different from albuterol alone ($p < 0.05$). (From Ueno *et al.*[4] Reprinted by permission from *Life Sciences.*)

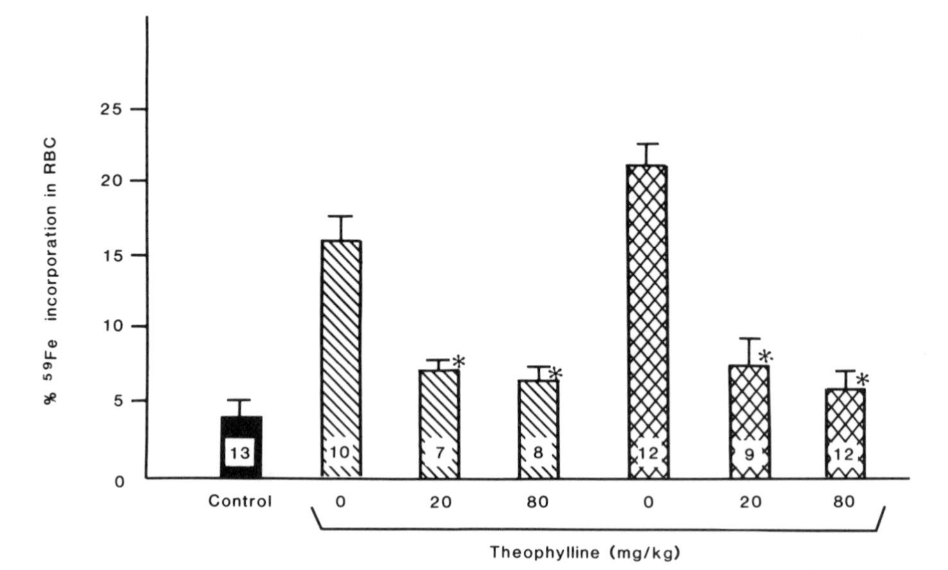

FIGURE 6. Inhibitory effects of theophylline (20 and 80 mg/kg/day, i.p.) on the enhancement of radioiron incorporation in red cells induced by ADE (▧: 1600 nmol/kg, i.v.) and NECA (▩: 100 nmol/kg/day, i.v.) in combination with 4 hr hypoxia. Each value represents the mean ± SEM of 7 to 13 mice. The number of mice is at the bottom of each bar. Mean percent ^{59}Fe incorporation in red cells when standard erythropoietin was administered at 200 mU/mouse and 800 mU/mouse in these experiments were 5.48 ± 0.73 and 25.07 ± 1.46, respectively. The *asterisk* over each bar indicates that this group is significantly different from ADE or NECA alone ($p < 0.05$). The mean hematocrit values ranged between 60.6 and 63.7%, and the hematocrit values were not significantly different in any of the groups. (From Ueno *et al.*[4] Reprinted by permission from *Life Sciences*.)

in Ep levels in the medium was seen indicating a selective inhibition of Ep secretion in this system. There was no cytotoxic effect of these concentrations of CHA on the cells as indicated by the ^{51}Cr release assay. We have not at this time completed the studies on the A_2 agonist NECA in this system, but in preliminary studies we have been unable to trigger this cell line to enhance erythropoietin with NECA.

The next criteria for proving that adenosine is involved in mediating Ep production is to determine the effects of adenosine receptor antagonists on Ep production in response to hypoxia. For this work we used the nonselective adenosine antagonist theophylline. As shown in FIGURE 6 a dosage of 1600 nmol/kg, i.v. adenosine hemisulfate produced a significantly greater increase in radioiron incorporation when compared with hypoxia alone, and both the 20 and 80 mg/kg doses of theophylline significantly inhibited this rise. In addition, note the much more significant elevation in radioiron incorporation seen with NECA at a dose of 100 nm/kg i.v. than with adenosine when combined with hypoxia which was again significantly inhibited by 20 and 80 mg/kg theophylline. Incidentally, theophylline is a phosphodiesterase inhibitor and actually increases cyclic AMP levels in tissues. The 80-mg/kg dose produces a slight increase in cyclic AMP in the kidney, but this is not apparently involved in this mechanism. Theophylline is a competitive adenosine receptor antagonist.

To continue this study with CHA we used a selective adenosine A_1 antagonist PACPX [1,3,-Dipropyl-8-(2-amino-4-chlorophenyl)xanthine] (to be published). The selective A_1 antagonist PACPX was evaluated at concentrations between 10^{-7} and 10^{-5} M on our

Ep-producing renal carcinoma cell line in which the medium contained 10^{-6} M adenosine. 10^{-7}–10^{-5} M PACPX actually produced a significant increase in Ep levels in the culture medium as assessed by our Ep radioimmunoassay.[18] A possible explanation for this increase in Ep levels is that the Ep-producing RC cells have predominantly A_1 receptors, and when these receptors are blocked with PACPX, the effects of the adenosine A_2 receptors, which may be slight to moderate on the cell surface of these cells, is unmasked enabling the adenosine in the culture medium to produce an A_2 receptor response to increase Ep production.

To continue these studies of our adenosine/oxygen free radical model, as mentioned earlier the hypoxanthine generated during hypoxia can be acted upon by xanthine oxidase to generate oxygen free radicals H_2O_2 and O_2^- ion. Grisham *et al.*[19] using an isolated perfused hypoxemic heart model to produce an oxygen deficit noted that the xanthine oxidase inhibitor allopurinol, the oxygen free radical scavengers catalase and superoxide dismutase protected the heart against hypoxic damage. Hypoxia is known to decrease ATP and increase adenosine, inosine and hypoxanthine in the kidney.[3] Hypoxanthine is converted nonenzymatically to xanthine which is acted upon by xanthine oxidase. Xanthine oxidase is produced by an action of proteases which are activated by calcium entering the cell during hypoxia to convert xanthine dehydrogenase to xanthine oxidase. Hypoxanthine, when acted upon by xanthine oxidase, produces primarily the oxygen free radicals H_2O_2 and O_2^- ion. O_2^- ion is actually dioxygen with one less electron, and under the influence of superoxide dismutase is metabolized to H_2O_2 which under the influence of catalase is further metabolized to water. For our studies, we generated the oxygen free radicals O_2^- and H_2O_2 in a xanthine-xanthine oxidase system and also used superoxide dismutase to scavenge the O_2^- ion generated and catalase to scavenge H_2O_2 produced.

Note in FIGURE 7 that using our erythropoietin-producing renal carcinoma cell line with xanthine at a concentration of 10^{-5} M and xanthine oxidase in concentrations from 8×10^{-7} up to 5×10^{-4} units/ml an exponential increase in Ep production in these RC cells used at early confluency was seen. We postulate that the oxygen free radicals may be very important in enhancing Ep production. This system was used as an *in vitro* system to assess whether or not these oxygen free radicals have a direct affect on Ep production. However, further *in vivo* studies are needed to determine if hypoxia and regional changes in kidney perfusion during hypoxia correlate with increases in oxygen free radicals and Ep production. We carried out additional studies with the incubation of xanthine-xanthine oxidase in the above concentrations in control medium for 24 hr in the presence of Ep. No significant change in the immunoreactivity of Ep was seen in the presence of xanthine-xanthine oxidase at concentrations of 10^{-5} M xanthine plus xanthine oxidase (8×10^{-7}–5×10^{-4} units/ml) when compared with the levels of Ep in control medium for 24 hr.

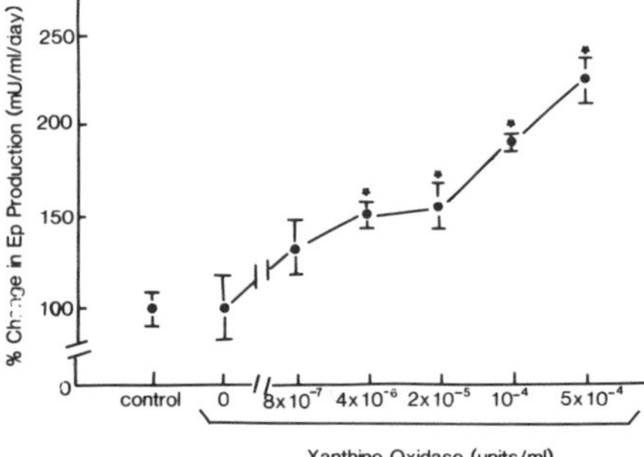

FIGURE 7. Effects of 10^{-5} M xanthine and increasing concentrations of xanthine oxidase on erythropoietin (Ep) production by renal carcinoma (RC) cells in culture. RC cells were incubated for 24 hr in culture media in "early confluency." After incubation, the viable cells were counted and the culture media assayed for Ep (RIA). The values are expressed as the percentage of the basal level of Ep which equals 100% and represents the mean ± SEM of 6 different samples. *Asterisk:* $p < 0.05$. (From Ueno *et al.*[6] Reprinted by permission from *Biochemical and Biophysical Research Communications*.)

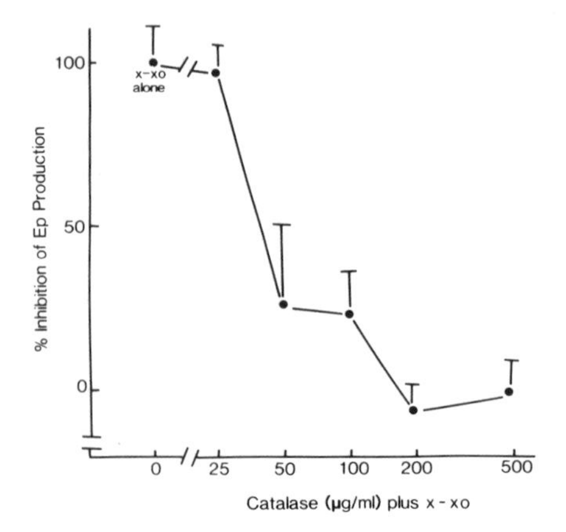

FIGURE 8. Inhibitory effects of catalase on the increased erythropoietin (Ep) production induced by 10^{-5} M xanthine plus 5×10^{-4} units/ml xanthine oxidase. The renal carcinoma (RC) cells were incubated with several concentrations of catalase (specific activity 17,000 U/mg) plus xanthine-xanthine oxidase (X-XO) for 24 hr in "early confluency." Values are expressed as percentage of the enhancement level of Ep production induced by X-XO and represent the mean ± SEM of 6–12 different samples. 0% represents the basal level of Ep production. (From Ueno *et al.*[6] Reprinted by permission from *Biochemical and Biophysical Research Communications.*)

We next turned to studies of catalase to determine whether catalase, which scavenges the H_2O_2 generated in these cells, by the addition of xanthine-xanthine oxidase. Note in FIGURE 8 that when H_2O_2 is scavenged in this xanthine-xanthine oxidase system with the use of catalase there is a very significant dose-response-related decline in Ep production when compared with xanthine-xanthine oxidase alone. This would indicate that the oxygen free radical generated which is important in enhancing Ep production is probably H_2O_2. On the other hand, as shown in FIGURE 9 when superoxide dismutase, which is a scavenger of superoxide ion, is added to the xanthine-xanthine oxidase system in the renal carcinoma cell line, there is actually an enhancement of Ep production. This indicates that perhaps the O_2^- ion generated could have an inhibitory effect on Ep production in this system which is probably overridden by the high level of production of H_2O_2.

We next carried out studies where we generated H_2O_2 using glucose oxidase in the RC cell line. FIGURE 10 shows that concentrations of .032 up to 4 mU/ml glucose oxidase produced a significant enhancement of Ep in the RC cell line which is further support for our hypothesis that H_2O_2 is the primary oxygen free radical which is selective in increasing the generation of Ep. To pursue this problem further, as seen in FIGURE 11, we added H_2O_2 directly to the Ep-producing RC cells in concentrations of 4×10^{-8}–5×10^{-4} M and observed that H_2O_2 itself produced a very marked and significant increase in Ep generation in these RC cells. These concentrations employed in the studies of H_2O_2 as well as the concentrations of glucose oxidase used did not produce a cytotoxic effect, as

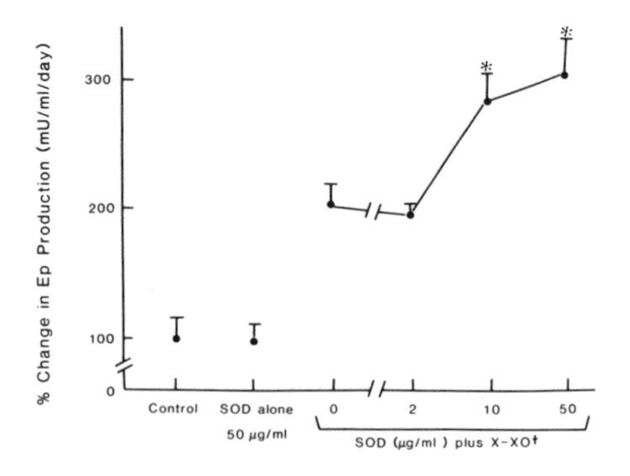

FIGURE 9. Effects of superoxide dismutase (SOD) on erythropoietin (Ep) production induced by X-XO in renal carcinoma cells in culture. Basal level (control) of Ep—19.04 ± 3.04 mU/ml/day. +X-XO = combination of xanthine (10^{-5} M) and xanthine oxidase (5×10^{-4} units/ml). *Asterisk:* significantly ($p < 0.05$) different from X-XO alone. N = 6. (From Ueno *et al.*[6] Reprinted by permission from *Biochemical and Biophysical Research Communications.*)

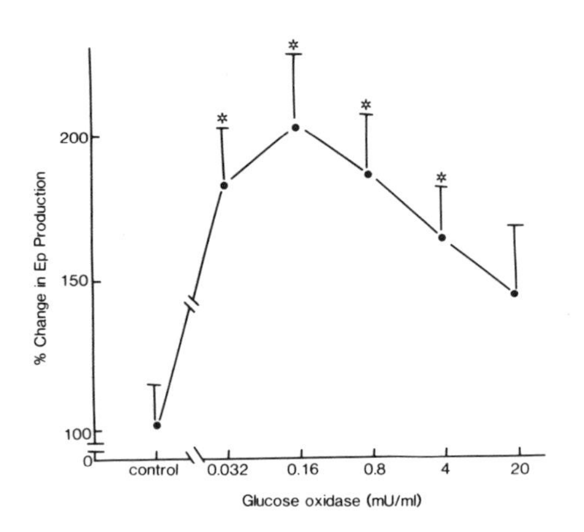

FIGURE 10. Effects of glucose oxidase on erythropoietin (Ep) production in renal carcinoma (RC) cell cultures. RC cells in "early confluency" were incubated for 24 hr. The values are expressed as the percentage of the basal level of Ep which equals 100% and represent the mean ± SEM of 6 different samples. *Asterisk:* $p < 0.05$. (From Ueno *et al.*[6] Reprinted by permission from *Biochemical and Biophysical Research Communications.*)

assessed by the ^{51}Cr release assay, nor did the xanthine-xanthine oxidase combination cause any change in Ep directly in the medium.

SUMMARY

We have presented a model for the role of external messenger substances in hypoxic stimulation of kidney production of erythropoietin. These autacoids probably act in concert to activate the adenylate cyclase system to enhance production and/or secretion of erythropoietin. The phosphoproteins generated in this system could act at the level of transcription and translation of erythropoietin as well as at the level of release of erythropoietin from the cell. Even though eicosanoids and beta-2-adrenergic agonists may be involved in mild to moderate hypoxia, it seems more likely that adenosine is more involved in erythropoietin production with increasing severity of hypoxia. Adenosine may play a very early role in hypoxia following the decrease in ATP to trigger erythropoietin production, and hydrogen peroxide may be generated from hypoxanthine, a metabolite of adenosine, during reoxygenation and regional changes in blood flow in the normal kidney and perhaps in certain renal and hepatic tumors. Further work is necessary *in vivo* to completely clarify the role of adenosine and oxygen free radicals in regulating kidney production of erythropoietin.

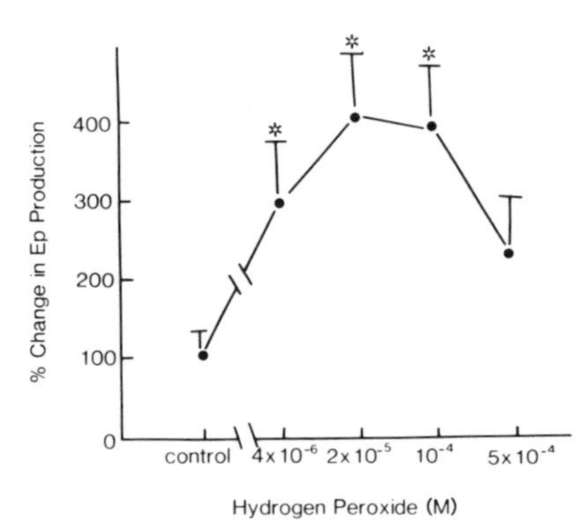

FIGURE 11. Effects of hydrogen peroxide on erythropoietin (Ep) production by renal carcinoma (RC) cells. RC cells in "early confluency" were incubated for 24 hr in culture medium. The values are expressed as the percentage of basal levels of Ep which equals to 100% and represents the mean ± SEM of 6 different samples. *Asterisk:* $p < 0.05$. (From Ueno *et al.*[6] Reprinted by permission from *Biochemical and Biophysical Research Communications.*)

REFERENCES

1. FISHER, J. W. 1989. Regulation of erythropoietin (Ep) production. Handbook of Physiology. (In press.)
2. FISHER, J. W. 1988. Pharmacologic modulation of erythropoietin production. Ann. Rev. Pharmacol. Toxicol. **28:** 101–122.
3. MILLER, W. L., R. A. THOMAS, R. M. BERNE & M. RUBIO. 1978. Adenosine production in the ischemic kidney. Circ. Res. **43:** 390.
4. UENO, J., J. BROOKINS, B. BECKMAN & J. W. FISHER. 1988. A_1 and A_2 adenosine receptor regulation of erythropoietin production. Life Sci. **43:** 229–237.
5. NELSON, P. K., J. BROOKINS & J. W. FISHER. 1983. Erythropoietic effects of prostacyclin (PGI_2) and its metabolite 6-keto-prostacyclin (PG) E_1. J. Pharmacol. Exp. Ther. **226:** 493.
6. UENO, M., J. BROOKINS, B. S. BECKMAN & J. W. FISHER. 1988. Effects of reactive oxygen metabolites on erythropoietin production in renal carcinoma cells. Biochem. Biophys. Res. Commun. **154:** 773–780.
7. MCCORD, J. M. 1985. Oxygen-derived free radicals in post-ischemic tissue injury. N. Engl. J. Med. **312:** 159.
8. FINK, G.D. & J. W. FISHER. 1977. Stimulation of erythropoiesis by beta adrenergic agonists. I. Characterization of activity in polycythemic mice. J. Pharmacol. Exp. Ther. **202:** 192–198.
9. FINK, G.D. & J. W. FISHER. 1977. Stimulation of erythropoiesis by beta adrenergic agonists. II. Mechanism of action. J. Pharmacol. Exp. Ther. **202:** 199–208.
10. JELKMANN, W., B. BECKMAN & J. W. FISHER. 1979. Enhanced effects of hypoxia on erythropoiesis in rabbits following beta-2 adrenergic activation with albuterol. J. Pharmacol. Exp. Ther. **211:** 99.
11. DALY, J. W. 1983. *In* Physiology and Pharmacology of Adenosine Derivatives. J. W. Daly, Y. Kuroda, J. W. Phillis, H. Shimizu & M. Ui, Eds.: 59–70. Raven Press, New York, NY.
12. SHIMIZU, H. 1983. *In* Physiology and Pharmacology of Adenosine Derivatives. ed. by J. W. Daly, Y. Kuroda, J. W. Phillis, H. Shimizu & M. Ui, Eds.: 31–40. Raven Press, New York, NY.
13. CHURCHILL, P. C. & M. C. CHURCHILL. 1985. A_1 and A_2 adenosine receptor activation inhibits and stimulates renin secretion of rat renal cortical slices. J. Pharmacol. Exp. Ther. **232:** 589–594.
14. BEAUCHAMP, C. & I. FRIDOVICH. 1970. A mechanism for the production of ethylene from methional: The generation of hydroxyl radicals by xanthine oxidase. J. Biol. Chem. **245:** 4641.
15. OSSWALD, H., H. J. SCHMITZ & R. KEMPER. 1977. Tissue content of adenosine, inosine and hypoxanthine in the rat kidney after ischemia and post-ischemic recirculation. Pflugers Arch. **371:** 45.
16. DIONISI, O., T. GALEOTTI, T. TERRANOVA & A. AZZI. 1975. Superoxide radicals and hydrogen peroxide formation in mitochondria from normal and neoplastic tissues. Biochim. Biophys. Acta **403:** 292.
17. HAGIWARA, M., I. CHEN, R. MCGONIGLE, B. BECKMAN, F. H. KASTEN & J. W. FISHER. 1984. Erythropoietin production in a primary culture of human renal carcinoma cells maintained in nude mice. Blood **63:** 838.
18. REGE, A. B., J. BROOKINS & J. W. FISHER. 1982. A radioimmunoassay for erythropoietin: Serum levels in normal human subjects and patients with hemopoietic disorders. J. Lab. Clin. Med. **100:** 829–843.
19. GRISHAM, M. B., W. J. RUSSEL, R. S. ROY & J. M. MCCORD. 1986. Reoxygenation injury in the isolated perfused working rat heart: Role of xanthine oxidase and transferrin. Superoxide and superoxide dismutase. *In* Chemistry, Biology and Medicine. G. Rotilio, Ed.: 571–575. Elsevier. Amsterdam.

Erythropoietin-Dependent and Erythropoietin-Producing Cell Lines

Implications for Research and for Leukemia Therapy

W. DAVID HANKINS,[a,b] KYUNG L. CHIN,[a] ROBERT DONS,[a]
AND GEORGE SIGOUNAS[a]

[a]*Armed Forces Radiobiology Research Institute*
NMC-NCR
Bethesda, Maryland 20814
and
[b]*National Institutes of Health*
Bethesda, Maryland 20205

INTRODUCTION

Many laboratories, including ours, have used primary *in vitro* cultures of bone marrow, spleen, and fetal liver cells to study the interactions of erythropoietin with its target cell, the so called "Epo responsive cell."[1,2] Recently cell purification procedures and models of drug-induced[3] and virus-induced[4] erythroid hyperplasia made it possible for investigators to study large numbers of relatively purified primary erythroid target cells. These *in vitro* studies along with *in vivo* studies led to extensive knowledge about erythropoiesis. Nevertheless, while it is clear that erythropoietin plays a primary role in the regulation of red cell production, the control of the production of this hormone, the nature of its receptor, and its mechanism of action have not been elucidated. In contrast to other endocrine systems, the study of erythropoietin has been significantly hindered by the absence of any established cell lines which produce significant quantities of the hormone as well as cell lines which respond to the hormone. Such lines would undoubtedly be useful for further studies on erythropoietin gene transcription and erythroid differentiation, proliferation, receptor status, and signal transduction pathways.

In this report, we describe two new classes of erythroleukemia cell lines which we have isolated from animals rendered leukemic by a particular isolate of murine leukemia virus. While our cell lines share some properties with the traditional, extensively studied murine erythroleukemia cells,[5] two important new properties have been observed. The first class of lines produces erythropoietin (Epo) and exhibits altered or rearranged genes for this hormone. In the second class the cells do not secrete the hormone but display an absolute requirement for the hormone for their viability. A third class, which is only mentioned here for comparison, neither secretes nor responds to erythropoietin (WDH and KLC, unpublished). The isolation of these lines has opened new experimental doors for analysis of the transcriptional regulation of the hormone and its mechanism of action. In addition the Epo-dependent lines suggest that Epo may play a physiologic role in survival or viability rather than act as a specific inducer of erythroid differentiation. Finally, because of this viability requirement, we suggest that antihormone therapy *in vivo* may be efficacious in controlling this, and perhaps other types of leukemia.

Almost two decades have elapsed since the late Dr. Charlotte Friend derived erythroleukemia cell lines which appeared to be arrested in maturation and independent of erythropoietin.[5] Dr. Friend's observation that dimethyl-sulfoxide induced hemoglobin synthesis spawned a whole new field of investigation and volumes of publications on the biochemistry of terminal erythroid differentiation in these cell lines. Studies on murine erythroleukemia cells (MELC) have produced an enormous database on the sequence of erythroid-specific events. Nevertheless, the induction of hemoglobin by DMSO, a nonphysiologic agent, has predictably revealed little about the relation of erythropoietin to erythroid differentiation. The many studies of MELC induction have been adequately reviewed[6] and will not be further discussed in this article.

Erythropoietin-Producing Erythroid Cell Lines

Although extrarenal production has been documented, it is generally held that the major *in vivo* site of production of Epo is the kidney.[1,2] During recent attempts to isolate new cell lines, observations were made by Dr. Tambourin and Dr. Varet in Paris and one of us (WDH) at the National Institutes of Health which led to the suspicion that Epo might be produced by erythroleukemia cells. (These observations are summarized in REFERENCES 7 and 8.) We therefore isolated a number of new lines and screened all the existing erythroleukemia lines which we could obtain for the production of erythroid growth factors. In our initial screen we tested the ability of conditioned media from more than 70 cell lines to stimulate hemoglobin synthesis by cultured fetal liver cells. These lines included erythroleukemia cells from mice, rats, and humans. Several lines were found to produce factors that stimulated hemoglobin synthesis. All of the positive lines were from leukemic mice that had been inoculated as newborns with a replication competent helper virus isolated from Friend virus stocks.[7,8] Most of our subsequent studies were performed on conditioned media from two cell lines, HFTC-5 and TP-3.[8,9]

Biologic Similarity of Erythroid Growth Factors and Erythropoietin

Since the ^{59}Fe assay used in the initial screen could have been influenced by factors other than Epo, we tested the media from HFTC-5 and TP-3 in a total of nine biological assays outlined in TABLE 1. These included six *in vitro* assays and three *in vivo* assays.[8] We observed that the media conditioned by both cell lines was active in all nine biological assays. We therefore concluded that while each of the individual assays may have deficiencies, the collective data provided compelling evidence that the biologic activity of these erythroid growth factors was identical to that of Epo.

Immunologic Similarity of Erythroid Growth Factors and Erythropoietin

TABLE 2 presents evidence that the erythropoietic activities in the media from HFTC-5 exhibited immunologic properties of Epo.[8] Conditioned media from HFTC-5 were mixed with a rabbit antiserum against erythropoietin. After 30 minutes of incubation, goat antirabbit gamma globulin was added and the immune precipitate removed by centrifugation. This approach was chosen instead of the alternative method of direct inoculation of the antiserum into the mouse without prior exposure to the hormone. Our rationale was that the erythropoietic agent in conditioned media might not be Epo but only an enhancer or inducer of endogenous Epo. If such were the case, *in vivo* inoculation of antiserum would still abolish the erythropoietic activity. Because neutralization was accomplished *in vitro* before inoculation, we concluded that the erythropoietic agent in conditioned medium was immunologically similar to Epo.

TABLE 1. Erythroid Stimulation by Conditioned Media from HFTC-5 and TP-3[a]

	Form of Data	No Addition	HFTC-5	TP-3
In vitro assays				
1. FVA cells	cpm/5 $\times$ 10^5 cells	56	4,350	3,414
2. DNA synthesis	cpm/4 $\times$ 10^5 cells	1,128	32,133	7,851
3. CFU-E	colonies/10^6 cells	0	885	308
4. 2-day ^{59}Fe	cpm/5 $\times$ 10^6 cells	224	9,593	2,840
5. 8-day BFU-E	bursts/10^6 cells	0	150	62
6. 8-day ^{59}Fe	cpm/10^6 cells	40	2,918	582
In vivo assays				
7. Polycythemic mouse (^{59}Fe)	% uptake	0.08	22.2	4.6
8. Appearance of nucleated rbc	yes/no	no	yes	yes
9. Nude mouse tumors	hematocrit	48	68	57

[a]In assays 1 through 4, standard Epo was added at a final concentration of 0.25 mL. HFTC-5 and TP-3 conditioned media (4-day harvest) were assayed at a final concentration of 10% vol/vol. In assays 5 and 6, the final concentration of standard Epo was 2.5 U/mL and conditioned media were concentrated tenfold (Amicon filter) before use in the assay at 10% vol/vol. In assays 7 and 8, 0.25 units of standard Epo or 1 mL of conditioned media was injected subcutaneously into plethoric mice. In assay 7, all samples were coded samples and were assayed "blind" by JS. In assay 8, we tested whether inoculation of 1.0 mL of media into polycythemic mice produced morphologically identifiable erythroblasts 48 hours later. In assay 9, we tested whether HFTC-5 and TP-3 were tumorigenic in nude mice and if so, whether tumor-bearing mice exhibited polycythemia. Cells (2 $\times$ 10^6) from each line were inoculated subcutaneously into three NIH nu/nu mice at 4 weeks of age. Controls were given a single injection of phosphate-buffered saline. Animals inoculated with HFTC-5 and TP-3 exhibited subcutaneous tumors eight weeks postinoculation. At this time, blood was collected from the tail and hematocrits (% packed cell volume) determined. The observed increase in hematocrits appeared to be due to increased red cell production since no changes in the thickness of the white blood cell layer (buffy coat) were observed.

Biochemical Similarity of Erythroid Growth Factors and Erythropoietin

The conditioned media was exposed to a variety of physical treatments such as heating, freeze-thawing, enzyme digestions, and others. In all cases, erythroid growth factors present in the conditioned media of HFTC-5 and TP-3 displayed properties similar

TABLE 2. Neutralization of Erythropoietic Activity Produced by HFTC-5 by Erythropoietin Antiserum[a]

Treatment	^{59}Fe Uptake in Polycythemic Mouse Assay (%)
Uninjected	0.08 = 0.01
Normal rabbit serum + HFTC-5 media + GARGG	15.00 = 1.47
Epo-antiserum + HFTC-5 media + GARGG	0.70 = 0.28
(Epo-antiserum + GARGG), then + HFTC-5 media	9.40 = 1.70

[a]Rabbit erythropoietin antiserum (0.2 ml) was added to 6 ml of test media and incubated at 37°C for 30 minutes after which 0.5 ml of goat antirabbit gamma globulin (GARGG) was added. After further incubation at 37°C for 30 minutes, precipitates were removed by centrifugation at 5,000 rpm for ten minutes. One milliliter of the supernatant was injected subcutaneously into plethoric mice. Two days later the mice received 1 μCi ^{59}Fe (ICN). Seventy-two hours subsequent to injection of the isotope, 0.5 ml of blood was removed and the percentage of ^{59}Fe uptake determined. Standard error of the mean, six mice per assay.

to those published for erythropoietin. To further assess the biochemical character of the erythroid growth factors, a five-step purification scheme was devised that allowed purification of this erythropoietic activity to apparent homogeneity.[9] The methods employed included lectin affinity chromatography (wheat germ agglutinin), gel filtration (ultra gel ACA44), ion exchange, hydroxylapatite, and high performance liquid chromatography (HPLC). Following the final purification procedure of HPLC, the recovered erythropoietic activity was electrophoresed on polyacrylamide gels and was detected by silver staining as a single band with an apparent molecular weight of 35,000 daltons. This molecular weight and all of the other biochemical properties, which we assessed, were similar for the erythroid growth factors and erythropoietin.

Summary and Implications of Erythropoietin-Producing Erythroid Cell Lines

All these properties provide substantial evidence that the growth factor produced by these erythroleukemia cells is indeed authentic mouse erythropoietin. In unpublished studies we have documented that when a murine erythropoietin gene is used to hybridize to cytoplasmic RNA from these Epo-producing cells, a 1.6-kb messenger RNA species is detected. In other erythroleukemia cells, which do not make erythropoietin, no such message was detected. This mRNA for Epo appears to have the same size as that reported for normal Epo.[10] Upon observing the apparent normality of the protein and the mRNA we recently began to examine the genomic structure of the erythropoietin gene in these erythroleukemia cells. Restriction enzyme digest of high molecular weight DNA prepared from these cells revealed that at least two of the Epo-producing cell lines have rearranged erythropoietin genes. The precise nature of these genomic alterations is currently under investigation.

The production of Epo by erythroid cells was surprising. Two explanations for this phenomenon may be suggested. First, since all of the positive lines were derived from virus-infected mice, it is possible that the inoculated virus or a recombinant virus may have integrated near the Epo gene and activated its transcription. Second, it is possible that normal erythroid precursors may transiently secrete Epo at a certain stage of differentiation and that these erythroleukemia cell lines may represent an expansion of erythroid progenitors "selected by the virus" at that stage. If such were the case, this would have the conceptual implication that autocrine production of erythroid growth factors by erythroid cells may be a part of normal red blood cell development.

Erythropoietin-Dependent Cell Lines

For many years our laboratory has been involved in one way or another in the study of virus-induced alteration of growth properties of blood-forming precursors and leukemic transformation. An *in vitro* transformation system was devised which permitted a direct analysis of putative transforming retroviruses with their hemopoietic targets.[11,12] Thus bone marrow and other hemopoietic tissues were directly infected by transforming retroviruses *in vitro*. When these infected cells were subsequently plated in semisolid methylcellulose, erythroid colonies developed under culture conditions which would not support colony formation by normal, uninfected erythroid precursors. The induction of colonies by retroviruses was referred to as virus-induced erythroid transformation.

Initial studies with the Friend spleen focus-forming virus indicated that the virus conveyed a growth/survival advantage to cells within a fairly narrow "window" in the erythroid pathway. Two unexpected observations emanated from our *in vitro* transformation studies with the Friend virus.[13] The transformed erythroid precursors were not

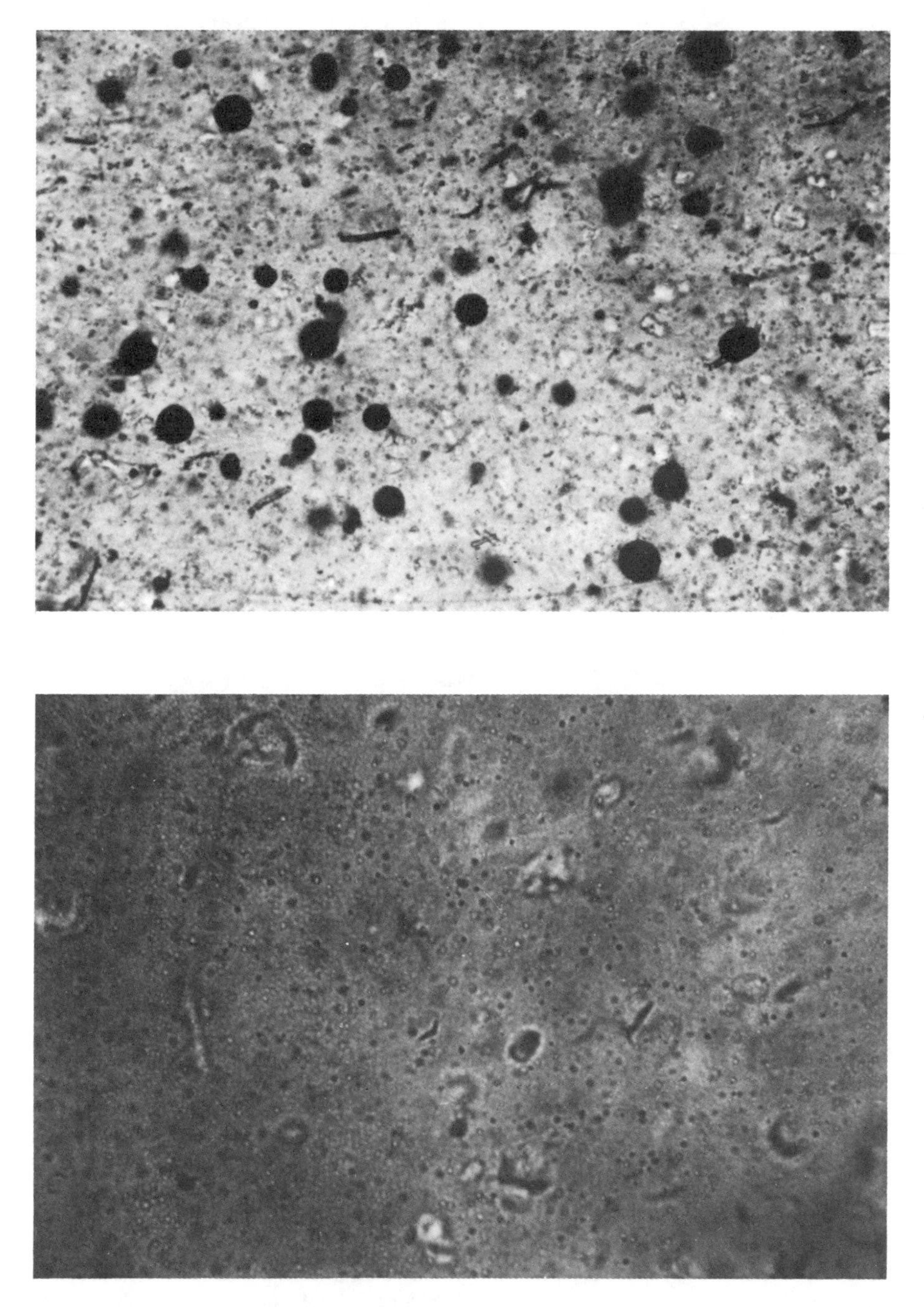

FIGURE 1. Effect of Epo on leukemia cell growth *in vitro*. Spleen cells from passage 5 of a transplantable erythroleukemia line were seeded at a concentration of 1 million cells/ml in methylcellulose. Connaught step III erythropoietin was added (0.5 U/ml) to half the cultures (*top*) and the remainder (*bottom*) received an equivalent amount of Hanks' balanced salt solution. Photographs were made *in situ* using a Leitz inverted microscope at a magnification of ×25.

blocked in differentiation and were very sensitive to erythropoietin (rather than being hormone independent). We initially thought that the retention of hormone dependence and an ability to differentiate was unusual for transformed cells and was probably unique to Friend virus transformants. Subsequent studies, however, demonstrated that virtually all the oncogene-containing retroviruses tested (including ras, abl, mos, src, fes, and others) would transform erythroid progenitors without causing a block in differentiation or a loss in hormone sensitivity.[14]

Because of the consistent properties of these transformed cells, we proposed an alternative model of transformation[13,14] which suggested that cancer cells, in general, could gain a heritable growth advantage—become transformed—while maintaining the ability to differentiate and a dependence on normal physiologic regulators, *i.e.*, hormones. These two properties, hormone dependence and the ability of the transformed cell to undergo terminal maturation, appeared to have significant implications for tumor progression and through the use of an antihormone approach offered a potential new therapy for leukemia. To test our hypothesis, we derived several transplantable F-MuLV-induced erythroleukemias according to Oliff *et al.*[15] Spleen cells from mice bearing such transplantable leukemias were placed *in vitro* in a variety of hormonal culture

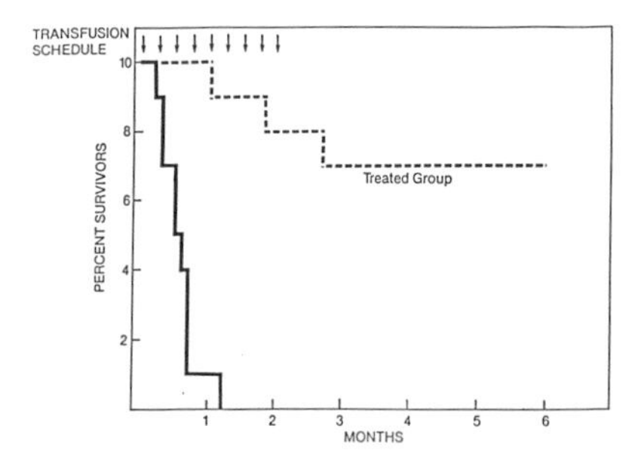

FIGURE 2. Effect of transfusion on the survival of mice during the early stages of leukemia development. Treated groups received transfusions at times indicated by the *arrows*.

conditions.[16] We found that the only additive which supported colony formation was erythropoietin. Even more impressive was that the requirement for Epo appeared absolute, since no colonies were formed in its absence (FIG. 1). This dramatic effect of Epo on the leukemic cells led us to (a) test the effects of Epo deprivation on erythroleukemia growth *in vivo;* and (b) attempt to establish continuous lines of Epo-dependent cells.

Antihormonal Approach to the Treatment of Leukemia

To study the effects of Epo deprivation we employed a physiological feedback procedure to lower endogenous Epo. Weekly transfusions of packed red blood cells were administered to mice bearing a transplanted leukemia. The mice which had received hypertransfusion survived two to three times as long as those leukemic mice which remained untreated. It was considered that the beneficial effects of hypertransfusion may have been unrelated to Epo. However, when transfused mice were administered exogenous Epo inoculations, these mice succumbed to leukemia and died even earlier than their untreated counterparts. For this reason, we believe that the slower growth of leukemia cells in hypertransfused mice was due to erythropoietin deprivation. In a subsequent study we observed an even more dramatic effect when the hypertransfusion was administered at an early stage of this leukemia[17] (FIG. 2).

TABLE 3. Dependence of Subcultured Erythroleukemia Colonies on Epo Cell/ml × 10^6

	Passage Number								
	1	2	3	4	5	6	7	8	9
Epo present	2.00	1.10	1.50	0.85	1.95	1.00	2.00	0.70	1.75
Epo absent	0.05	0.05	0.05	0.10	0.25	0.70	0.30	0.10	0.10

The absolute dependence of the leukemic cells on erythropoietin was of interest since the leukemic cells, when examined microscopically, did not appear to be hemoglobinized. Historically, erythropoietin has been thought of as an agent which induces terminal erythroid differentiation. We, and others, have attempted to elucidate how erythropoietin induced the expression of erythroid-specific genes such as hemoglobin, glycophorin, etc. Taken at face value, the result in FIGURE 1 suggests that erythropoietin may perform a viability role rather than an inductive role. Thus, in a sense, erythropoietin may simply be a permissive viability factor and allow committed erythroid precursors to execute their developmental potential—a part of which is the expression of erythroid-specific genes, hemoglobin, and glycophorin. This would further suggest that the erythroid lineage, and perhaps other developmental lineages, arise as the result of selection rather than induction. In any case, we reasoned that the sensitivity of these cells to erythropoietin should render them useful for studying the mechanism of action of this developmental hormone. Therefore, we attempted to establish permanent cell lines from these primary cultures.

In the experiment shown in TABLE 3, colonies were dispersed in culture medium and subcultivated with or without erythropoietin every three days for nine passages. At each passage, the cells growing with erythropoietin were washed and resuspended at 1/4 the concentration. The results indicated that erythropoietin was essential throughout the experiments and without the hormone these erythroleukemia cells quickly died. This provided further evidence of the viability role of the Epo. We have subsequently carried these cells in culture for more than one year and have derived several established lines. These lines can be stored frozen, and when thawed they retain their absolute dependence on erythropoietin. As evidenced by the dose-response data in TABLE 4, the cells are extremely sensitive to the addition of Epo.

In experiments not shown, we have detected B-globin mRNA in these cells although hemoglobin was not detectable. Dr. Sigounas has observed that addition of hemin to the culture medium results in an increase in benzidine positivity of the cells from less than 1% to about 80% within four days. During this same time period, he has observed a ten- to fiftyfold increase in the globin mRNA.

In summary, the sensitivity, and indeed dependence, of these erythroleukemia cells on erythropoietin have far-reaching implications for the therapy of erythroleukemias and perhaps other leukemias as well. Second, the Epo-dependent lines provide a useful model

TABLE 4. Dose-Response Effects of Epo on Cell Growth

Erythropoietin	Cell/ml × 10^6
3 U/ml	0.916
1 U/ml	0.860
0.3 U/ml	0.612
0.1 U/ml	0.450
0.03 U/ml	0.283
0.01 U/ml	0.138
0 U/ml	0.013

in which to investigate the interaction of the hormone with its receptor. Finally, derivation of Epo-dependent cell lines led us to reconsider whether the role of erythropoietin in red blood cell development is inductive or permissive. Experiments aimed at distinguishing these two alternatives are in progress.

REFERENCES

1. GRABER, S. & S. B. KRANTZ. 1981. Erythropoietin and the control of red blood cell production. Annu. Rev. Med. **29:** 51.
2. SPIVAK, J. L. 1986. The mechanism of action of erythropoietin. Int. J. Cell Cloning **4:** 139–166.
3. DICKERMAN, H. W., T. C. CHENG, H. H. KAZAZIAN, JR. & J. L. SPIVAK. 1976. The erythropoietic mouse spleen—a model system of development. Arch. Biochem. Biophys. **177:** 1–9.
4. KOURY, M. J., M. C. BONDURANT, D. T. DUNCAN, S. B. KRANTZ & W. D. HANKINS. 1982. Specific differentiation events induced by erythropoietin in cells infected in vitro with the anemia strain of Friend virus. Proc. Natl. Acad. Sci. USA **79:** 635–639.
5. FRIEND, C., W. SCHER, J. G. HOLLAND & T. SATO. 1971. Hemoglobin synthesis in murine virus-induced leukemic cells in vitro: Stimulation of erythroid differentiation by dimethyl sulfoxide. Proc. Natl. Acad. Sci. USA **68:** 378.
6. MARKS, P. A., M. SHEFFERY & R. A. RIFKIND. 1987. Induction of transformed cells to terminal differentiation and the modulation of gene expression. Cancer Res. **47:** 659–666.
7. TAMBOURIN, P., N. CASADEVALL, J. CHOPPIN, C. LACOMBE, J. HEARD, S. FICHELSON, F. WENDLIN, W. D. HANKINS & B. VARET. 1983. Production of erythropoietin-like activity by a murine erythroleukemia cell line. Proc. Natl. Acad. Sci. USA **80:** 6269.
8. HANKINS, W. D., C. EASTMENT & J. SCHOOLEY. 1986. Erythropoietin, an autocrine regulator? Serum-free production off erythropoietin by cloned erythroid cell lines. Blood **68:** 263–268.
9. QIAN, R. L., K. CHIN, J. K. KIM, H. M. CHIN, J. CONE & W. D. HANKINS. 1986. Purification of murine erythropoietin produced in serum-free cultures of erythroleukemia cells. Blood **68:** 258–262.
10. SHOEMAKER, C. & L. D. MITSOCK. 1986. Murine erythropoietin gene: Cloning, expression, and human gene homology. Mol. Cell Biol. **6:** 849–853.
11. HANKINS, W. D., T. A. KOST, M. J. KOURY & S. B. KRANTZ. 1978. Erythroid bursts produced by Friend leukemia virus in vitro. Nature **276:** 506–508.
12. HANKINS, W. D., S. B. KRANTZ, T. A. KOST & M. J. KOURY. 1980. In vitro transformation of mouse hematopoietic cells by the spleen focus-forming virus. Cold Spring Harbor Symp. Quant. Biol. **44** Part 2: 1211–1214.
13. HANKINS, W. D. 1983. Increased erythropoietin sensitivity after in vitro transformation of hematopoietic precursors by RNA tumor viruses. J. Natl. Cancer Inst. **70:** 725–734.
14. HANKINS, W. D. & J. KAMINCHIK. 1984. Modification of erythropoiesis and hormone sensitivity by RNA tumor viruses. Prog. Clin. Biol. Res. **148:** 141–152.
15. OLIFF, A., S. RUSCETTI, E. C. DOUGLAS & E. M. SCOLNICK. 1981. Isolation of transplantable erythroleukemia cells from mice infected with helper-independent Friend murine leukemia virus. Blood **58:** 244.
16. HOSSAIN, A., J. K. KIM & W. D. HANKINS. 1986. Treatment of a fatal transplantable erythroleukemia by procedures which lower endogenous erythropoietin. J. Cell. Biochem. **30:** 311–318.
17. HANKINS, W. D., K. CHIN & G. SIGOUNAS. 1987. Hormone associated therapy of leukemia: Reflections. Prog. Clin. Biol. Res. **262:** 257–267.

Studies of the Constitutive Expression of the Mouse Erythropoietin Gene[a]

NEGA BERU, JEFFREY MCDONALD, AND
EUGENE GOLDWASSER

The University of Chicago
Department of Biochemistry and Molecular Biology
920 East 58th Street
Chicago, Illinois 60637

INTRODUCTION

The IW32 and NN10 cell lines are erythroleukemic cell lines isolated from the spleens of mice infected with two different biologically cloned viruses, the former with a virus derived from a helper-independent ecotropic Friend murine leukemia virus and the latter with a virus derived from a complete Friend virus (anemia strain).[1–3] The cell lines were obtained from Drs. Bruno Varet and Pierre Tambourin of Hôpital Cochin, Paris, France. Both cell lines constitutively secrete erythropoietin (epo),[1–3] the glycoprotein hormone that induces red cell formation in mammals.[4,5] The gross properties of epo from IW32 cells are essentially the same as those of normal mouse plasma epo.[6]

Northern blot analysis using a cloned mouse genomic probe[7,8] showed that these cells contain a normal-size epo mRNA (FIG. 1). A control cell line, IW201, which was derived in the same manner as NN10, but does not secrete epo,[2] showed no detectable message. The studies presented in this report were undertaken to understand the mechanism of constitutive expression of the rearranged mouse epo gene in these cells. We felt that understanding the mechanism of epo gene expression in these cells might shed light on the normal regulation of epo gene expression in response to hypoxia.

The NN10 Cell Line Has a Normal Epo Gene While the IW32 Cell Line Has the Normal As Well As a Rearranged and Amplified Epo Gene

Southern blot analysis (FIG. 2) using DNA isolated from IW32, NN10 and IW201 showed that the epo gene in IW32 cells has undergone rearrangement and amplification while the normal locus was retained,[9] as shown in the *Eco RI, Pvu II* and *Hind III* digested samples. Finding no unique band due to the rearranged epo gene in the *Pst I* and *Bam HI* digested samples suggests that the fragments obtained by these enzymes do not span the rearrangement breakpoint. The NN10 and IW201 cell lines contain only the normal epo locus which is found in normal mouse liver.[9]

Both the normal as well as the rearranged epo genes of IW32 have been cloned,[9] and their restriction maps are shown in FIGURE 3. The rearrangement in IW32 consists of a breakpoint approximately 1 kb upstream of an otherwise structurally unchanged epo gene as well as downstream region.

[a]Supported in part by Grant HL 30121 from the National Institutes of Health.

DNase I Hypersensitivity Studies

Since it was important to ascertain which of the two genes in IW32 is transcriptionally active, we conducted DNase I hypersensitivity studies[9] which showed that the normal gene is in a chromatin structure resistant to DNase I while the rearranged epo gene was in a configuration which was hypersensitive. Because active genes generally have DNase I-hypersensitive chromatin,[10,11] we tentatively concluded that, in IW32, it must be the rearranged gene which is transcriptionally active.

Unlike the normal gene in IW32, the normal gene in NN10 cell line is in a DNase I-hypersensitive state (FIG. 4) in accordance with its transcriptional activity. The epo gene is contained in a 17-kb *Hind III* fragment (FIG. 2). The fragment generated by DNase I treatment is 5.2 kb indicating that the DNase I hypersensitive site is about 1.2 kb upstream of the Xba I site immediately 5′ to the epo gene (FIG. 3). Since it is the rearranged epo gene that is transcriptionally active in IW32 cells and not the normal gene as in NN10 cells, we infer that the mechanism of constitutive epo gene expression is different in these cell lines.

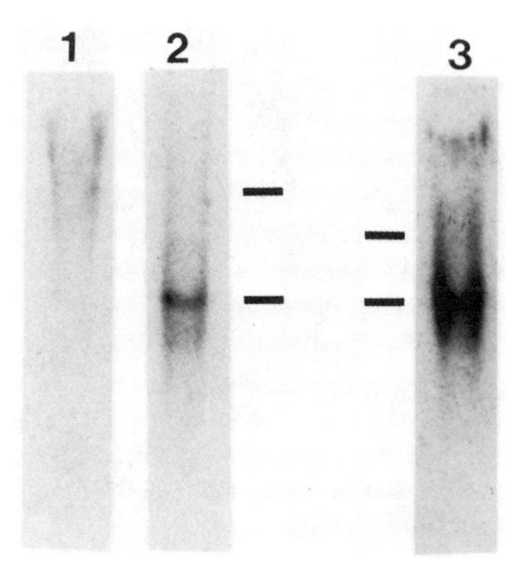

FIGURE 1. Northern blot analysis of poly(A)$^+$ RNA from IW201, IW32 and NN10 cell lines. Twenty μg of poly(A)$^+$ RNA from each cell line was electrophoresed and probed with a ^{32}P-labeled epo probe[8] after being transferred to GeneScreenPlus. After being washed, the filters were exposed to X-ray film for 4 days (*lanes 1, 2*) and 8 days (*lane 3*) at −70°C with intensifying screens. **Lane 1,** cell line IW201; *lane 2,* cell line IW32; **lane 3,** cell line NN10. The *markers* are 18S and 28S ribosomal RNA.

The Viruses Are Not Directly Involved in the Activation of the Epo Gene

Since these cells were virally transformed, it was possible that the virus was directly involved in the epo gene expression observed in IW32 and NN10 cells. One possibility was that the provirus may have inserted near the epo gene thereby bringing the epo gene under the control of viral long terminal repeat, transcriptional regulatory elements. Such activation has been observed for a number of oncogenes.[12–16] This was an especially attractive hypothesis because it could also explain the rearrangement observed 1 kb upstream of the epo gene in IW32 cells. This does not appear to be the case. Probing the λ clone containing the rearranged epo gene with a full length FMuLV viral probe showed no evidence of viral sequences.[9] It should be noted, however, that this λ clone contains only 4.5 kb of DNA upstream and 9.5 kb of DNA downstream of the epo gene (see FIG. 3) and therefore, this experiment does not rule out the possibility of viral enhancers acting from locations further upstream or downstream.

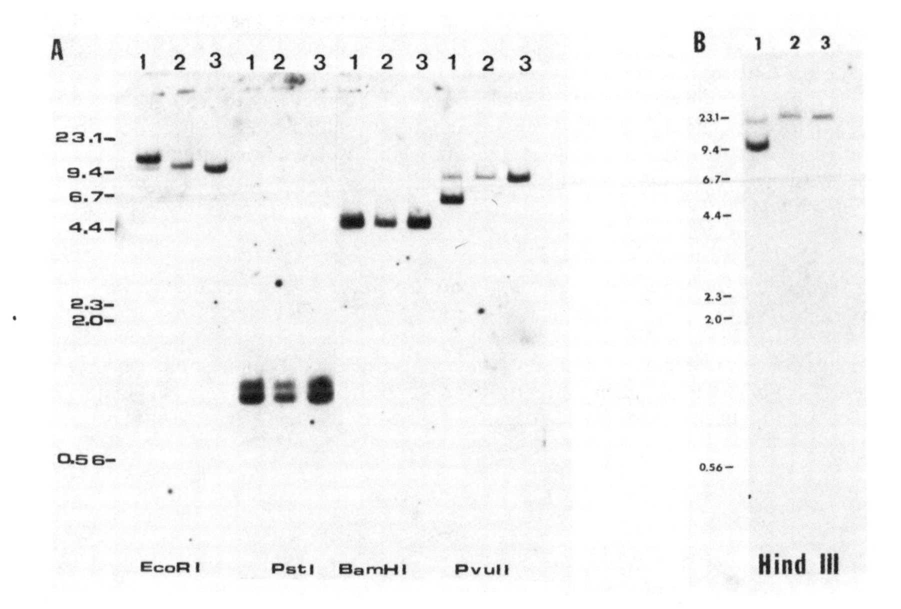

FIGURE 2. Southern blot analysis of DNA isolated from IW201, IW32 and NN10 cell lines. Approximately 10 μg of high molecular weight DNA isolated from the three cell lines was digested with the indicated restriction enzymes. After electrophoresis and blotting onto GeneScreenPlus, the filters were probed with a ^{32}P-labeled epo probe[8] (**A**) or a ^{32}P-labeled monkey cDNA probe (**B**). *Lane 1,* cell line IW32, *lane 2,* cell line IW201; *lane 3,* cell line NN10. There was some variation in the amount of DNA loaded per lane. The difference is most notable in the *Pvu II*-digested samples between lanes 1 and 2 on the one hand and lane 3 on the other. Similarly, in the *Eco RI*-digested samples, lane 1 had considerably less DNA than lanes 2 and 3. The molecular weight *markers* are in kilobases.

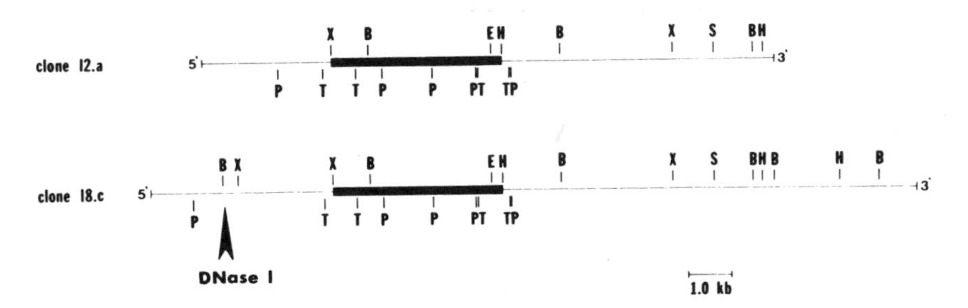

FIGURE 3. The restriction maps of the two epo loci of IW32 cell line. Clone 12.a represents the normal epo gene while clone 18.c represents the rearranged and amplified epo gene. The entire coding region lies within the *heavy line* which has been sequenced.[7,23] B, *Bam HI;* E, *Eco RI;* H, *Hind III;* P, *Pst I;* S, *Sal I;* T, *BstX I;* X, *Xba I.* The DNase I hypersensitive site upstream of the rearranged and amplified gene is indicated.

The mechanism of epo gene activation in NN10 cells also probably does not involve proviral integration near the epo gene. We have cloned the epo gene from NN10 cells and we do not find viral sequences within 19 kb upstream or 5 kb downstream of the epo gene using the full length viral probe. The second possibility was that the epo gene may have been transduced into the retroviral genome and thereby activated as has been described for C-myc.[17–19] However, probing of viral genomic RNA isolated from virions pelleted from the IW32 cell culture medium with an epo probe failed to show any epo sequences in the viral genome.[9]

Sequencing and Sequence Analysis of the Upstream Regions of the Normal and Rearranged Epo Genes of IW32 Cell Line

It is conceivable that in IW32, a translocation or deletion event near the epo gene may have brought the gene under formerly distant, positive transcriptional, regulatory elements or may have removed local negative regulatory elements. In order to answer this question, the upstream regions of both epo loci (normal and rearranged) were subcloned into the single-stranded phage M13 and sequenced by the chain termination method of Sanger[20] after a sequential series of overlapping clones was obtained.[21] We used the specific-primer-directed DNA sequencing method[22] to resolve ambiguous sequences. Sequence compilation and analysis as well as data base searches were done using software obtained from IntelliGenetics, Inc., Mountainview, CA.

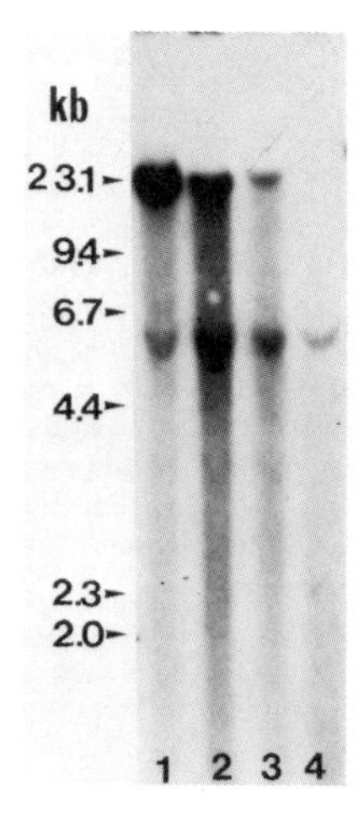

FIGURE 4. DNase I hypersensitivity of the epo gene in the NN10 cell line. Nuclei were isolated from the NN10 cell line and treated with increasing amounts of DNase I for 2 minutes at room temperature, after which the reaction was stopped and DNA isolated. Ten μg of DNA was digested with *Hind III* and electrophoresed on a 0.6% agarose gel and blotted onto GeneScreenPlus. The probe used was a mouse genomic probe encompassing exons 2, 3, 4 and part of 5.[8] DNase I concentrations used were 2 (*lane 1*), 4 (*lane 2*), 6 (*lane 3*) and 10 (*lane 4*) μg/ml. The fragment generated due to DNase I digestion is 5.2 kb in length.

The rearrangement breakpoint is 530 bp upstream of the *Xba I* site immediately 5′ to the epo gene (see FIG. 5) or 930 bp upstream of the proposed start site of transcription.[23] Sequence analysis shows that the upstream region of the normal gene contains two B1 repeat elements, the murine homologue of the human *Alu* type 1 family, 1.8 and 2.5 kb upstream of the *Xba I* site and one B2 repeat element 1 kb upstream of the *Xba I* site. These B repeat elements are missing in the upstream region of the rearranged gene; instead, the new translocated sequence contains one B1 repeat element 3.6 kb upstream of the *Xba I* site. Comparison of these B repeat elements with the consensus B1 and B2 repeat elements[24,25] shows a high degree of sequence conservation.

It is noteworthy that the 3-kb upstream region of the normal epo gene contains two B1 repeat elements. Together with a third B1 repeat element in the 3′ untranslated region of the mouse gene,[7] these represent an unusually high concentration of B1 repeat elements around the epo gene, considering that since there are 1–5 × 10^5 copies of B1 repeat elements per haploid genome in the mouse, the expected distribution is one repeat element per 6–10 kb of DNA.[25] Clustering of *Alu* repeats has been implicated in genome rearrangements and amplification of rearranged fragments.[26] The clustering of B1 repeat elements around the epo gene could therefore be a contributing factor to the presence of the rearranged and amplified locus in the hypotetraploid genome[2] of the IW32 cell line.

Possible Mechanisms of Constitutive Expression of the Epo Gene

In an attempt to explain the DNase I hypersensitive site upstream of the rearranged epo gene[9] and activation of the epo gene, we subjected this sequence to a promoter search by computer. Indeed, a region in the far upstream sequence of the rearranged gene was found which contained sequences (*i.e.*, "CCAAT" and "TATA" boxes, start site of transcription, ribosome binding site as well as an open reading frame) needed for a gene. This gene was in the complementary strand to that coding for the epo gene (FIG. 5).

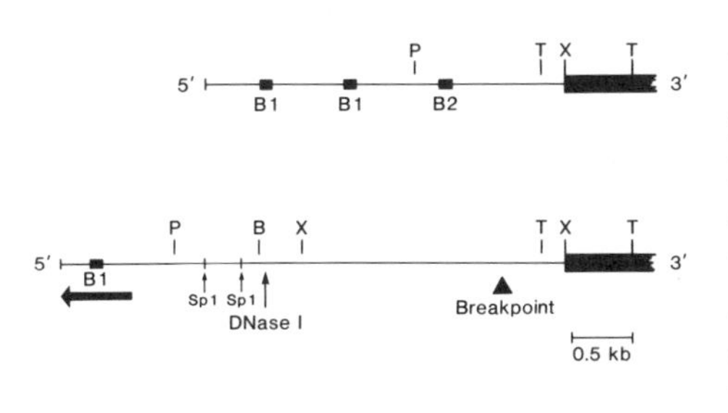

FIGURE 5. The upstream regions of the normal (*top*) and rearranged (*bottom*) epo genes of IW32 cells showing the results of sequence analysis. The *triangle* shows the rearrangement breakpoint and the DNase I hypersensitive site in the upstream region of the rearranged gene[9] is indicated. B1, B1 repeat element; B2, B2 repeat element; SP1, human transcriptional factor *Sp1* binding hexanucleotide. *Leftward arrow* shows the location of the new gene starting at the computer-predicted start site of transcription. Restriction enzyme abbreviations are as in FIGURE 3.

It would appear then that the epo gene at the rearranged locus may have been activated, at least in part, because a translocation event has brought a new gene on the complementary strand with its DNase I hypersensitive site into close proximity to it. This notion is supported by the presence of two *Sp1* transcriptional factor binding hexanucleotides (GGGCGG) in the upstream region of the "new gene" (590 and 900 bp upstream of the start site of transcription as predicted by computer). As few as one of these elements has been shown to be sufficient to specify a strong binding site for the human transcriptional factor *Sp1*.[27,28]

Only the 5′ end of the new gene, containing the code for about 30 amino acid residues, is contained in the λ clone. Using the available sequence we have been unable to find homology to any known gene in the rodent files of Genbank (NIH DNA sequence library) as well as the MBL library (European Molecular Biology Laboratories). Similarly, a search of the NBRF (National Biomedical Research Laboratories) or Swiss Prot (European Molecular Biology Laboratories) protein data banks using the translated product of the open reading frame failed to find any identity. Efforts are underway to identify messenger RNA transcribed from this gene and to clone the complete gene in order to fully characterize it.

Understanding the mechanism of constitutive epo gene expression in IW32 cell line may not tell us much about how expression of the epo gene is normally regulated. However, elucidation of how the NN10 cell line epo gene has been turned on toward constitutive expression may do so. The mechanism of constitutive epo gene expression in the NN10 cell line must be different because in this cell line there is no translocation-mediated activation. There is no rearranged and amplified epo gene as in IW32 cell line. One possibility is that in the NN10 cell line a transacting factor has been expressed which, in turn, has activated the epo gene. We are currently studying this possibility by DNA footprinting and retardation assays using the promoter region of the gene and NN10 nuclear extracts.

SUMMARY

The IW32 and NN10 cell lines are erythroleukemic cell lines which secrete erythropoietin (epo) into the culture medium constitutively. IW32 cells have a rearranged and amplified epo gene in addition to the normal gene. NN10 cells contain only the normal epo gene. DNase I hypersenitivity studies suggested that, in IW32 cells, it is the rearranged gene which is transcriptionally active. Sequence analysis of the upstream region of the rearranged epo gene suggests that the rearrangement has served to introduce a transcriptionally active gene close to the epo gene. This juxtaposition may explain the transcriptional activation of the epo gene at the rearranged locus.

REFERENCES

1. TAMBOURIN, P., N. CASADEVALL, J. CHOPPIN, C. LACOMBE, J. M. HEARD, S. FICHELSON, F. WENDLING & B. VARET. 1983. Proc. Natl. Acad. Sci. USA **80:** 6269–6273.
2. CHOPPIN, J. C., N. CASADEVALL, C. LACOMBE, F. WENDLING, E. GOLDWASSER, R. BERGER, P. TAMBOURIN & B. VARET. 1985. Exp. Hematol. **13:** 610–615.
3. CHOPPIN, J. C., C. LACOMBE, N. CASADEVALL, O. MULLER, P. TAMBOURIN & B. VARET. 1984. Blood **64:** 341–347.
4. GOLDWASSER, E. 1976. Fed. Proc. **34:** 2285–2292.
5. GOLDWASSER, E., S. KRANTZ & F. F. WANG. 1985. *In* Mediators in Cell Growth and Differentiation. R. J. Ford & A. L. Maizel, Eds.: 103–107. Raven Press. New York, NY.
6. CHOPIN, J., J. EGRIE, N. CASADEVALL, C. LACOMBE, O. MULLER & B. VARET. 1987. Exp. Hematol. **15:** 171–176.
7. MCDONALD, J., F.-K. LIN & E. GOLDWASSER. 1986. Mol. Cell. Biol. **6:** 842–848.
8. BERU, N., J. MCDONALD, C. LACOMBE & E. GOLDWASSER. 1986. Mol. Cell. Biol. **6:** 2571–2575.
9. MCDONALD, J., N. BERU & E. GOLDWASSER. 1987. Mol. Cell. Biol. **7:** 365–370.
10. WEINTRAUB, H. & H. GRONDINE. 1976. Science **193:** 848–853.
11. AFFARA, N., J. FLEMMING, P. S. GOLDFARB, E. BLACK, B. THIELE & P. R. HARRISON. 1985. Nucleic Acids Res. **13:** 5629–5644.
12. HAYWARD, W. S., B. G. NEAL & S. M. AUSTIN. 1981. Nature(London) **290:** 475–480.
13. NOORI-DALOII, M. R., R. A. SWIFT, H. J. KUNG, L. B. CRITTENDEN & R. L. WITTER. 1981. Nature(London) **294:** 574–576.
14. NUSSE, R. & H. E. VARMUS. 1982. Cell **31:** 99–109.
15. FUNG, Y. K. T., W. G. LEWIS, L. B. CRITTENDEN & H. J. KUNG. 1983. Cell **33:** 357–368.
16. VAN OOYEN, A. & R. NUSSE. 1984. Cell **39:** 233–240.
17. LEVY, L. S., M. B. GARDNER & J. W. CASEY. 1984. Nature(London) **308:** 853–856.
18. MULLINS, J. I., D. S. BRODY, R. C. BINARI, JR. & S. M. COTTER. 1984. Nature(London) **308:** 856–858.

19. NEIL, J. C., D. HUGHES, R. MCFARLANE, N. M. WILKIE, D. E. ONIONS, G. LEES & O. JARRETT. 1984. Nature(London) **308:** 814–820.
20. SANGER, F., S. NICKLEN & A. R. COULSON. 1977. Proc. Natl. Acad. Sci. USA **74:** 5463–5467.
21. DALE, R. M. K., B. A. MCCLURE & J. P. HOUCHINS. 1985. Plasmid **13:** 31–40.
22. STRAUSS, E. C., J. A. KOBORI, G. SIU & L. E. HOOD. 1986. Anal. Biochem. **154:** 353–360.
23. SHOEMAKER, C. B. & L. D. MISTOCK. 1986. Mol. Cell. Biol. **6:** 849–858.
24. KRAYEV, A. S., T. V. MARKUSHEVA, D. A. KRAMEROV, A. P. RYSKOV, K. G. SKRYABIN, A. A. BAYER & G. P. GEORGIEV. 1982. Nucleic Acids Res. **10:** 7461–7475.
25. PROSPT, F. & G. F. V. WOUDE. 1984. Nucleic Acids Res. **12:** 8381–8392.
26. CALABRETTA, B., D. C. ROBBERSON, H. A. BARRERA-SALDANA, T. P. LAMBRON & G. F. SAUNDERS. 1982. Nature(London) **296:** 219–225.
27. GIDONI, D., S. D. WILLIAM & R. TIJAN. 1984. Nature **312:** 409–413.
28. GIDONI, D., J. T. KADONAGA, H. BARRERA-SALDANA, K. TAKAHASHI, P. CHAMBON & R. TIJAN. 1985. Science **230:** 511–517.

Molecular Mechanism for the Inhibitory Action of Interferon on Hematopoiesis

DONALD ORLIC, ROBERT GILL, ROXANNE FELDSCHUH, FEDERICO QUAINI, ANDREA MALICE, AND CLAUDIO SANDOVAL

Department of Cell Biology and Anatomy
New York Medical College
Basic Sciences Building
Valhalla, New York 10595

INTRODUCTION

Several recent papers[1,2] demonstrate a link between aplastic anemia and high titers of interferon (IFN) in the circulation and bone marrow. Endogenous IFNs are produced by fibroblasts (IFNβ) and leukocytes (IFNα) including virus-activated lymphocytes (IFNγ), and when present in bone marrow sera they constitute a microenvironmental factor. Interferons function as biological response modifiers that inhibit virus reproduction and cellular DNA and protein synthesis. Their inhibitory effect on protein synthesis[3–6] and cell proliferation[4,5,7,8] occurs through the induction of several enzyme systems that require dsRNA for activation and use ATP as the sole substrate.[4,6,9–11] One pathway involves 2′-5′adenylate synthetase (2-5AS) which generates 2-5A, a potent translational inhibitor that activates endonuclease. This pathway, referred to collectively as the 2-5AS system, attacks both free and polysome-bound mRNA.[12–15]

Several studies have shown that IFN inhibits Friend erythroleukemic cells[16] and mouse marrow CFU-E and human BFU-E growth.[17] Our earlier findings suggested that the increase in erythropoiesis in phenylhydrazine-injected rats correlated directly with an increase in 2-5AS activity in marrow and spleen of adult rats.[18] The level of 2-5AS activity was determined for bone marrow and spleen cells by assaying cell lysates for the conversion of labeled ATP to 2-5A. By contrast, rats made polycythemic by hypertransfusion showed a decrease in erythropoiesis and a decrease in 2-5AS activity. In another study we utilized hypobaric hypoxia, a physiologic stimulus for erythropoiesis, and observed the highest 2-5AS activity at the time of peak erythropoiesis and the lowest 2-5AS activity during the period of decreased erythropoiesis that occurred in the posthypoxic, polycythemic mice.[19]

Early stages of erythropoiesis are characterized by extensive cell proliferation and hemoglobin synthesis and it seems unlikely that there would normally be any 2-5AS activity in proerythroblasts and basophilic erythroblasts. However, in late erythroblasts there are decreases in RNA content, protein synthesis and cell division. For this reason, we proposed that the normal onset of 2-5AS activity occurs during late erythroblast maturation. According to this hypothesis, 2-5A-activated endonuclease degrades erythroblast mRNA thus initiating a block to further globin synthesis and cell division. As the number of late erythroblasts increases with hypoxia and decreases with polycythemia, the rate of 2-5AS activity increases and decreases proportionally.

The red cell series seems well-suited to studies on the mechanism of IFN action.

Exposure of proerythroblasts and/or basophilic erythroblasts to IFN may induce early activation of the 2-5AS system and subsequent inhibition of erythropoiesis. The data reported in this paper support this hypothesis.

In recent *in vitro* studies, human leukocyte IFNα (HuIFNα) was tested for its effect on growth of human marrow cells. Both the colony forming unit-erythroid and -megakaryocyte (CFU-E and CFU-Meg) were inhibited.[20] No data were presented for the mechanism by which HuIFNα inhibited human hematopoietic colony growth. However, others were able to demonstrate that HuIFNα induced HeLa cells to produce a 33-kD 2-5AS and a 110-kD 2-5AS.[21] These two forms of 2-5AS, but especially the larger molecule, are dependent on dsRNA (poly I:poly C) for activation and expression of their inhibitory property.

In recent clinical trials, HuIFNα was found to induce 2-5AS in peripheral mononuclear cells[22] and to have a cytotoxic effect on hematopoiesis as seen by an induced leukopenia, thrombocytopenia and anemia.[23] Daily injections of HuIFNα resulted in a normocytic, normochromic anemia which required weeks to months for recovery after discontinuation of injections. This suggested an interference with erythropoiesis.[23,24]

Recombinant human leukocyte IFNαA and αD (rHuIFNαA and αD) subtypes are both derived from the human myeloblast cell line KG-1. The rHuIFNαA subtype which is in extensive clinical trials in several oncologic disorders is not active in the murine system and rHuIFNαD is only moderately active in mice. However, the hybrid form, rHuIFNαA/D, is highly active in mice. This hybrid gene constructed using plasmid cleavage with the restriction endonuclease Bgl II and recombination, is expressed in *E. coli*. Analysis of rHuIFNαA/D indicates that the N-terminus 1 through 61 amino acids are derived from the alpha A gene and the subsequent 62 through 165 amino acids from the alpha D gene according to the following scheme:

1. IFNαA and IFNαD are derived from a human myeloblast cell line KG-1.
2. IFNαA/D gene is expressed in *E. coli* and active in murine cells.
3. IFNαA/D amino acid sequence. 1 61 62 165 A D
4. IFNαA/D is active in murine system.

Pharmacokinetic parameters and tissue distribution of rHuIFNαA/D were assessed in CD-1 mice after an intravenous injection of 1–2 × 10^6 units. Serum concentration-time data indicated a mean elimination half-life of 33 minutes and the highest units rHuIFNαA/Dper gram tissue were found in the kidney.[25] Additional studies using CD-1 mice indicated that a single injection of 1 μg of protein of rHuIFNαA/D conferred protection against encephalomyocarditis virus infection[26] and in a dose-dependent manner protected against Semliki Forest virus and Herpes Simplex Type 2 virus.[27]

From the above it is clear that rHuIFNαA/D is effective in CD-1 mouse cells and, because of the availability of rHuIFNαA/D in large quantities, we used this type IFN as well as murine IFNα for the *in vivo* and *in vitro* studies reported here on the regulation of hematopoiesis.

MATERIALS AND METHODS

Percoll Separation of Spleen Cells

Adult BALB/c mice were injected with 60 mg/kg phenylhydrazine (PHZ) on days 1, 2 and 4 and spleens were removed on day 5. Spleen cell suspensions were prepared as outlined below. The cell suspensions from 3 spleens were pooled for each experiment and were diluted in PBS to a concentration of approximately 1 × 10^9 cells/2.0 ml solution.

Percoll solutions were prepared according to published procedures.[28] Stock solutions of 10 times the standard strength of PBS (10 × PBS) and of 1% (w/v)$CaCl_2$/1% (w/v) $MgCl_2$ (Ca/Mg) were stored at −20°C. The osmolarity of the Percoll supplied was adjusted by the following procedure. For each gradient, 20 ml of Percoll + 1.5 ml of the 10 × PBS + 0.2 ml of Ca/Mg solution was adjusted to pH 7.2 by the dropwise addition of 0.01 M-HCl with rapid stirring. Then the osmolarity was adjusted to 295–310 mosmol/lby the dropwise addition of water or NaCl. This solution is by definition 100% Percoll and was diluted to make 10 ml gradient fractions of 30%, 40%, 50% and 60% Percoll by using PBS solution (10 ml of 10 × PBS + 1 ml of Ca/Mg solution per 100 ml).

Spleen cells were separated on discontinuous Percoll gradient in the following manner. A five-step Percoll gradient was produced by sequentially layering 2.0 ml of each gradient fraction into the bottom of a 15-ml centrifuge tube from a 5-ml syringe, under gravity flow. Six tubes were prepared for each experiment. Spleen cells were loaded, after thorough resuspension, onto the top of the gradient in 2.0 ml of PBS medium and the gradient centrifuged at 3000 g for 30 min. Fractions (2 ml) were collected by hand from the top of the gradient by using a fired Pasteur pipette. The positions of the erythroblast and myeloid blast cells at the interfaces of the gradient can generally be readily determined visually and the appropriate 6 fractions pooled. To remove Percoll, the combined fractions were diluted 20-fold with PBS medium and the cells pelleted at 300 g for 5 min. The cell fractions were resuspended in media and counted. A suspension of 20 × 10^6 cells from each layer were spun down and resuspended in buffer A (200 μl). The fractions were frozen at −80°C until the 2-5AS enzyme assay was done (see below for methodology).

In Vitro *Interferon Studies on BALB/c Spleen Cells*

Mouse interferon α and β (2.8 × 10^5 units/mg protein, Sigma Chemical Co., St. Louis, Mo.) was stored at −20°C in NCTC at a dilution of 22,000 U/0.1 ml. A final dilution of 10,000 U in 0.4 ml NCTC was used in the cell cultures described below.

Adult male BALB/c mice were injected with PHZ on days 1, 2 and 4 and killed on day 5. A spleen cell suspension was prepared from 3 pooled spleens for each experiment. From the cell suspension a total of 30 × 10^6 nucleated cells were maintained in liquid culture medium for 8 hr at 37°C in 5% CO_2. Each 1 ml culture medium contained 0.4 ml NCTC, 0.1 ml cells, 0.1 ml FCS and 0.4 ml IFN in NCTC (controls contained 0.4 ml NCTC without IFN). The cells were collected and lysed and the lysate was applied to a poly I:poly C agarose column. The absorbed 2-5AS was assayed for conversion of ATP to 2-5A (see below for methodology).

In another series of experiments, spleen cell suspensions were centrifuged on Ficoll-Hypaque at 400 g for 30 min in order to separate the lymphocytes from the erythroblasts and reticulocytes. A total of 20 × 10^6 nucleated cells from the pellet were incubated in liquid culture medium as described above and assayed for 2-5AS activity.

2-5AS Assay

Single cell suspensions were prepared from spleens of mice. An aliquot of the cell suspension in the test tube, equivalent to 100 × 10^6 nucleated cells, was transferred to a 1.5-ml microcentrifuge tube and centrifuged at 15,600 g for one min and the supernatant was removed.

A modified version of the method of Justesen *et al.*[11] for determining 2-5AS activity was employed on the cells in the pellet. To each microcentrifuge tube of 100 × 10^6 cells was added 1 ml of Buffer A (10 mM Hepes, 90 mM KCl, 1.5 mM Mg $(OAc)_2$, 1 mM dithiothreitol (DTT), 10% glycerol and 0.5% NP-40) followed by vortexing. Each microcentrifuge tube was submersed in liquid nitrogen for 20 sec followed by incubation in a 37°C water bath for 10 min. This regimen of vortexing, freezing and thawing was done 3 times to insure breakup of cells into a lysate.

An aliquot of lysate equivalent to 10 $\times$ 10^6 cells (100 μl) was transferred to a microcentrifuge tube. The control consisted of 100 μl Buffer B (20 mM Hepes, 2 mM Mg $(OAc)_2$, 1mM DTT, 10% glycerol, 100 mM KCL). A 40 μl volume of poly I:poly C-agarose (PL Biochemical, Piscataway, NJ) was added to each microcentrifuge tube followed by vortexing and a 15 min incubation at 20°C. This incubation allowed the poly I:poly C-agarose to specifically bind 2-5A synthetase if it was present in the lysate.

The poly I:poly C-agarose lysate suspension was washed 4 times with 1 ml aliquots of Buffer B. Following the last wash and removal of the supernatant, 40 μl of a master mix consisting of 20 μl Buffer C (80 mM Tris HCL, 80 mM Mg $(OAc)_2$, 4 mM DTT, 0.4 mg/ml bovine serum albumin), 4 μl (8^{-14}C) ATP, specific activity 51 mCi/mM, ICN Lot #1497113, 8 μl 30 μg/ml poly I:poly C in Buffer B solution and 8 μl 50 mM ATP was added to each microcentrifuge tube which was vortexed, centrifuged, and incubated for 3 hr at 37°C in a water bath.

After 3 hr of incubation the tubes were vortexed and centrifuged, and 30 μl of supernatant from each was transferred to another set of microcentrifuge tubes. A 7-μl volume of a 300-mM glucose/300 units/ml hexokinase solution was added to each tube. They were then vortexed and incubated 5 min at 37°C in a water bath.

After centrifugation, 10 μl of supernatant from each microcentrifuge tube was placed on a PEI cellulose plate (Sybron Brinkman, Westbury, NY) and allowed to dry. Migration in 1.0 M LiCl took approximately 2 hr. When the migration front reached 2 cm from the top of the PEI cellulose plate, the plate was removed from the chromatography chamber and examined under UV light. The radioactive 2-5A and ADP spots were quantitated by scintillation counting in Dimiscent (National Diagnostics, Somerville, NJ).

The average cpm in our experiments were approximately 20,000. Thus 0.196 nM (8^{-14}C) ATP was used per tube. The amount of nonradioactive ATP per tube in 8 μl of 50 mM ATP was 400 nM. Since the radioactive ATP (0.196 nM) was negligible in comparison to the nonradioactive ATP (400 nM), for our calculations the amount of ATP used at the start of the experiment was considered to be 400 nM. The percent conversion of ATP to 2-5A during the 3-hr incubation was calculated as follows: cpm 2-5A/(cpm 2-5A + cpm ADP) $\times$ 100 = percent conversion of ATP to 2-5A in 3 hr. To determine the nM of ATP converted to 2-5A during a 3 = hr incubation, it is necessary to obtain the product of the percent ATP converted to 2-5A during a 3-hr incubation (from above) and the quantity of ATP at the start of the experiment, which was 400 nM. The 2-5AS activity can thus be determined for a 10 $\times$ 10^6 cell concentration.

Animal Model for In Vivo *Studies*

Four-week-old female CD-1 mice, 18-19g, (Charles River Labs, Kingston, NY) were used in each experiment. The mice were housed six per cage in a viral-free environment. Water and food pellets were provided *ad libitum*. Mean body weights were determined prior to injection and on day of sacrifice.

Interferon Procedures

Recombinant human interferon alpha A/D (generously supplied by Hoffman-LaRoche, Nutley, NJ) was stored in concentrated form (50 $\times$ 10^6 units (U)/0.5 ml) in

0.25 M ammonium acetate. The interferon was diluted to aliquots of 1×10^6 U/ml in normal saline and frozen until used. On the day of injection 0.2 ml of 1×10^6 U/ml solution was injected intraperitoneally (IP) into "Hi Dose IFN" mice (corresponding to 10^7 U/kg or approximately 2×10^5 U/mouse). An aliquot of this solution was further diluted to 1×10^5 U/ml in normal saline and 0.2 ml of this solution was injected IP into "Lo Dose IFN" mice (10^6 U/kg or 2×10^4 U/mouse). Control mice were injected IP with 0.2 ml normal saline. Mice were injected for 7 or 14 days and sacrificed on day 8 or 15, respectively.

Cardiac Puncture and Blood Cell Procedures

Cardiac puncture was performed under ether anesthesia. Between 0.5 and 1.0 ml of blood was obtained routinely, using a heparinized syringe fitted with a 25-gauge needle. Aliquots of blood were used for determination of platelet counts and volume. Hematocrits were determined for each mouse.

Morphology Procedures

Mice were killed by cervical dislocation under ether anesthesia following cardiac puncture. The spleen was removed and weighed.

Light Microscopy. Half of the spleen was cut into two or three pieces and used to prepare sections for light microscopy as follows. Tissues were fixed in 3% glutaraldehyde in phosphate buffer, dehydrated in a graded ethyl alcohol series: 50%, 70%, 95%, 100% (2×)—1 hr each, embedded for JB-4 (methylmethacrylate), sectioned at 4 μmeters, and stained with Gills hematoxylin, triple strength.

Electron Microscopy. Four to five 1-mm^3 pieces were used to prepare sections for electron microscopy as follows. Tissues were fixed in 3% glutaraldehyde in phosphate buffer, followed by 1% osmium tetroxide in 0.1 N phosphate buffer, dehydrated in a graded ethyl alcohol series, followed by a graded propylene oxide (Prop Ox) series consisting of 50%, 70%, 95% EtOH (30 min each), 100% EtOH (3×10 min each), 1:1 EtOH:Prop Ox (15 min), 1:3 EtOH:Prop Ox (15 min), Prop Ox (2×20 min each), infiltrated in Embed 812 (Electron Microscopy Sciences, Fort Washington, PA) as a mixture of 1:1 Prop Ox: Embed 812 for 2 hr, followed by 1:3 Prop Ox: Embed 812 overnight, then embedded in pure Embed 812 at 60°C. Sections were stained with uranyl acetate and lead citrate. Sections were photographed using a JEOL 100C electron microscope.

Spleen Cell Suspensions

Half of the spleen was passed through a stainless steel 60 mesh (Becton-Dickinson, Rutherford, NJ) into 5 ml alpha minimal essential medium (αMEM) without nucleotides (Gibco, Grand Island, NY). Cells were passed through a transfer pipette 20 times, then through an 18-gauge needle 20 times to break up cell clumps. One hundred μl of cell suspension was diluted with 300 μl αMEM, and cell counts of nucleated spleen cells were performed using a hemocytometer.

RESULTS AND DISCUSSION

In Vitro *Studies Using Percoll Separated Erythroblasts*

2-5AS activity was present in Percoll separated erythroblasts from spleen cell suspensions. Adult male BALB/c mice were injected with PHZ and their highly erythropoietic spleens were used to obtain cell suspensions for Percoll separation. Bands were formed at the interfaces of 30, 40, 50, 60 and 100% Percoll and a pellet was also obtained. The cells of these 6 samples were assayed in 6 separate experiments and results are seen in FIGURE 1.

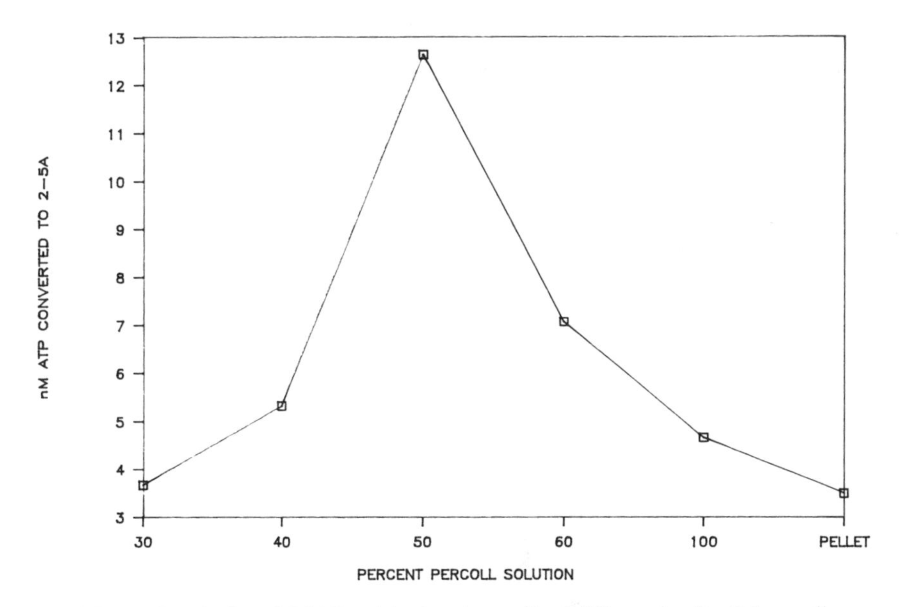

FIGURE 1. Quantitation of 2-5AS activity in spleen cells of different density. Spleen cells were separated into 5 bands and a pellet using discontinuous Percoll solutions. For the enzyme assay, 20×10^6 cells were used from each band and pellet. 2-5AS activity was measured on the basis of nM ATP converted to oligomers of 2-5A during a 3-hr incubation. There were 4 to 6 values for each point on the curve. The same trend occurred in each of the 6 experiments but with large variations. Because of these large variations and the small number of samples, no statistical analysis was done.

The density of polychromatophilic and orthochromatic erythroblasts is thought to be between 1.070 and 1.081 g/ml.[29] This suggests that these cells would be stopped at the 50 and 60% Percoll bands. In these preparations the highest number of nucleated cells were found in the 50% Percoll band and the highest enzyme activity occurred in the 50 and 60% Percoll bands. Reticulocytes and erythrocytes (density of 1.084 to 1.092 g/ml) were found mostly in the 100% band and pellet.

The levels of 2-5AS activity present in the erythroblast fractions obtained above were not induced by exogenous interferon. We consider this "native" 2-5AS activity to be a normal regulator of late erythroblast maturation by blocking cell division and hemoglobin synthesis.

In Vitro *Studies Using Mouse IFNα and β*

Our studies (TABLE 1) with BALB/c mice show an increase in 2-5AS activity in spleen cells cultured with IFN. When 30 × 10^6 nucleated cells from unseparated cell suspensions obtained from highly erythropoietic spleens of PHZ-injected mice were cultured with 10,000 U of mouse fibroblast IFNα and β for 8 hr, the 2-5AS activity in these highly erythropoietic cell populations increased 70%. In 11 separate experiments the nM ATP converted to 2-5A increased from 3.96 ± 0.30 (mean ± SEM) in control samples to 6.73 ± 0.58 (mean ± SEM) in IFN-treated samples. The paired Student t test on these 11 experiments indicates a value of $p < 0.0001$.

In another study using PHZ-injected BALB/c mice, spleen cell suspensions were separated using Ficoll-Hypaque density gradient centrifugation. The cells of the pellet were cultured at a ratio of 20 × 10^6 per 1 ml liquid culture medium with or without murine IFNα and β. Exposure of the cells to IFN for 8 hr resulted in an increase in 2-5AS activity of 38.1% above control values. This increase in 2-5AS activity seen in TABLE 2 is thought to be the result of IFN induction of the enzyme in erythroblasts. Lymphocytes were removed from the samples by Ficoll separation and any granulocytes present would not respond to IFN with synthetase activity.

TABLE 1. Effect of Murine IFNα and β on 2-5AS Activity in Unseparated Mouse Spleen Cell Suspensions[a]

Exp. No.	No. of cells per Sample	nM ATP to 2-5A: Control	nM ATP to 2-5A: 10,000 U IFN	IFN-Induced Increase in 2-5A Synthesis
1	30 × 10^6	2.72	8.56	215%
2	30 × 10^6	4.08	5.28	29%
3	30 × 10^6	5.95	10.72	80%
4	30 × 10^6	3.22	6.87	113%
5	30 × 10^6	3.26	6.50	99%
6	30 × 10^6	3.94	6.09	55%
7	30 × 10^6	5.35	7.95	49%
8	30 × 10^6	3.27	6.12	87%
9	30 × 10^6	4.01	4.95	23%
10	30 × 10^6	4.55	7.52	65%
11	30 × 10^6	3.21	3.52	10%
Mean		3.96[b]	6.73	70%
± SEM		± 0.30	± 0.58	—

[a]The spleens were highly erythropoietic following induction of anemia by 3 injections of phenylhydrazine. Cells were incubated for 8 hr in liquid culture medium.

[b]Student t test (paired) indicated $p < 0.0001$.

In Vivo *Studies Using rHuIFNαA/D*

Effects of rHuIFNαA/D on body weight, spleen weight and spleen cellularity were observed. Daily injections of Lo Dose (10^6 U/kg) or Hi Dose (10^7 U/kg) rHuIFNαA/D did not alter the growth pattern of these young adult mice during the 2 week experimental period suggesting that there was no systemic toxicity. As seen in FIGURE 2, saline-injected and Lo Dose IFN-injected mice were virtually the same weight at weeks 1 and 2. Mice receiving Hi Dose IFN injections for 2 weeks gained 3.8 grams and although they

TABLE 2. Effect of Murine IFNα and β on 2-5AS Activity in Ficoll-Separated High Density Cells from Pellets of Mouse Spleen Cell Suspensions[a]

Exp. No.	No. of Cells per Sample	nM ATP to 2-5A		IFN-Induced Increase in 2-5A Synthesis
		Control	10,000 U IFN	
1	20×10^6	4.2	6.2	52%
2	20×10^6	6.9	14.0	103%
3	20×10^6	19.9	25.8	30%
4	20×10^6	11.6	10.8	−7%
5	20×10^6	23.7	31.4	32%
6	20×10^6	9.2	16.2	76%
Mean		12.6[b]	17.4	48%
± SEM		± 3.1	± 3.8	

[a]Cells were incubated for 8 hours in liquid culture medium.
[b]Student *t* test (paired) indicated $p < 0.02$.

weighed nearly 1 gram less than controls, this was not a significant difference. However, there was a major increase in spleen wet weight after 7 and 14 days of rHuIFNαA/D injections as seen in FIGURE 3. After 14 daily injections, the mean splenic weight in mg was 100 ± 0.005 SEM for control mice, 126 ± 0.007 SEM for Lo Dose IFN-injected mice, and 172 ± 0.015 SEM for Hi Dose IFN-injected mice. The Lo and Hi Dose IFN groups were significantly different from each other and from the control group (Scheffe Multiple Comparison Test). The 72% increase in spleen weight in Hi Dose IFN mice versus the control mice was not due to any change in splenic hemodynamics such as blood retention. Instead, this 72% increase was the result of an increase in the absolute number of nucleated cells per spleen (TABLE 3). The total number of nucleated spleen cells at 1 week increased from 186.3×10^6 in control mice to 245.6×10^6 in Hi Dose IFN-injected mice. This was a 32% increase. After 2 weeks of daily injections the total number of nucleated cells increased from 154.5×10^6 per control spleen to 225.1×10^6 per

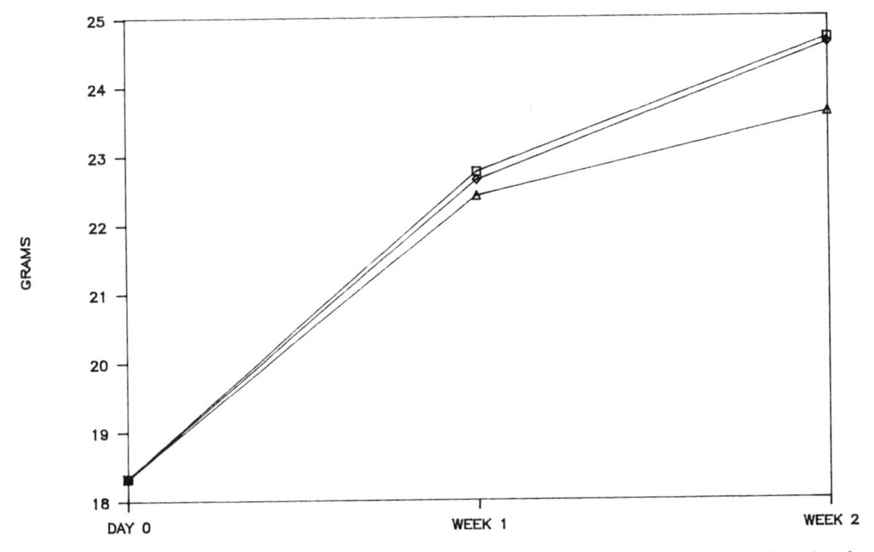

FIGURE 2. Body weights. The body weight of adult CD-1 mice did not change significantly after 7 to 14 daily injections of rHuIFNαA/D. Saline □, Lo IFN ◇, Hi IFN △; n = 12.

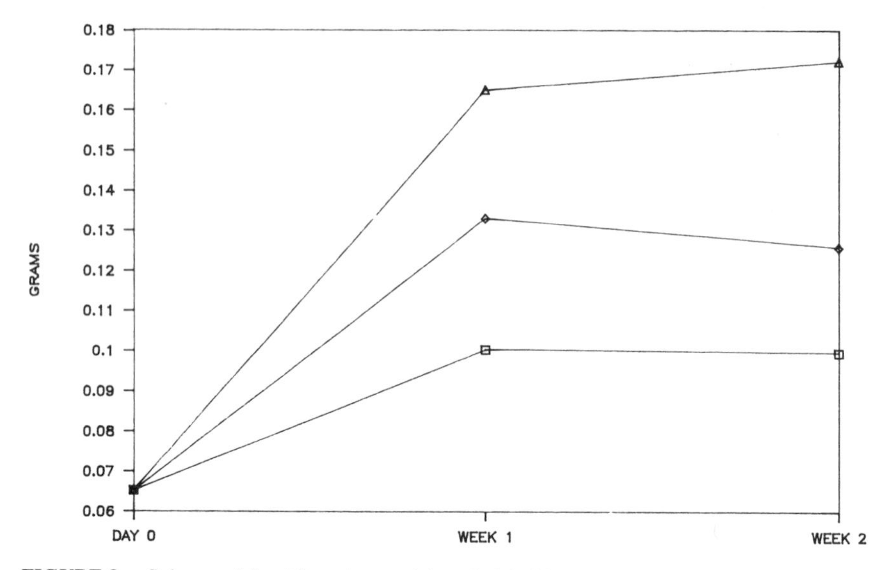

FIGURE 3. Spleen weights. The spleen weights of adult CD-1 mice injected daily for 7 to 14 days increased significantly over control values. The largest increase in spleen weight occurred when Hi Dose IFN was injected and this was associated with the largest increase in nucleated (developing) hematopoietic cells. Saline □, Lo IFN ◇, Hi IFN △; n = 12.

spleen in Hi Dose IFN-injected mice. This was a 46% increase in total spleen cell number. The largest increase occurred in the nucleated erythroid compartment accompanied by no apparent change in absolute number of lymphocytes (data not included).

In Vivo Effects of rHuIFNαA/D on Red Cells

At week 1 and week 2 of Lo and Hi Dose IFN treatment, there were increases in percent erythroblasts in both marrow and spleen. In the spleen, a large increase in percent erythroblasts occurred after 7 and 14 daily injections of Hi Dose IFN (data not shown). These large increases in percent splenic erythroblasts are further enhanced by the increase in spleen size (see FIG. 3) to yield a markedly increased number of splenic as well as marrow erythroblasts. Nevertheless, with all of this apparently enhanced erythropoiesis, there occurred a 17.1% decrease in hematocrit after 1 week of daily Hi Dose IFN injections (significantly different; $p < 0.0003$). The 6.4% and 2.9% decreases in hematocrit seen after 1 and 2 weeks, respectively, of daily Lo Dose IFN injections were not significantly different (TABLE 4).

TABLE 3. Estimated Absolute Number of Nucleated Cells per Spleen

Treatment[a]	Week 1	Week 2
Saline	186.3×10^6	154.5×10^6
LoIFN	199.8×10^6	206.9×10^6
HiIFN	245.6×10^6	225.1×10^6

[a]Mice received 7 or 14 daily injections of saline or 10^6 U/Kg or 10^7 U/Kg rHuIFNαA/D

Since mouse erythrocytes have a circulating life-span of approximately 40 to 43 days, the decrease in hematocrit may be directly related to normal red cell senescence and death and ineffective erythrocytic renewal. The latter may be the result of a block to maturation at the erythroblast stage.

In Vivo *Effects of rHuIFNαA/D on Platelets*

There occurs a dose dependent and significant decrease in the number of circulating platelets and an increase in platelet volume after injecting rHuIFNαA/D (data not shown). These 2 platelet parameters, number times volume, can be used to obtain the platelet hematocrit which is lower after both 1 and 2 weeks of daily IFN injections. In ongoing studies, we are investigating these effects of rHuIFNαA/D on platelet production. These

TABLE 4. Hematocrit Values in Mice Injected Daily with rHuIFNαA/D[a]

	Week 1			Week 2		
	Saline	LoIFN	HiIFN	Saline	LoIFN	HiIFN
	48.6	44.8	36.8	44.9	35.8	38.0
	51.6	38.2	37.0	44.3	41.1	44.4
	50.7	41.5	39.0	46.4	41.7	47.2
	48.4	40.7	32.7	44.8	46.2	36.7
	51.2	43.1	36.7	41.5	44.0	24.1
	44.2	45.6	36.8	45.5	49.0	33.3
	47.0	44.1	40.0	51.9	43.4	38.3
	48.5	42.2	44.6	42.9	40.1	39.0
	43.9	41.8	40.0	45.5	43.4	44.2
	41.5	41.8	39.6	42.9	41.9	36.2
	37.5	35.3	36.6	40.7	40.0	37.0
	37.9	42.3	35.3	42.0	43.8	40.4
	35.6	47.6	31.6	41.5	47.8	
Mean	45.1	42.2	37.4[b]	44.2	42.9	38.2[b]
± SEM	±1.5	±0.9	±0.9	±0.8	±1.0	±1.7

[a]Mice were injected daily with saline or 10^6 U/Kg or 10^7 U/Kg rHuIFNαA/D.

[b]Significantly different values from control and Lo IFN values as established by the Scheffle Multiple Comparison Analysis.

studies involve morphometric analysis of individual megakaryocyte volume and number of megakaryocytes per unit volume of splenic red pulp.

Microscopy of Spleens of rHuIFNαA/D-Injected Mice

Although the spleen did enlarge after mice received daily injections of rHuIFNαA/D for 1 or 2 weeks, the white pulp did not appear to increase in volume and morphometric studies are currently underway to quantitate this observation. The increase in spleen weight seems to be entirely due to an increase in red pulp. Spleen cell differentials indicate a large increase in erythroblasts (data not shown) and light microscopic examination of spleen sections shows large groups of erythroblasts (FIG. 4). These cells are structurally normal when viewed by light and electron microscopy, however, their maturation may be blocked or prolonged by interferon-induced 2-5AS activity as suggested below in FIGURE 5.

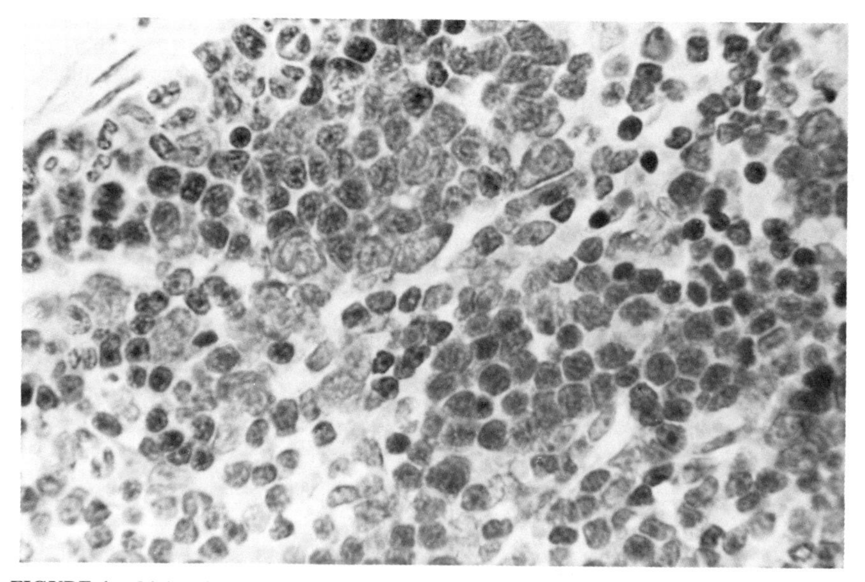

FIGURE 4. Light micrograph of spenic red pulp from a mouse injected with 7 daily doses of 10^7 U/Kg rHuIFNαA/D. The field illustrates a virtually continuous sheet of erythroblasts. Although present in what appeared to be a greatly increased number, there were no observed changes in structural characteristics of these erythroblasts. 640X.

Hypothesis for IFN-Inhibition of Hematopoiesis.

Our interpretation of the anemia and thrombocytopenia that develop in mice injected with rHuIFNαA/D is that IFN induces a (transient or long-term) block to erythroblast and megakaryocyte maturation through an early activation of the 2-5AS system. This enzyme system includes endonuclease which can degrade mRNA and thus interfere with protein synthesis and cell division. This may lead to a failure to complete maturation and a subsequent "backing-up" of blasts in the spleen. This hypothesis is expressed in FIGURE 5.

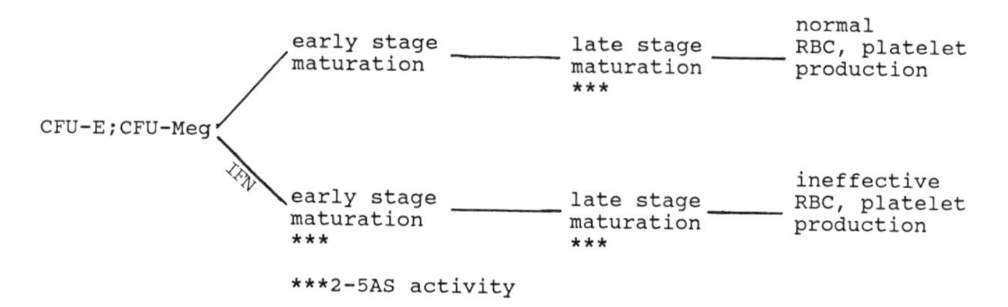

FIGURE 5. Proposed effect of IFN on blood cell production. Interferon induces an early activation of the 2-5AS system in developing "blasts" of the erythroid and megakaryocytic series. The 2-5AS block to maturation is interpreted as a form of ineffective hematopoiesis.

SUMMARY

Spleen cell suspensions obtained from adult mice were separated by Ficoll/Hypaque and Percoll density gradient centrifugation. The enriched erythroblast populations were maintained in liquid culture medium for 8 hours with 10,000 units of murine interferon (IFN) α and β. Exposure of these cell cultures to murine IFNα and β resulted in a 48% to 70% increase in 2-5adenylate synthetase (2-5AS) activity.

In parallel studies, adult mice were injected daily for 1 or 2 weeks with recombinant human IFNαA/D (rHuIFNαA/D) at a dose of 10^6 or 10^7 units/kg body weight. This treatment did not significantly affect body weight but did produce a mean 70% increase in spleen wet weight and a mean 46% increase in number of nucleated cells per spleen. This increase in number of splenic hematopoietic cells did not result in a corresponding increase in number of circulating cells. In fact, during this 1 to 2 week period the hematocrit dropped from 45% to 38% in mice injected with high dose rHuIFNαA/D.

From these findings we propose that IFN induces an early 2-5AS activity in erythroblasts and megakaryocytes. This 2-5AS activity, which is known to inhibit protein synthesis in other cell systems, is thought to be responsible for the block or prolongation in blood cell maturation observed in the present studies.

REFERENCES

1. Zoumbos, N. C., P. Gascon, J. Y. Djeu & N. S. Young. 1985. Proc. Natl. Acad. Sci. USA **82:** 188–192.
2. Zoumbos, N. C., P. Gascon, J. Y. Djeu, S. R. Trost & N. S. Young. 1985. N.Engl.J. Med. **312:** 257–265.
3. Ball, L. A. 1979. Virology **94:** 282–296.
4. Baglioni, C., S. Bevin, P. A. Maroney, M. A. Minks, T. Nilsen & D. West. 1980. Ann. N.Y. Acad. Sci. **350:** 497–509.
5. Revel, M., A. Kimchi, L. Shulman, A. Fradin, R. Shuster, E. Yakobson, Y. Chernajovsky, A. Schmidt, A. Shure & R. Bendori. 1980. Ann. N.Y. Acad. Sci. **350:** 459–472.
6. Kerr, I. M., D. H. Wreschner, R. H. Silverman, P. J. Cayley & M. Knight. 1981. Adv. Cyclic Nucleotide Res. **14:** 469–478.
7. Schmidt, A., Y. Chernajovsky, L. Shulman, P. Federman, H. Berissi & M. Revel. 1979. Proc. Natl. Acad. Sci. USA **76:** 4788–4792.
8. Ball, L.A. 1980. Ann. N.Y. Acad. Sci. **350:** 486–496.
9. Roberts, W. K., A. Hovanessian, R. E. Brown, M. J. Clemens & I. M. Kerr. 1976. Nature **264:** 477–480.
10. Minks, M. A., S. Benvin, P. A. Maroney & C. Baglioni. 1979. J. Biol. Chem. **254:** 5058–5064.
11. Justesen, J., D. Ferbus & M. N. Thang. 1980. Proc. Natl. Acad. Sci. USA **77:** 4618–4622.
12. Baglioni, C., M. A. Minks & P. A. Maroney. 1978. Nature **273:** 684–686.
13. Clemens, M. J. & C. M. Vaquero. 1978. Biochem. Biophys. Res. Commun. **83:**59–67.
14. Schmidt, A., A. Zilbertstein, L. Shulman, P. Federman, H. Berissi & M. Revel. 1978. FEBS Letts. **95:** 257–263.
15. Wreschner, D. H., McCauley, J. W., J. J. Skehel & I. M. Kerr. 1981. Nature **289:** 414–417.
16. Matarese, G. P. & G. B. Rossi. 1977. J. Cell Biol. **75:** 344–354.
17. Ortega, J. A., J. Ma, N. A. Shore, P. P. Dukes & T. C. Merigan. 1979. Exp. Hematol. **7:** 145–150.
18. Orlic, D., J. M. Wu, R. D. Carmichael, F. Quaini, M. Kobylack & A. S. Gordon. 1982. Exp. Hematol. **10:** 478–485.
19. Orlic, D., E. Kirk & F. Quaini. 1985. Exp. Hematol. **13:** 821–826.

20. MAZUR, E. R., W. J. RICHTSMEIER & K. SOUTH. 1986. J. Interferon Res. **6:** 199–206.
21. ILSON, D. H., P. F. TORRENCE & J. VILCEK. 1986. J. Interferon Res. **6:** 5–12.
22. MERRITT, J. A., L. A. BALL, K. M. SIELAFF, D. M. MELTZER & E. C. BORDEN. 1986. J. Interferon Res. **6:** 189–198.
23. QUESADA, J. R., M. TALPAZ, A. RIOS, R. KURZROCK & J. U. GUTTERMAN. 1986. J. Clin. Oncol. **4:** 234–243.
24. INGIMARSSON, S., K. BERGSTROM, L. A. BROSTROM, K. CANTELL & H. STRANDER. 1980. Acta Med. Scand. **208:** 155–159.
25. BOHOSLAWEC, O., P. W. TROWN & R. J. WILLS. 1986. J. Interferon Res. **6:** 207–213.
26. WECK, P. K., E. RINDERKNECHT, D. A. ESTELL & N. STEBBING. 1982. Infect. Immun. **35:** 660–665.
27. KRAMER, M. J., R. DENNIN, C. KRAMER, G. JONES, E. CONNELL, N. ROLON, A. GRUARIN, R. KALE & P. W. TROWN. 1983. J. Interferon Res. **3:** 425–435.
28. HARRISON, F. L., T. M. BESWICK & C. J. CHESTERTON. 1981. Biochem. J. **194:** 789–796.
29. SHORTMAN, K. & K. SELIGMAN. 1969. J. Cell Biol. **42:** 783–793.

The Role of Interleukin-3 and Heme in the Induction of Erythropoiesis[a]

FRANCIS C. MONETTE

Department of Biology
Boston University
2 Cummington Street
Boston, Massachusetts 02215

INTRODUCTION

Recent research has made it abundantly clear that hemopoiesis is governed by protein growth factors which are produced extrinsically to their cellular targets. Such factors appear to fall into two broad categories, multipotential growth factors and lineage-restricted growth factors.[1,2] The lymphokine interleukin-3 (IL-3) is a prime example of a multilineage growth factor (or multi-CSF) since it stimulates the proliferation and development of virtually all myeloid cells whereas erythropoietin (EPO), monocyte-colony-stimulating factor (M-CSF), etc. are examples of lineage-specific growth factors. An understanding of hematopoietic regulation relies on a detailed accounting of the biology of each of these growth factors as well as their possible interactions with each other. In addition, it will be necessary to delineate the role of other types of factors such as interferons, prostaglandins, etc. in the modulation of hemopoiesis, not all of which are necessarily protein in nature. An example of a nonprotein factor is heme which has been shown to exert a multitude of effects at several stages of hematopoietic cell development. It will be the purpose of this paper to review the basic biological properties of IL-3 and heme and describe their specific role(s) and possible interactions in fostering hemopoietic cell development in the mammal.

Interleukin-3

Biological Properties of Interleukin-3

By virtue of its effects on early multipotential stem cells, IL-3 may influence lymphoid cell development as well as that of myeloid cells. The reader is referred to reviews on the former subject by Ihle *et al.*[3] and Schrader.[4] IL-3 is clearly a product of activated T-helper cells[4,5] which have been stimulated by antigen or lectins. Many reports ascribe a cell surface phenotype of Lyt-1 +; Lyt-2-; Ia-; and Ig- to the lymphoid IL-3-producing cells. The lymphokine can also be obtained from cloned T-cell lines.[5] In addition, certain myeloid leukemias (*e.g.*, WEHI-3B) represent pathological sources of the lymphokine. Both the human[6] as well as the murine[7–9] IL-3 genes have been cloned as has the gene for gibbon IL-3.[10] Expression vectors for the IL-3 cDNA have consisted of both prokaryotic[11,12] as well as mammalian cell lines.[6,8,9,13,14] The observations that both recombinant IL-3 (rIL-3) produced by bacteria as well as naturally-produced IL-3 (nIL-3)

[a] The work described herein was made possible by Public Health Service Grants DK 35325 and DK 37366 from the National Institutes of Health.

exhibit similar biological activities[14,15] strongly suggests that the carbohydrate moiety present on the molecule, although extensive, is not required for its biological activity *in vivo* or *in vitro*. Ihle's group[16] reported that the IL-3 gene maps to chromosome 11 of the mouse whereas LeBeau *et al.*[17] demonstrated that the human IL-3 gene is located in a region of chromosome 5 which also contains the genes for a number of other growth factors.

Molecular and Biochemical Properties of Interleukin-3

IL-3 is a heavily glycosylated monomeric glycoprotein consisting of 166 amino acids with a signal peptide of 27 a.a..[7,18] A biologically active IL-3 preparation has been chemically synthesized by Clark-Lewis *et al.*[19] who described a preparation consisting of 134 a.a. Ihle's group[20] purified murine nIL-3 to apparent homogeneity on SDS-PAGE and obtained an Mr of 28,000. Their preparation was characterized by a specific activity of 0.05 ng/unit. The successful iodination of IL-3 was reported by Palaszynski & Ihle[21] who estimated the IL-3 receptor density to vary between 2000 and 5000 per cell for different factor-dependent cell lines, whereas Crapper *et al.*[23] obtained estimates of only ~1000 per cell. These same two groups also determined that IL-3 bound to its specific cell receptor with an equilibrium dissociation constant (Kd) between 10^{-11} M[21] and 10^{-12} M[22] suggesting the existence of high-affinity receptors on IL-3-dependent cells. The work of Park et al.[23] suggested that maximal biological activity can be obtained with only 50% receptor saturation; however, the full extent of receptor internalization on their estimate of saturation was not made clear.

By employing an iodinated IL-3 preparation, Nicola and Metcalf[24] have shown specific and equal binding to target cells including myeloid cells morphologically recognizable as granulocytic, monocytic, and eosinophilic and that binding of IL-3 decreased with the cell ^{125}IL-3 maturation level. In contrast, ^{125}IL-3 bound poorly to morphologically recognizable erythroid and lymphoid cells. One group[25] has suggested that binding of the IL-3 receptor results in the down-modulation of other CSF receptors with specific consequences on cellular differentiation pathways. Several groups have provided data suggesting that the activation of protein kinase C is involved in the signal transduction of IL-3 on its target cells,[26,27] whereas fluctuations in cytoplasmic ATP,[28] glucose transport,[29] and calcium[30] have also been implicated.

Biological Activities of Interleukin-3

Schrader[4] has argued for considering IL-3 as a "panspecific hemopoietin" since it clearly functions as a growth factor with a characteristically broad range of cell targets, at least in the *in vitro* setting. Its *in vivo* role in day-to-day hemopoiesis, however, remains obscure due to the inability to detect its presence in normal body fluids such as serum.[31] In fact, some groups[15,32] have speculated that IL-3 functions *in vivo* only under conditions of severe immunological challenge and thus does not play a significant role as a regulator of hemopoiesis in normal adult life.[33] Nevertheless, substantial evidence has accumulated which indicates that IL-3 not only promotes the survival of early hematopoietic cell progenitors *in vitro* but that it specifically stimulates the *proliferation* of early progenitors committed to the granulocytic, monocytic/macrophagic, megakaryocytic, erythroid, and eosinophilic cell differentiation pathways.[9,34] Thus, IL-3 has been shown to support the *in vitro* proliferation of early hematopoietic stem cells as identified by both *in vivo* (*i.e.*, CFU-S)[35] and *in vitro* (CFU-GEMM and "blast" cell colonies)[36,37] methodologies. The work of Suda *et al.*[37] suggests that IL-3 supports multipotential cell colony growth in "serum-free" medium, the only factor described to-date to do so. In addition, their work suggests a direct action of IL-3 on cellular targets. Fabian *et al.*[38] recently demonstrated that a brief exposure of marrow cells to IL-3 *in vitro* significantly augmented their ability to reconstitute the marrows of irradiated recipients.

The broad proliferative action of IL-3 revealed by the *in vitro* studies described above strongly suggest that this molecule should have many observable effects on hemopoiesis when injected *in vivo*. However, the effectiveness of *in vivo* infusions of IL-3 has only recently been demonstrated. Several groups[14,32,39,40] have reported that IL-3 administration to normal animals greatly augments CFU-S, CFU-E, and mast cell numbers in the spleen while having little effect on bone marrow progenitors. In contrast, Broxmeyer's group[41,42] recently demonstrated a medullary as well as splenic effect on progenitor cell number and cycling levels in lactoferrin-suppressed animals. By utilizing this assay system they were able to show an effective IL-3 dose level at least several orders of magnitude *lower* than previously reported.

Interleukin-3 Effects on Cellular Function

Although there have been few reports of IL-3 effects on mature cell functions, Crapper *et al.*[43] demonstrated augmentation of phagocytosis by marrow and peritoneal macrophages exposed to IL-3 *in vitro*. Likewise, activation of cellular function by IL-3 was reported by Lopez *et al.*[10] who showed that the lymphokine stimulated cell-mediated cytotoxicity and phagocytosis by eosinophils, with little effect, however, on neutrophils.

Synergism of Interleukin-3 with Other Growth Factors

Recent studies have made clear that hematopoietic growth factors are capable of interaction with one another so that much lower concentrations prove optimally effective when combined than when assayed separately. To-date, at least four different growth factors have been shown to act synergistically with IL-3 to promote cell proliferation either *in vitro* or *in vivo* including M-CSF (CSF-1), EPO, GM-CSF, and Interleukin-4 (TABLE 1). Just how growth factors interact in a synergistic fashion is not certain, but it has been suggested that such interaction may induce proliferation and receptor acquisition in early stem cells which then become responsive to terminal growth factors in an orderly, hierarchical manner.[51]

Heme

Action of Heme in Hemopoiesis

Most, if not all, cells require iron for their normal metabolism and employ the 80,000 dalton B_1-globulin transferrin (TF) molecule to direct iron to specific membrane receptors.[52,53] However, some cells may employ alternate pathways in addition to TF for obtaining iron from their environment. Recent evidence supports the notion that cells may incorporate heme either by binding iron protoporphyrin directly[54] or alternatively, by employing a hemopexin-mediated mechanism.[55] Over the last decade numerous studies have served to establish a central role for heme on a variety of cellular activities. The effects of heme on globin synthesis and protein synthesis in general have been reviewed elsewhere.[56] The observations that heme (in the form of hemin, ferric chloride proto-

TABLE 1. Synergistic Activities of Interleukin-3[a]

IL-3 Source	Factor Synergized	Cell Target	Reference
Hemopoietin-2	nCSF-1	CFU-M	44,45
nmIL-3	nCSF-1	blood monocyte; peritoneal macrophage	46
nmIL-3	nCSF-1	CFU-GM	47
rmIL-3	nCSF-1	*in vivo:* CFU-GM BFU-E CFU-GEMM	48
nmIL-3	nhEPO	CFU-GEMM	37
rmIL-3	rmGM-CSF	*in vivo:* CFU-GM BFU-E CFU-GEMM	48
nmIL-3	rmIL-4	mast cell line (SN-1)	49
rmIL-3	rmIL-4	BFU-E CFU-G CFU-M CFU-mast	50

[a]Abbreviations: n, natural; r, recombinant; m, murine; h, human.

porphyrin IX) greatly stimulates primitive erythroid colonies *in vitro*[57,58] as well as *in vivo*[59] suggested that it may also modulate erythropoiesis in a very fundamental way.

Biochemical Properties of Hemin

Hemin (ferric chloride protoporphyrin IX), defined by its lipophillic protoporphyrin ring structure, exhibits an M_r of 652 daltons. In normal serum unbound heme is not usually detected; however, serum levels exceeding 10 μM have been reported for certain hemolytic disease states.[60] Hemopexin is a serum glycoprotein which binds heme with high affinity and plays a major role in transporting heme in the circulation.[61] The successful iodination of hemopexin has permitted the characterization of heme-binding receptors on a variety of cells. Smith and Morgan[55] described a hemopexin receptor on hepatic membranes having a K_d of 6.8×10^{-7} M and an M_r of 115,000 daltons. However, hemopexin binding is not limited to liver cells, as Taketani *et al.* have demonstrated specific binding to surface receptors of both erythroleukemic (K-562) cells[62] and myeloid leukemic (HL-60) cells.[63] Hemopexin receptors in both hematopoietic cell lines exhibited a slightly higher binding affinity than was demonstrated for hepatic cells (*i.e.*, $K_d = 1$ to 5×10^{-9} M). However, the receptor distribution was nearly fivefold greater in HL-60 cells than in K-562 cells (42,000 and 8400 per cell, respectively).

In addition to transferrin and hemopexin, some cells are able to obtain iron by binding heme directly. Several groups have described specific heme-binding receptors on K562 cells,[62,64,65] HL-60 cells,[66] and HeLa cells.[67] Galbraith *et al.*[54] have identified heme-binding proteins with very high affinity ($K_d = \sim 10^{-11}$ M) on murine erythroleukemic (MEL DS-19) cells. Heme binding on these cells was clearly saturable, reversible, trypsin-sensitive, and stereospecific. They estimated the heme receptor density at ~10,000 per cell. Thus, in addition to the transferrin and hemopexin pathways of iron transport, a number of cells appear capable of obtaining heme iron from their environment directly. This immediately raises the question of the reason for multiple and overlapping iron incorporation pathways at the cellular level. Could heme be providing another function other than serving simply as an iron source?

The Role of Heme in Hemopoiesis

One possibility is that heme serves as an accessory factor in hematopoietic cell growth and differentiation. The results of numerous studies by my laboratory[56–59] are consistent with this interpretation. The addition of hemin to plasma clot or methyl cellulose clonal cell cultures produces a substantial enhancement in erythroid colony formation by normal bone marrow (FIG. 1) and the effect appears greater on earlier stages of erythroid cell differentiation. These findings have been recently confirmed in "serum-free" marrow cultures[68] which established: (1) hemin's ability in conjunction with IL-3 to support multipotential stem cell growth in the absence of other exogenous growth factors; and (2) a substantial reduction in growth factor requirements when marrow cells are cultured in the presence of hemin, resulting in a "shift-to-the-left" in the growth factor titration curves.[69]

Further work[70] has suggested that hemin may interact synergistically with IL-3 in promoting colony formation in "serum-free" marrow cell cultures in that the combination of IL-3 plus hemin yeilds >20-fold more colonies than does either IL-3 or hemin alone (TABLE 2). Interestingly, hemin has additional effects on primary marrow cell cultures when normal marrow cells are cultured in semisolid medium and re-fed every 10 to 14 days with a small volume of culture medium containing all the necessary growth factors, the clonal development of individual CFU-GEMM continues for periods of up to 60 to 90 days.[71] This suggests that hemin, along with IL-3 and EPO, may provide the growth

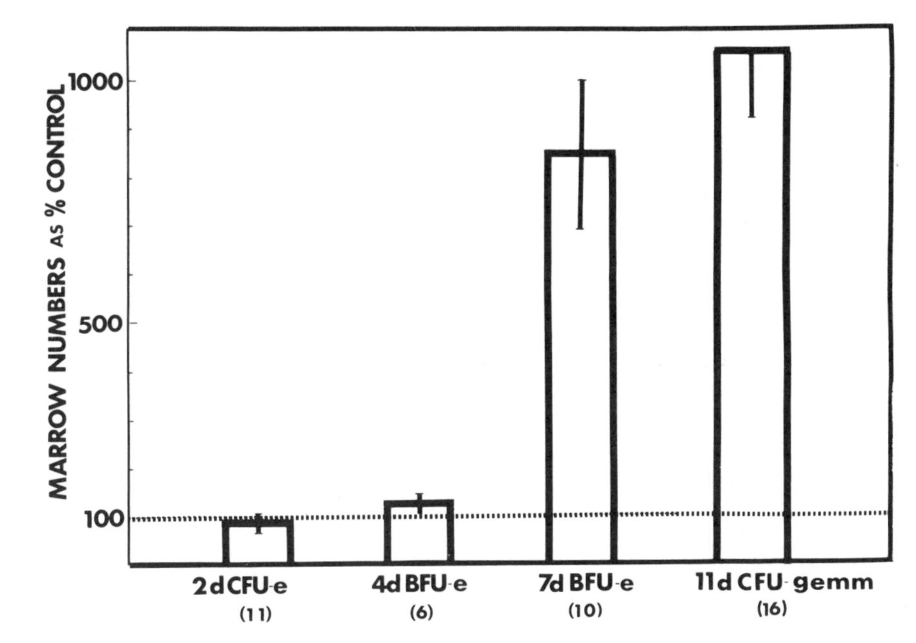

FIGURE 1: The relative effects of hemin on bone marrow-derived erythroid colonies.

TABLE 2. Synergistic Interaction Between Hemin and Interleukin-3 on Multipotential Stem Cell Colony Formation in "Serum-Free" Marrow Cultures[a]

Factor Added	Colony Number per 10^5 Cells
Interleukin-3	±
Hemin	±
Interleukin-3 plus hemin	+ + + + + + + + + + + +

[a]From REFERENCE 70. Highly purified recombinant erythropoietin was added to all cultures.

factor environment necessary for the continued proliferation and terminal differentiation of committed erythroid stem cells.

CONCLUSIONS

The lymphokine interleukin-3 appears to be a necessary growth factor for the development of a broad spectrum of hematopoietic cells *in vitro*. The fact that a gene coding for IL-3 production has been isolated from the rodent and primate genomes argues persuasively for the likelihood of a biological role *in situ*. Many hematopoietic growth factors greatly augment the biological effects of IL-3 via their synergistic interactions with the lymphokine. Since this synergism has been shown to occur *in vivo* as well as *in vitro*, and highly purified recombinant preparations have been utilized throughout these investigations, thus precluding the effects of possible contaminants, it is most likely that synergism of growth factors is a common biological phenomenon *in situ*. Synergism of growth factors can be expected to substantially alter the biologically-effective concentrations of such factors, perhaps by several or more orders of magnitude (FIG. 2). This would be especially true if a particularly sensitive parameter is chosen as an end point. As Broxmeyer and his colleagues have demonstrated,[41,42,48] the entry of cells into the S-phase of cell cycle may be a good example of an experimental end point to demonstrate synergism.

Another factor which appears capable of synergizing with IL-3 is the protoporphyrin hemin. Although hemin is quite distinct from other molecules already defined as hematopoietic growth factors (it is not a protein!), its role in modulating erythroid cell growth and differentiation may be no less important. This is particularly true since specific heme-binding proteins have now been isolated from a variety of cell types. Such findings provide support for the notion of an even wider biological role for this metalloporphyrin. With the availability of highly purified recombinant materials and the ability to grow cells

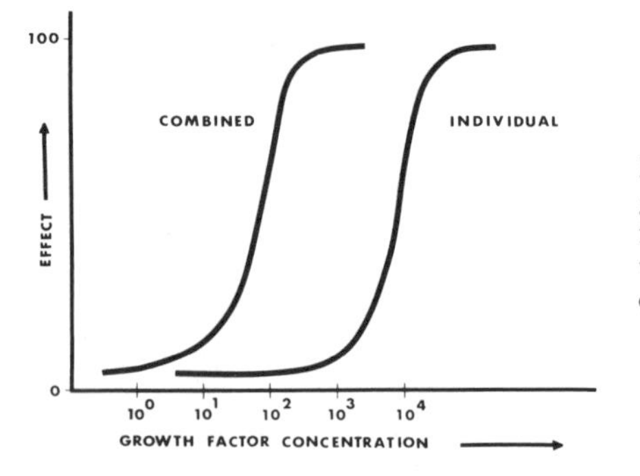

FIGURE 2: Effect of synergism between growth factors: comparison of titration curves for individual growth factors with that observed when factors are combined. Growth factor concentration is in arbitrary units.

in almost totally-defined culture systems, it should soon be possible to approach the experimental problem of hematopoietic cell developement in a highly critical manner.

REFERENCES

1. Whetton, A. D. & T. M. Dexter. 1986. Haemopoietic growth factors. Trends Biochem. Sci. **11:** 207–211.
2. Clark, S. C. & R. Kamen. 1987. The human hematopoietic colony-stimulating factors. Science **236:** 1229–1237.
3. Ihle, J. N., L. Rebar, J. Keller, J. C. Lee & A. J. Hapel. 1982. Interleukin 3: Possible roles in the regulation of lymphocyte differentiation and growth. Immunol. Rev. **63:** 5–32.
4. Schrader, J. W. 1986. The panspecific hemopoietin of activated T lymphocytes (Interleukin-3). Ann. Rev. Immunol. **4:** 205–230.
5. Prystowsky, M. B., J. M. Ely, D. I. Beller, L. Eisenberg, J. Goldman, M. Goldman, E. Goldwasser, J. Ihle, J. Quintans, H. Remold, S. N. Vogel & F. W. Fitch. 1982. Alloreactive cloned T cell lines: VI. Multiple lymphokine activities secreted by helper and cytolytic cloned T lymphocytes. J. Immunol. **129:** 2337–2344.
6. Yang, Y.-C., A. B. Ciarletta, P. A. Temple, M. P. Chung, S. Kovacic, J. S. Witek-Giannotti, A. C. Leary, R. Kriz, R. E. Donahue, G. G. Wong & S. C. Clark. 1986. Human IL-3 (Multi-CSF): Identification by expression cloning of a novel hematopoietic growth factor related to murine IL-3. Cell **47:** 3–10.
7. Fung, M. C., A. J. Hapel, S. Ymer, D. R. Cohen, R. N. Johnson, H. D. Campbell & I. G. Young. 1984. Molecular cloning of cDNA for mouse interleukin-3. Nature **307:** 233–237.
8. Yokota, T., F. Lee, D. Rennick, C. Hall, N. Arai, T. Mosmann, G. Nabel, H. Cantor & K. Arai. 1984. Isolation and characterization of a mouse cDNA clone that expresses mast-cell growth-factor activity in monkey cells. Proc. Natl. Acad. Sci. USA **81:** 1070–1074.
9. Hapel, A. J., M. C. Fung, R. M. Johnson, I. G. Young, G. Johnson & D. Metcalf. 1985. Biologic properties of molecularly cloned and expressed murine interleukin-3. Blood **65:** 1453–1459.
10. Lopez, A. F., L. B. To, Y.-C. Yang, J. R. Gamble, M. F. Shannon, G. F. Burns, P. G. Dyson, C. A. Juttner, S. Clark & M. A. Vadas. 1987. Stimulation of proliferation, differentiation, and function of human cells by primate interleukin 3. Proc. Natl. Acad. Sci. USA **84:** 2761–2765.
11. Campbell, H. D., S. Ymer, M. C. Fung & I. G. Young. 1985. Cloning and nucleotide sequence of the murine interleukin-3 gene. Eur. J. Biochem. **150:** 297–303.
12. Kindler, V., B. Thorens, S. De Kossodo, B. Allet, J. F. Eliason, D. Thatcher, N. Farber & P. Vassalli. 1986. Stimulation of hematopoiesis in vivo by recombinant bacterial murine interleukin 3. Proc. Natl. Acad. Sci. USA **83:** 1001–1005.
13. Gough, N. M., J. Gough, D. Metcalf, A. Kelso, D. Grail, N. A. Nicola, A. W. Burgess & A. R. Dunn. 1984. Molecular cloning of cDNA encoding a murine haematopoietic growth regulator, granulocyte-macrophage colony stimulating factor. Nature **309:** 763–767.
14. Cohen, D. R., A. J. Hapel & I. G. Young. 1986. Cloning and expression of the rat interleukin-3 gene. Nucleic Acids Res. **14:** 3641–3658.
15. Metcalf, D., C. G. Begley, G. R. Johnson, N. A. Nicola, A. F. Lopez & D. J. Williamson. 1986. Effects of purified bacterially synthesized murine multi-CSF (IL-3) on hematopoiesis in normal adult mice. Blood **68:** 46–57.
16. Ihle, J. N., J. Silver & C. A. Kozak. 1987. Genetic mapping of the mouse interleukin 3 gene to chromosome 11. J. Immunol. **138:** 3051–3054.
17. LeBeau, M. M., N. D. Epstein, S. J. O'Brien, A. W. Nienhuis, Y.-C. Yang, S. C. Clark & J. D. Rowley. 1987. The interleukin 3 gene is located on human chromosome 5 and is deleted in myeloid leukemias with a deletion of 5q. Proc. Natl. Acad. Sci. USA **84:** 5913–5917.
18. Ihle, J. N., J. Keller, S. Oroszlan, L. E. Henderson, T. D. Copeland, F. Fitch, M. B.

PRYSTOWSKY, E. GODWASSER, J. W. SCHRADER, E. PALASZYNSKI, M. DY & B. LEBEL. 1983. Biologic properties of homogeneous interleukin 3. I. Demonstration of WEHI-3 growth factor activity, mast cell growth factor activity, P cell stimulating factor activity, colony-stimulating factor activity, and histamine-producing cell-stimulating factor activity. J. Immunol. **131:** 282–287.

19. CLARK-LEWIS, I., R. REBERSOLD, H. J. ZILTENER, J. W. SCHRADER, L. E. HOOD & S. B. H. KENT. 1986. Automated chemical synthesis of a protein growth factor for hemopoietic cells, interleukin 3. Science **231:** 134–139.
20. IHLE, J. N., J. KELLER, L. HENDERSON, F. KLEIN & E. W. PALASZYNSKI. 1982. Procedures for the purification of interleukin 3 to homogeneity. J. Immunol. **129:** 2431–2436.
21. PALASZYNSKI, E. W. & J. N. IHLE. 1984. Evidence for specific receptors for interleukin 3 on lymphokine-dependent cell lines established from long-term bone marrow cultures. J. Immunol. **132:** 1872–1879.
22. CRAPPER, R. M., I. CLARK-LEWIS & J. W. SCHRADER. 1985. Analysis of the binding of a hemopoietic growth factor, P-cell-stimulating factor, to a cell surface receptor using quantitative absorption of bioactivity. Exp. Hematol. **13:** 941–947.
23. PARK, L. S., D. FRIEND, S. GILLIS & D. L. URDAL. 1986. Characterization of the cell surface receptor for a multi-lineage colony-stimulating factor (CSF-2α). J. Biol. Chem. **261:** 205–210.
24. NICOLA, N. A. & D. METCALF. 1986. Binding of iodinated multipotential colony-stimulating factor (Interleukin-3) to murine bone marrow cells. J. Cell. Physiol. **128:** 180–188.
25. WALKER, F., N. A. NICOLA, D. METCALF & A. W. BURGESS. 1985. Hierarchical down-modulation of hemopoietic growth factor receptors. Cell **43:** 269–276.
26. FARRAR, W. L., T. P. THOMAS & W. B. ANDERSON. 1985. Altered cytosol/membrane enzyme redistribution on interleukin-3 activation of protein kinase C. Nature **315:** 235–237.
27. EVANS, S. W., D. RENNICK & W. L. FARRAR. 1986. Multilineage hematopoietic growth factor interleukin-3 and direct activators of protein kinase C stimulate phosphorylation of common substrates. Blood **68:** 906–913.
28. WHETTON, A. D. & T. M. DEXTER. 1983. Effect of haematopoietic cell growth factor on intracellular ATP levels. Nature **303:** 629–631.
29. DEXTER, T. M., A. D. WHETTON & G. W. BAZILL. 1984. Haemopoietic cell growth factor and glucose transport. Its role in cell survival and the relevance of this in normal haemopoiesis and leukaemia. Differentiation **27:** 163–167.
30. ROSSIO, J. L., F. W. RUSCETTI & W. L. FARRAR. 1986. Ligand-specific calcium mobilization in IL-2- and IL-3-dependent cell lines. Lymphokine Res. **5:** 163–172.
31. CRAPPER, R. M., I. CLARK-LEWIS & J. W. SCHRADER. 1984. The in vivo functions and properties of persisting cell-stimulating factor. Immunology **53:** 33–44.
32. METCALF, D., C. G. BEGLEY, N. A. NICOLA & G. R. JOHNSON. 1987. Quantitative responsiveness of murine hemopoietic populations in vitro and in vivo to recombinant multi-CSF (IL-3). Exp. Hematol. **15:** 288–295.
33. NATHAN, D. 1988. This volume.
34. IHLE, J. N., J. KELLER, A. REIN, J. CLEVELAND & U. RAPP. 1985. Interleukin-3 regulation of the growth of normal and transformed hematopoietic cells. *In* Growth Factors and Transformation. J. Feramisco, B. Ozanne & C. Stiles, Eds.: 211–219. Cold Spring Harbor Laboratory, Cold Spring Harbor, NY.
35. SPIVAK, J. L., R. R. L. SMITH & J. N. IHLE. 1985. Interleukin 3 promotes the in vitro proliferation of murine pluripotent hematopoietic stem cells. J. Clin. Invest. **76:** 1613–1621.
36. SCHRADER, J. W. & I. CLARK-LEWIS. 1982. A T cell-derived factor stimulating multipotential hemopoietic stem cells: Molecular weight and distinction from T cell growth factor and T cell-derived granulocyte-macrophage colony-stimulating factor. J. Immunol. **129:** 30–35.
37. SUDA, J., T. SUDA, K. KUBOTA, J. N. IHLE, M. SAITO & Y. MIRUA. 1986. Purified interleukin-3 and erythropoietin support the terminal differentiation of hemopoietic progenitors in serum-free culture. Blood **67:** 1002–1006.
38. FABIAN, I., I. BLEIBERG, I. RIKLIS & Y. KLETTER. 1987. Enhanced reconstitution of hematopoietic organs in irradiated mice, following their transplantation with bone marrow cells pretreated with recombinant interleukin 3. Exp. Hematol. **15:** 1140–1144.
39. LORD, B. I., G. MOLINEUX, N. G. TESTA, M. KELLY, E. SPOONCER & T. M. DEXTER. 1986.

The kinetic response of haemopoietic precursor cells, in vivo, to highly purified, recombinant interleukin-3. Lymphokine Res. **5:** 97–104.
40. NAPARSTEK, E., K. WAGNER, D. HARRISON & J. S. GREENBERGER. 1987. Hematopoietic effects of continuous intravenous infusion of mice with growth factors produced by the WEHI-3 cell line. Acta Haematol. **77:** 1–5.
41. BROXMEYER, H. E., D. E. WILLIAMS, S. COOPER, R. K. SHADDUCK, S. GILLIS, A. WAHEED, D. L. URDAL & D. C. BICKNELL. 1987. The comparative effects in vivo of recombinant murine interleukin-3, natural murine colony stimulating factor-1 and recombinant murine granulocyte-macrophage colony stimulating factor on myelopoiesis in mice. J. Clin. Invest. **79:** 721–730.
42. BROXMEYER, H. E., D. E. WILLIAMS & S. COOPER. 1987. The influence in vivo of natural murine interleukin-3 on the proliferation of myeloid progenitor cells in mice recovering from sublethal dosages of cyclophosphamide. Leuk. Res. **11:** 201–205.
43. CRAPPER, R. M., G. VAIRO, J. A. HAMILTON, I. CLARK-LEWIS & J. W. SCHRADER. 1985. Stimulation of bone marrow-derived and peritoneal macrophages by a T lymphocyte-derived hemopoietic growth factor, persisting cell-stimulating factor. Blood **66:** 859–866.
44. STANLEY, R. R., A. BARTOCCI, D. PATINKIN, M. ROSENDAAL & T. R. BRADLEY. 1986. Regulation of very primitive, multipotent hemopoietic cells by hemopoietin 1. Cell **45:** 667–674.
45. BARTELMEZ, S. H. & R. Stanley. 1985. Synergism between hemopoietic growth factors (HGFs) detected by their effects on cells bearing receptors for a lineage specific HGF: Assay of hemopoietin 1. J. Cell. Physiol. **122:** 370–378.
46. CHEN, B. D.-M. & C. R. CLARK. 1986. Interleukin 3 (IL 3) regulates the in vitro proliferation of both blood monocytes and peritoneal exudate macrophages: Synergism between a macrophage lineage-specific colony-stimulating factor (CSF-1) and IL 3. J. Immunol. **137:** 563–570.
47. KOIKE, K., E. R. STANLEY, J. N. IHLE & M. OGAWA. 1986. Macrophage colony formation supported by purified CSF-1 and/or interleukin 3 in serum-free culture: Evidence for hierarchical difference in macrophage colony-forming cells. Blood **67:** 859–864.
48. BROXMEYER, H. E., D. E. WILLIAMS, G. HANGOC, S. COOPER, S. GILLIS, R. K. SHADDUCK & D. C. BICKNELL. 1987. Synergistic myelopoietic actions in vivo after administration to mice of combinations of purified natural murine colony-stimulating factor 1, recombinant murine interleukin 3, and recombinant murine granulocyte/macrophage colony-stimulating factor. Proc. Natl. Acad. Sci. USA **84:** 3871–3875.
49. SCHMITT, E., B. FASSBENDER, K., BEYREUTHER, E. SPAETH, R. SCHWARZKOPF & E. RUDE. 1987. Characterization of a T cell-derived lymphokine that acts synergistically with IL 3 on the growth of murine mast cells and is identical with IL 4. Immunobiology **174:** 406–419.
50. RENNICK, D., G. YANG, C. MULLER-SIEBURG, C. SMITH, N. ARAI, Y. TAKABE & L. GEMMELL. 1987. Interleukin 4 (B-cell stimulatory factor 1) can enhance or antagonize the factor-dependent growth of hemopoietic progenitor cells. Proc. Natl. Acad. Sci. USA **84:** 6889–6893.
51. QUESENBERRY, P. J. 1986. Synergistic hematopoietic growth factors. Int. J. Cell Cloning **4:** 3–15.
52. AISEN, P. & I. LISTOWSKY. 1980. Iron transport and storage proteins. Annu. Rev. Biochem. **49:** 357–393.
53. OCTAVE, J.-N., Y.-J. SCHNEIDER, A. TROUET & R. R. CRICHTON. 1983. Iron uptake and utilization by mammalian cells. I: Cellular uptake of transferrin and iron. Trends Biochem. Sci. **8:** 217–220.
54. GALBRAITH, R. A., S. SASSA & A. KAPPAS. 1985. Heme binding to murine erythroleukemia cells. Evidence for a heme receptor. J. Biol. Chem. **260:** 12198–12202.
55. SMITH, A. & W. T. MORGAN. 1985. Hemopexin-mediate heme transport to the liver. Evidence for a heme-binding protein in liver plasma membranes. J. Biol. Chem. **260:** 8325–8329.
56. MONETTE, F. C. & G. SIGOUNAS. 1984. Factors affecting the proliferation and differentiation of clonogenic hematopoietic stem cells in vitro. Blood Cells **10:** 261–286.
57. MONETTE, F. C. & S. A. HOLDEN. 1982. Hemin enhances the in vitro growth of primitive erythroid progenitor cells. Blood **60:** 527–530.
58. MONETTE, F.C. & S. A. HOLDEN. 1982. Hemin enhancement of primitive erythroid progen-

itors in vitro: Relationship to burst-promoting activity (BPA). Exp. Hematol. **10** (Suppl. 12): 281–294.
59. Monette, F. C., S. A. Holden, M. J. Sheehy & E. A. Matzinger. 1984. Specificity of hemin action in vivo at early stages of hematopoietic cell differentiation. Exp. Hematol. **12:** 782–787.
60. Muller-Eberhard, V., J. Javid, H. H. Liem, A. Hanstein & M. Hanna. 1968. Plasma concentration of hemopexin, haptoglobin and heme in patients with various hemolytic diseases. Blood **32:** 811–815.
61. Muller-Eberhard, Y. & W. T. Morgan. 1975. Porphyrin-binding proteins in serum. Ann. N. Y. Acad. Sci. **244:** 624–649.
62. Taketani, S., H. Kohno & R. Tokunaga. 1986. Receptor-mediated heme uptake from hemopexin by human erythroleukemia K562 cells. Biochem. Int. **13:** 307–312.
63. Taketani, S., H. Kohno & R. Tokunaga. 1987. Cell surface receptor for hemopexin in human leukemia HL 60 cells. J. Biol. Chem. **262:** 4639–4643.
64. Louache, F., U. Testa, P. Pelicci, P. Thomopoulos, M. Titeux & H. Rocant. 1984. Regulation of transferrin receptor in human hematopoietic cell lines. J. Biol. Chem. **259:** 11576–11582.
65. Majuri, R. & R. Grasbeck. 1987. A rosette receptor assay with haem-microbeads. Demonstration of a haem receptor on K562 cells. Eur. J. Haematol. **38:** 21–25.
66. Taketani, S., H. Kohno, Y. Naitoh & R. Tokunaga. 1987. Isolation of the hemopexin receptor from human placenta. J. Biol. Chem. **262:** 8668–8671.
67. Ward, J. H., I. Jordan, J. P. Kushner & J. Kaplan. 1984. Heme regulation of HeLa cell transferrin receptor number. J. Biol. Chem. **259:** 13235–13240.
68. Monette, F. C. & G. Sigounas. 1988. Growth of murine multipotent stem cells in a simple "serum-free" culture system: Role of interleukin-3, erythropoietin, and hemin. Exp. Hematol. **16:** 250–255.
69. Monette, F. C. & G. Sigounas. 1988. Molecularly characterized factors governing the growth of murine multipotent stem cells in serum-depleted marrow cultures. *In* Molecular Biology of Hemopoiesis. Plenum Press. New York, NY. In press.
70. Monette, F.C. & G. Sigounas. 1988. Hemin acts synergistically with interleukin-3 to promote the growth of multipotent stem cells (CFU-GEMM) in "serum-free" cultures of normal murine bone marrow. Exp. Hematol. **16:** 727–729.
71. Sigounas, G. & F. C. Monette. 1988. Long-term clonal development of marrow-derived multipotent stem cells in semi-solid cultures supplemented with interleukin-3, erythropoietin, and hemin. Submitted.

The Effects of Interleukin-3, Bryostatin and Thymocytes on Erythropoiesis

SAUL J. SHARKIS,[a] JOHN P. LEONARD,[a] JAMES N. IHLE,[b] AND W. STRATFORD MAY[a]

[a]*The Johns Hopkins Oncology Center*
600 North Wolfe Street
Baltimore, Maryland 21205
and
St. Jude Children's Research Hospital
Department of Biochemistry
332 North Lauderdale
P.O. Box 318
Memphis, Tennessee 38101

INTRODUCTION

Growth factors have become important for the analysis of the growth potential of hematopoietic progenitor cells both *in vitro* and *in vivo*. Several factors have been shown to have effects on the cells of the erythroid series. These include erythropoietin (EPO), interleukin-3 (IL-3), granulocyte-macrophage colony stimulating factor (GM-CSF), etc. IL-3 (and GM-CSF) has been shown to be released by activated T lymphocytes[1] as well as a myelogenous leukemic tumor cell line.[2] We have demonstrated that cells in the thymus (spleen and bone marrow as well) can either enhance or suppress the growth of erythroid progenitors *in vivo* and *in vitro*.[3,4] We have further recently demonstrated that a low molecular weight (approximately 900) macrocylic lactone extracted from a marine invertebrate (*bugula neritina*) has similar effects as the glycoproteins such as IL-3 and GM-CSF on human hematopoietic progenitors. It is the purpose of this present report to compare the effects of this product, bryostatin 1, and IL-3 on murine erythropoiesis with or without the influence of thymic regulatory accessory cells. We will utilize the W/W^v mouse which has been used in numerous studies of regulation of hematopoiesis because of an inherent hematopoietic stem cell (HSC) defect and profound erythropoietic deficits. We have also observed[6] an abnormal thymocyte subset in this mouse which in normal mice regulates HSC self-renewal and modulates the differentiation of erythroid progenitors. Since the disorder in W/W^v anemic mice involves T cells (which might produce IL-3) the effects of adding IL-3 or bryostatin and T cells to cultures of W/W^v bone marrow for the growth of erythroid progenitors may allow us to study the *in vitro* abnormality which characterizes this defect.

METHODS

Animals

Six to 8 week old male and female $WBB6F_1$ mice were purchased from the Jackson Laboratories (Bar Harbor, ME). These F_1 strain littermates, $+/+$ (hematopoietically normal) and W/W^v (HSC-lymphocyte defective) were offspring of WB/REJ ($W/+$) and C57B1/6J ($W^v/+$) parents.

TABLE 1. Effect of IL-3 on BFU-E from +/+ and *W/W* v Mice (Cultured in 3% FBS)

Type of Mouse	Concentration of IL-3 (Units/Culture)	BFU-E/5 × 10^4 BM Cells Plated Mean ± SEM[a]
+/+	none	0.0 ± 0.0
	1.2	1.0 ± 0.3
	2.5	0.0 ± 0.0
	5.0	1.5 ± 0.4
	10.0	8.3 ± 0.7
	20.0	10.7 ± 1.1
W/W v	none	0.0 ± 0.0
	1.2	4.6 ± 1.7
	2.5	3.0 ± 1.1
	5.0	5.3 ± 1.1
	10.0	6.9 ± 1.9
	20.0	10.8 ± 2.5

[a]Results represent the mean ± SEM for 3 separate experiments done at different times.

Growth Factors

Homogeneously purified interleukin-3 was obtained as previously described.[2] Bryostatin 1 was a generous gift from G. R. Pettit of Arizona State University. Both factors were prepared at appropriate concentrations in McCoy's media (Gibco, NY) and sterilized by filter. Recombinant human erythropoietin was purchased from Amgen (Thousand Oaks, CA).

Cell Suspensions

Bone marrow cells and thymocytes were harvested from mice sacrificed by cervical dislocation. Single cell suspension of these tissues were prepared as described.[3,4]

TABLE 2. Effect of IL-3 on CFU-E from +/+ and *W/W* v Mice (Cultured in 20% FBS)

Type of Mouse	Concentration of IL-3 (Units/Culture)	CFU-E/5 × 10^4 BM Cells Plated Mean ± SEM[a]
+/+	none	85.8 ± 7.8
	1.2	73.6 ± 7.8
	2.5	75.9 ± 8.1
	5.0	57.3 ± 5.2
	10.0	62.8 ± 4.8
	20.0	49.6 ± 4.9
W/W v	none	77.3 ± 6.0
	1.2	122.3 ± 15.5
	2.5	152.0 ± 11.4
	5.0	149.0 ± 11.0
	10.0	165.6 ± 14.0
	20.0	167.4 ± 19.2

[a]Results represent the mean ± SEM for 3 separate experiments done at different times.

Progenitor Cell Assays

We utilized the plasma clot method[7,8] for growth of CFU-E and BFU-E. We replaced or diluted the fetal calf serum with McCoy's media to study the effects of IL-3 and bryostatin on the progenitor cell growth in "low serum". Added IL-3, bryostatin and thymocytes in McCoy's replaced part or all of the McCoy's media in the assays. Clots in micro test plates were incubated at 37°C in 5% CO_2, 95% air for 2 days (CFU-E) or 7 days (BFU-E). Clots were fixed to microscope slides with 5% gluteraldehyde (Fisher Scientific, Silver Spring, MD), washed in distilled water and stained with benzidine. Erythroid progenitors were scored as ≥8 cells for colonies and ≥50 cells for bursts. When culturing bone marrow only 5×10^4 cells were plated, in co-cultures 10^6 thymocytes were added.

TABLE 3. Effect of Bryostatin on BFU-E from +/+ and *W/W* ^{v} Mice (Cultured in 3% FBS)

Type of Mouse	Concentration of Bryostatin (M/ml)	BFU-E/5 × 10^4 BM Cells Plated Mean ± SEM[a]
+/+	none	0.0 ± 0.0
	10^{-7}	0.5 ± 0.3
	10^{-8}	0.0 ± 0.0
	10^{-9}	0.3 ± 0.2
	10^{-10}	0.0 ± 0.0
	10^{-11}	0.0 ± 0.0
	10^{-12}	0.4 ± 0.2
	10^{-13}	0.0 ± 0.0
	10^{-14}	0.1 ± 0.1
	10^{-15}	0.0 ± 0.0
W/W ^{v}	none	6.3 ± 2.3
	10^{-7}	46.4 ± 16.4
	10^{-8}	26.5 ± 9.1
	10^{-9}	5.4 ± 2.5
	10^{-10}	2.8 ± 1.0
	10^{-11}	0.9 ± 0.4
	10^{-12}	2.8 ± 0.6
	10^{-13}	0.1 ± 0.1
	10^{-14}	1.1 ± 0.4
	10^{-15}	1.1 ± 0.1

[a]Results represent the mean ± SEM for 3 separate experiments done at different times.

Treatment of Cells with Anti-IL-3 Antibody

Anti-IL-3 antibody was obtained as described previously[9] by immunizing rabbits with IL-3. This antiserum neutralized IL-3 activity. Three hundred μg/ml was added to either 200 units of IL-3 or 10^{-11} M bryostatin. The results were compared to the effect of normal rabbit serum which was heat inactivated (30 minutes at 60°C).

RESULTS

Effects of IL-3 on Erythroid Colony Formation

In order to determine effects of IL-3 on erythroid progenitors of both normal (+/+) and anemic (*W/W* ^{v}) mice we examined colony formation (BFU-E and CFU-E) from both animals. BFU-E were grown in 3% fetal calf serum to reduce significantly any endogenous burst promoting activity in the serum. As can be seen in TABLE 1 both normal and anemic bone marrow BFU-E responded to increasing doses of IL-3; however, the response in the W/W^v mouse appears at lower concentrations than in the +/+ mouse. The effect of IL-3 on the later progenitor CFU-E is seen in TABLE 2. IL-3 added to CFU-E grown in 20% fetal calf serum from normal +/+ mice was not stimulatory and at 20.0 units per culture was inhibitory. On the other hand W/W^v CFU-E numbers increased in the presence of IL-3. Taken together this data suggests that IL-3 regulates erythropoiesis in the W/W^v mouse by stimulation of the early and late progenitors and is a more sensitive regulator for early progenitors in the anemic recipient.

TABLE 4. Effect of Bryostatin on CFU-E from +/+ and W/W^v Mice (Cultured in 20% FBS)

Type of Mouse	Concentration of Bryostatin (M/ml)	CFU-E/5 × 10^4 BM Cells Plated Mean ± SEM[a]
+/+	none	58.1 ± 2.5
	10^{-7}	57.9 ± 2.0
	10^{-8}	61.0 ± 5.1
	10^{-9}	56.8 ± 4.2
	10^{-10}	120.9 ± 8.4
	10^{-11}	294.3 ± 4.4
	10^{-12}	110.9 ± 7.5
	10^{-13}	90.4 ± 3.5
	10^{-14}	59.4 ± 2.8
	10^{-15}	60.1 ± 4.1
W/W^v	none	110.9 ± 11.8
	10^{-7}	236.9 ± 24.4
	10^{-8}	181.5 ± 24.4
	10^{-9}	185.9 ± 17.9
	10^{-10}	163.8 ± 10.5
	10^{-11}	145.1 ± 24.0
	10^{-12}	168.5 ± 8.0
	10^{-13}	147.8 ± 21.7
	10^{-14}	185.4 ± 24.4
	10^{-15}	125.9 ± 16.1

[a]Results represent the mean ± SEM for 3 separate experiments done at different times.

Effects of Bryostatin on Erythroid Colony Formation

Since we previously observed and reported a dose responsive modulation of bryostatin 1 on human erythroid progenitor cell growth we set up experiments to examine the effect of bryostatin on murine progenitors. We found (TABLE 3) that BFU-E from W/W^v but not +/+ marrow was stimulated by bryostatin 1. More interestingly (TABLE 4) bryostatin 1 stimulated CFU-E from W/W^v marrow over a wide concentration range but +/+ stimulation of CFU-E was confined to the same range which stimulated human CFU-E.[5] Since the W/W^v mouse responded to both IL-3 and bryostatin we tested whether the bryostatin effect could be blocked with an antibody to IL-3. As can be seen in TABLE 5 antibody to IL-3 has the capacity to block the bryostatin stimulatory effect on W/W^v erythroid colonies. Heat inactivated rabbit serum did not block the stimulatory effect of bryostatin

TABLE 5. Effect of Anti-IL-3 Neutralizing Antibody on CFU-E from *W/W* ^{v} in the Presence of Either IL-3 or Bryostatin

Treatment of Marrow	CFU-E/5 × 10^4 BM Cells Plated
None	41
300 μg/ml ∝-IL-3	29
20 units/ml IL-3	84
20 units IL-3 + 300 μg ∝-IL-3	16
10^{-11} M bryostatin	88
10^{11} M bryostatin + 300 μg ∝-IL-3	23
Heat inactivated normal rabbit serum	38

on erythropoiesis. Finally, lymphocytes from *W/W* ^{v} mice are defective in their ability in co-culture to stimulate CFU-E.[6] In order to determine if bryostatin could induce *W/W* ^{v} thymocytes to stimulate erythroid colony growth we co-cultured *W/W* ^{v} thymocytes with either *W/W* ^{v} or +/+ bone marrow in the absence or presence of added bryostatin. We observed (TABLE 6) that thymocytes from *W/W* ^{v} mice exposed to bryostatin now had the capacity to stimulate *W/W* ^{v} marrow to produce more CFU-E. In addition *W/W* ^{v} thymocytes inhibited +/+ bone marrow in co-culture when exposed to bryostatin suggesting the elaboration of an IL-3-like molecule from *W/W* ^{v} thymocytes when exposed to bryostatin.

DISCUSSION

IL-3 and bryostatin both stimulate human hematopoietic progenitors.[5] Our experiments were designed to examine in an erythropoietically compromised host if *in vitro* both these molecules could stimulate growth of erythroid progenitors and if the mechanism of action was similar for both growth factors. The glycoprotein IL-3 and the macrocylic (small molecular weight—approximately 900) lactone, bryostatin 1 both stimulate growth of early and late erythroid progenitors from the *W/W* ^{v} mouse. This suggested a possible

TABLE 6. Effect of 10^6 *W/W* ^{v} Thymocytes on the Action of Bryostatin 1 on CFU-E Growth from *W/W* ^{v} or +/+ Bone Marrow

Bone Marrow Donor	Bryostatin 1 Concentration (M)	Number of Added W/W^{v}Thymocytes	CFU-E/5 × 10^4 BM Cells Plated[a]
W/W ^{v}	—	—	147.25 ± 7.56
	10^{-11}	—	174.38 ± 8.89
	10^{-7}	—	208.88 ± 6.35[b]
	—	10^6	176.88 ± 5.68
	10^{-11}	10^6	305.00 ± 7.22[b]
	10^{-7}	10^6	414.63 ± 7.33[b]
+/+	—	—	149.50 ± 10.47
	10^{-11}	—	215.50 ± 7.58[b]
	10^{-7}	—	65.88 ± 9.18
	—	10^6	184.25 ± 6.70
	10^{-11}	10^6	187.75 ± 10.19
	10^{-7}	10^6	92.50 ± 8.45

[a]Mean ± SEM for 3 experiments.

[b]$p < 0.05$ when compared with control.

common pathway of action of the two molecules. Our antibody data further implicates bryostatin as a potential erythropoietic stimulatory factor with an IL-3-like action. W/W^v thymocytes in co-culture with W/W^v bone marrow in the presence of bryostatin results in stimulation of erythroid colonies suggesting a release by the thymocytes of a growth potentiator. We would like to believe that since the action of bryostatin is similar to that of IL-3 and the bryostatin effect can be blocked by neutralizing antibody to IL-3 that bryostatin is acting via release of IL-3 from otherwise functionally defective W/W^v thymic accessory cells. However, the production of IL-3 by W/W^v thymocytes has not been demonstrated. In fact we are unable to condition media from either thymocytes or bone marrow exposed to bryostatin for 48 hours (data not shown). Therefore, we cannot rule out the possibility that bryostatin acts directly on a defective erythroid progenitor in the W/W^v mice by altering the cell target (BFU-E and CFU-E) response to other growth factors or modulation of the signal transduction pathway of these cells. Bryostatin has been shown[10–12] to activate protein-kinase-C (PKC). This enzyme has been involved in the growth related signal transduction pathway in cells,[13] and it has been demonstrated to exist in a variety of forms. It is possible that W/W^v and +/+ cells have different forms of PKC and the action of bryostatin on these two cell types results in different erythropotentiating activities. Regardless of the definitive mechanism of action of this molecule, potentially it may prove useful as a biological response modifier for clinical use.

REFERENCES

1. CUTLER, R. L., D. METCALF, N. A. NICOLA & G. R. JOHNSON. 1985. Purification a multipotential colony stimulating factor from pokeweed mitogen-stimulated mouse spleen cell conditioned medium. J. Biol. Chem. **260:** 6579–6587.
2. IHLE, J. N., J. KELLER, L. HENDERSON, F. KLEIN & E. PALASYNSKI. 1982. Procedures for the purification of Interleukin-3 to homogeneity. J. Immunol. **129:** 2431–2436.
3. SHARKIS, S. J., W. W. JEDRZEJCZAK, A. AHMED, G. W. SANTOS, A. MCKEE & K. W. SELL. 1978. Anti-theta sensitive regulatory cell (TSRC) and hematopoiesis: Regulation of transplanted stem cells in W/W^v anemic and normal mice. Blood **52:** 802–817.
4. SHARKIS, S. J., J. L. SPIVAK, A. AHMED, J. MYSITI, R. K. STUART, W. WIKTOR-JEDRZEJCZAK, K. W. SELL & L. L. SENSENBRENNER. 1980. Regulation of hematopoiesis: Helper and suppressor influences of the thymus. Blood **55:** 524–527.
5. MAY, W. S., S. J. SHARKIS, A. H. ESA, V. GEBBIA, A. S. KRAFT, G. R. PETTIT & L. L. SENSENBRENNER. 1987. The antineoplastic bryostatins are multipotential stimulators of human hematopoietic progenitor cells. Proc. Natl. Acad. Sci. USA **84:** 8483–8487.
6. SHARKIS, S. J. 1987. Cell to cell interactions in erythropoiesis. The role of the thymus. *In* Molecular and Cellular Aspects of Erythropoietin and Erythropoiesis. I. N. Rich, Ed. NATO ASI Series H. Cell Biology Vol. **8:** 229–236. Springer-Verlag. Berlin.
7. STEVENSON, J. R., A. AXELRAD, D. L. MCLEOD & M. M. SHREEVE. 1971. Induction of colonies of hemoglobin synthesizing cells by erythropoietin *in vitro*. Proc. Natl. Acad. Sci. USA **68:** 1542–1546.
8. MCLEOD, D. L., M. M. SHREEVE & A. AXELRAD. 1974. Improved plasma culture system for production of erythrocytic colonies *in vitro*. Quantitative assay method for CFU-E. Blood **44:** 517–534.
9. BOWLIN, T. L., A. N. SCOTT & J. N. IHLE. 1984. Biologic preparation of Interleukin-3. II. Serologic comparison of 20-α-SD4 inducing activity, colony stimulating activity, and WEHI-3 growth factor activity by using an antiserum against IL-3. J. Immunol. **133:** 2001–2006.
10. KRAFT, A. S., V. V. BAKER & W. S. MAY. 1987. Bryostatin induces a rapid loss of calcium phospholipid dependent protein kinase activity from human promyelocytic leukemic cells (HL-60). Oncogene **1:** 111–118.
11. KRAFT, A. S., J. B. SMITH & R. L. BERKOW. 1986. Bryostatin, an activator of the calcium phospholipid-dependent protein kinase, blocks phorbol ester-induced differentiation of human promyelocytic leukemia cells HL-60. Proc. Natl. Acad. Sci. USA **83:** 1334–1338.

12. RAMSDELL, J. S., G. R. PETTIT & A. H. TASHTIAN. 1986. Three activators of protein kinase C, bryostatins, dioleins, and phorbol esters, show differing specificities of action on GH_4 pituitary cells. J. Biol. Chem. **261:** 17073–17080.
13. NISHIZUKA, Y. 1986. Studies and perspectives of protein kinase C. Science **233:** 305–312.

Nutrition Requirements for Mammalian Cells and Hematopoietic Growth Factor Production

MARCIA J. ARMSTRONG,[a] HENRY B. WARREN,[b]
PETER F. DAVIES,[b] AND NICHOLAS DAINIAK[a]

[a]*Departments of Medicine and Biomedical Research*
St. Elizabeth's Hospital of Boston
Boston, Massachusetts 02135

[a]*Department of Medicine*
Tufts University School of Medicine
Boston, Massachusetts 02111

[b]*Department of Pathology*
Brigham and Women's Hospital
Harvard Medical School
Boston, Massachusetts 02115

INTRODUCTION

The proliferation of populations of self-renewing mammalian bone marrow stem cells *in vitro* requires a variety of growth factors and nutrients. These factors include specific endocrine and paracrine hormones (such as erythropoietin, interleukin-3, granulocyte-macrophage colony-stimulating factor, erythroid burst-promoting activity, etc.), proteins (such as albumin and transferrin), trace elements (such as selenium and iron), vitamins, carbohydrates, amino acids, lipoproteins and lipids.[1,2] Several of the cell types found in bone marrow, such as lymphocytes and endothelial cells, have been found to release hematopoietic growth-promoting molecules into liquid culture medium.[3–5] Since the marrow culture milieu is highly complex, consisting of numerous growth-promoting substances, essential nutrients and multiple cell types, it is possible that interactions among nutritional factors and cells may alter growth factor production, resulting in net enhancement or suppression of hematopoietic proliferation and/or differentiation. Examples of such interactions that have recently been elucidated include augmentation of granulocyte-macrophage colony-stimulating factor and granulocyte colony-stimulating factor production by the monokine interleukin-1,[6–8] and enhancement of erythroid burst-promoting activity release by thyroid hormone.[9] Here, we review data suggesting that although purified phospholipids support hematopoietic colony formation under serum-free conditions, lipoproteins negatively regulate the release of hematopoietic growth factors from lymphoid tissue and endothelium.

Support of Hematopoietic Colony Formation by Exogenous Phospholipids and Cell-Derived Growth Factors

Because serum as well as crude plasma erythropoietin contains an abundant amount of lipoproteins, it was highly desirable to develop a serum-free culture system using as many highly purified products as possible in order to valuate phospholipid effects on hematopoietic colony formation. Therefore, a marrow culture system containing serum substitute (consisting of iron-saturated transferrin and Iscove's modified Dulbecco's medium), delipidated, highly purified human serum albumin and recombinant human erythropoietin was employed.[10] Albumin migrated as a single polypeptide with a broad isoelectric point after electrophoresis in the second dimension.[10] Under these conditions, we observed a dose-related enhancement of erythroid burst formation by purified phosphatidylcholine (see FIG. 1). In contrast, phosphatidylserine did not alter the number of erythroid bursts observed in culture of nonadherent bone marrow mononuclear cells. This observation was of particular interest to us since phosphatidylcholine is located in the outer half of the lipid bilayer of cell membranes and because we have previously observed that plasma membranes and shed membrane-derived vesicles express erythroid burst-promoting activity.[11]

Application of serum-free culture conditions to the preparation of medium conditioned by lymphocytes or endothelial cells was important not only for characterizing growth factors released by these cells but also for determining whether the presence of lipoproteins during the conditioning process altered growth factor release. Accordingly, lymphocyte-conditioned medium (LCM) was prepared by incubating 5 × 10^6 lymphocytes/ml in alpha medium supplemented with L-glutamine, streptomycin and penicillin. The conditioned medium was harvested 14 h later and added to serum-free bone marrow cultures. For the preparation of endothelial cell-conditioned medium (ECCM), bovine aortic endothelial cells or passaged human endothelial cells were grown in confluent layers at a density of 10^6/ml of nonmitogenic base medium containing DMEM, HEPES, and 5% lipoprotein-deficient plasma. For biochemical analysis, endothelial cells were grown to confluence under serum-free conditions. ECCM was harvested after two days of incubation, and added to bone marrow culture. Both LCM and ECCM were concentrated by filtration on an Amicon apparatus, and added to culture over a wide range of concentrations. In some cases, ECCM and LCM were added to culture together.

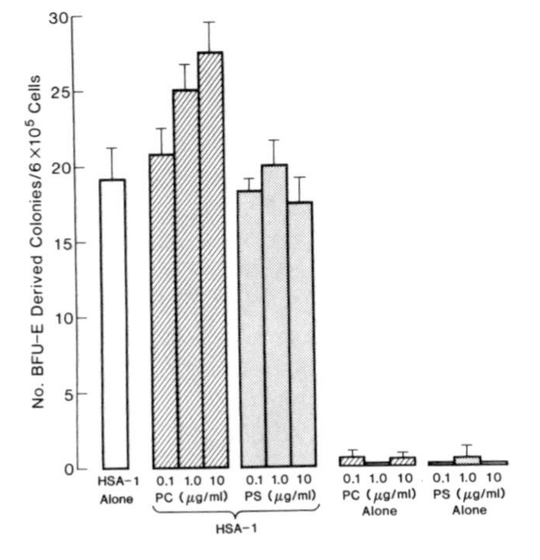

FIGURE 1. Stimulation of erythroid burst formation by purified phospholipids. Marrow cultures were established with NCTC-109 (*open bars*), phosphatidylcholine (PC, *hatched bars*) or phosphatidylserine (PS, *stippled bars*) in the presence and absence of albumin (HSA-1) freshly purified from whole human plasma by affinity chromatography using Cibacron blue F-3-GA conjugated to Sepharose. HSA-1 was delipidated prior to deionization and addition to culture. Mean ± SE bursts are displayed. (From Dainiak *et al.*[10] Reprinted by permission from *Experimental Hematology*.)

As shown in FIGURE 2, both LCM and ECCM augmented erythroid burst formation in a saturable fashion. Furthermore, when a single concentration of 1 × ECCM was added together with various amounts of LCM, an additive stimulatory effect was observed even in the plateau portion of the dose-response curve (see FIG. 2). This was also observed

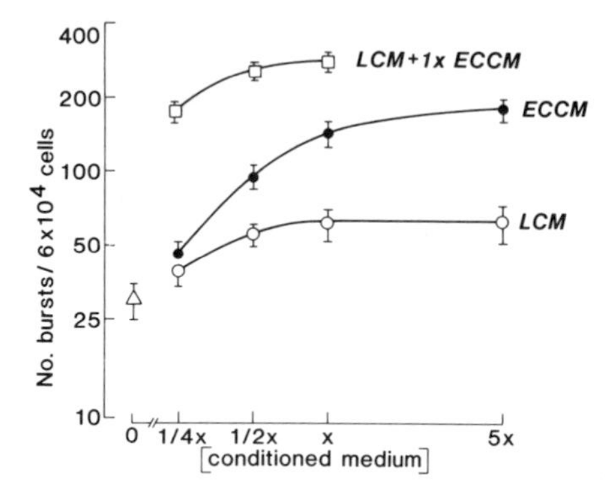

FIGURE 2. Support of erythroid burst formation by conditioned media. Lymphocyte conditioned medium (LCM) and medium conditioned by bovine aortic endothelial cells (ECCM) were added to cultures of marrow mononuclear cells containing nonmitogenic base medium (0) or 10% (v/v) LCM (*open circles*), ECCM (*closed circles*) or LCM plus 1 × ECCM (*squares*) in the presence of 2.0 U/ml recombinant erythropoietin. Mean ± SE bursts are displayed. Note that an additive effect is observed in cultures containing both ECCM plus LCM relative to those with LCM alone. Similar results were observed in cultures containing 1× LCM plus various amounts of ECCM (not shown).

in cultures plated with 1 × LCM plus various ECCM concentrations (data not shown). In addition, ECCM stimulated the proliferation of erythroid CFU-E-derived colonies, granulocyte/macrophage colonies and mixed hematopoietic colonies over a full range of erythropoietin, colony-stimulating factor and LCM concentrations, respectively.[12]

These data suggest that the growth factor(s) released from endothelial cells is distinct from that released from lymphocytes. This possibility is supported by the finding that in contrast to LCM, ECCM augments the size and number of Day 7 CFU-E derived colonies as well as Day 12 erythroid bursts (see FIG. 3). Moreover, the colony stimulatory activity was partially removed by preabsorption of ECCM with a recombinant GM-CSF antibody.[12] Finally, the approximate size of the active erythroid enhancing molecules in ECCM was assessed on preparative SDS gels. Unlike the active factor in LCM which was found to have an Mr of approximately 28,000 daltons,[13] the active moiety in ECCM comigrated with molecules having a molecular mass of 75,000 to 160,000 daltons.[12] These data strongly suggest that growth factors released by lymphocytes are biochemically and immunologically distinct from those released by endothelial cells.

Lipid Inhibition of Growth Factor Release

Little is known concerning the nutritional requirements for the production of growth factors by mammalian cells. However, recently Fox and Dicorleto have suggested that acetylated low density lipoproteins suppress the production of platelet-derived growth factor (PDGF) by endothelium.[14] This effect may be due to oxidation of modified lipoproteins.[15] Because we have previously shown that PDGF is a determinant of optimal erythroid progenitor cell proliferation,[16] we reasoned that exposure of endothelium or lymphocytes to low density lipoproteins (LDL) might impair the release of hematopoietic growth-promoting molecules. Therefore, LCM and ECCM were prepared in the presence of 0–400 μg LDL, acetylated LDL (AcLDL) or very low density lipoproteins (VLDL) per ml. Incubation of endothelial cells with lipoproteins markedly increased the cellular cholesterol content. Phase contrast microscopy of cholesterol-loaded cells showed no significant morphological alterations even with an average elevation in free cholesterol level of approximately 36%. Lymphocyte viability was greater than 99% following a one-day incubation with lipoprotein. As shown in FIGURE 4, incubation of lymphocytes with VLDL had no significant effect on the expression of growth factor in the spent medium. In contrast, incubation with LDL or AcLDL showed a profound inhibitory effect on growth factor expression. Similar results were observed following cholesterol-loading of endothelial cells.[12]

To determine whether the inhibitory effects of LDL and AcLDL were indeed due to the inhibition of growth factor release rather than a direct suppressive effect of the lipoproteins on hematopoietic progenitor cell growth, conditioned media were prepared in the presence and absence of lipid and, prior to addition to marrow culture, the media were delipidated by ultracentrifugation with solid potassium bromide. FIGURE 5 shows that delipidation of ECCM prepared with LDL unmasks stimulatory activity in the infranate. In contrast, no stimulatory activity is found in infranates of ECCM prepared with AcLDL.

Furthermore, dilution of ECCM prepared with 100–400 μg AcLDL per ml resulted in reversal of growth factor inhibition and a downward displacement of the dose-response curve. On the other hand, dilution of ECCM prepared with LDL resulted in leftward displacement of the curve, consistent with a kinetically distinct response.[12] Together, the data suggest that while LDL may directly suppress hematopoietic proliferation under serum-free conditions (*i.e.*, when endogenous cholesterol synthesis takes place), acetylated LDL impairs the production and/or release of hematopoietic growth factors from accessory cell populations.

Acetylated Lipoproteins and PDGF Production

We have previously shown that PDGF increases the number and size of bone marrow CFU-E-derived and BFU-E-derived colonies.[16] To probe the relationship of endothelial

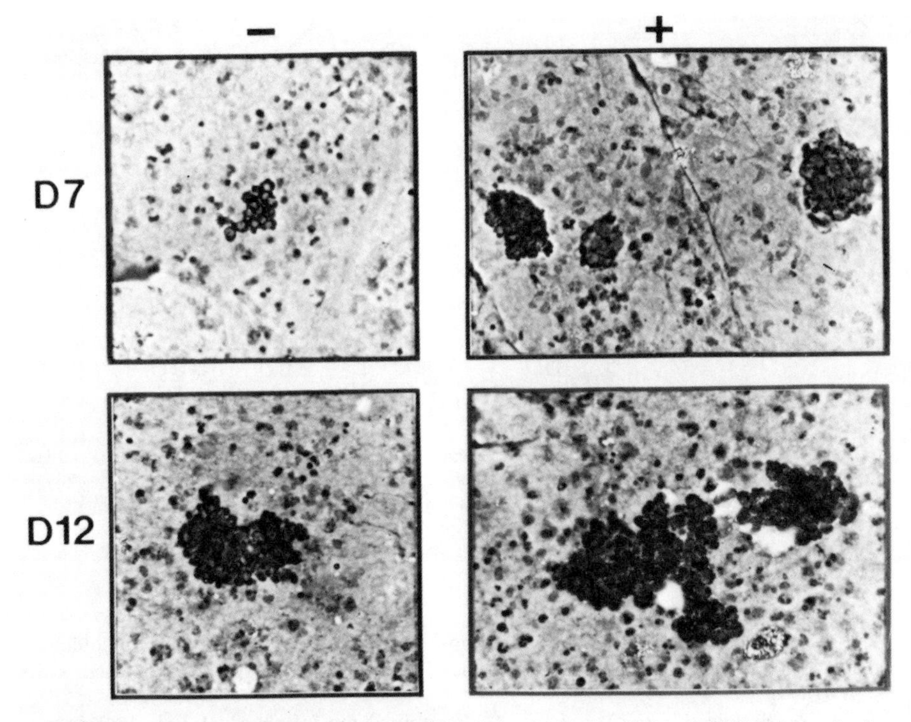

FIGURE 3. Photomicrograph of typical CFU-E-derived colonies (D7) and BFU-E-derived colonies (D12) formed in serum-free medium in the presence (+) and absence (−) of 10% ECCM. In addition to enhancing proliferation, ECCM increases colony size.

cell-derived erythroid-enhancing factors and PDGF-like proteins, we next examined whether suppression of erythroid growth factor release by AcLDL correlates with suppression of PDGF production. Thymidine labeling of smooth muscle cells was used to assay for PDGF in ECCM. In this assay, baseline and control levels are represented by labeling of cells incubated with 0.4% and 10% calf serum, respectively. FIGURE 6 shows the effects of ECCM prepared with serum or with each of two different concentrations of

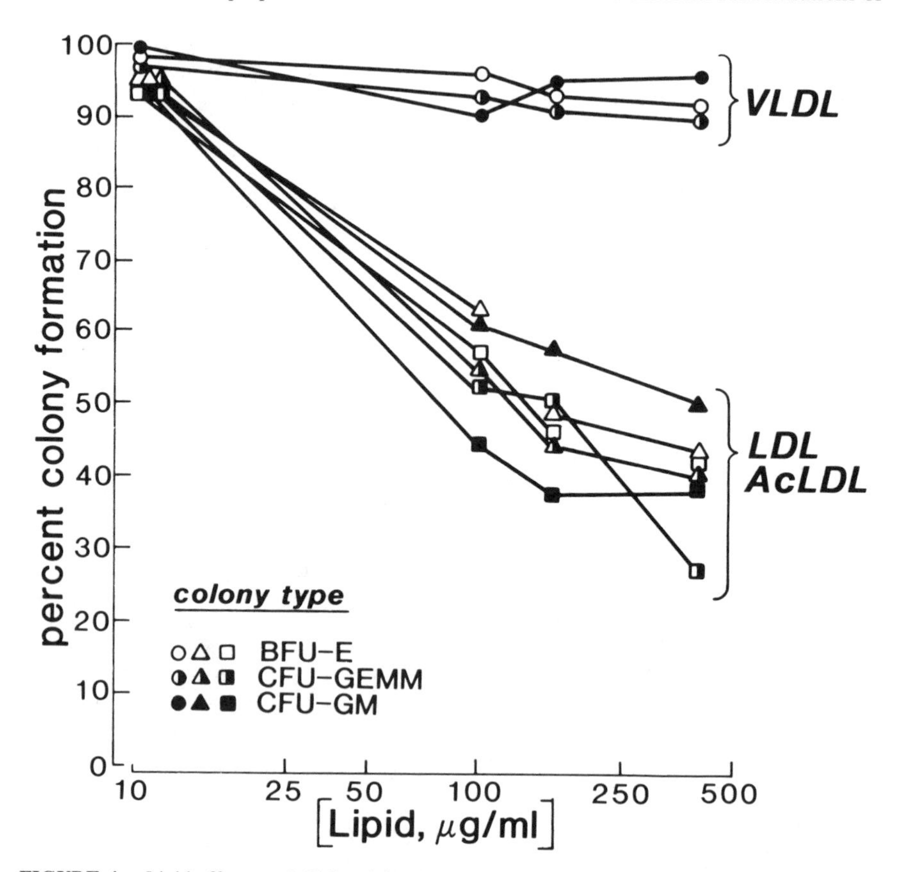

FIGURE 4. Lipid effects on LCM activity. Mean ± SE BFU-E-derived (*closed symbols*), CFU-GM-derived (*open symbols*) and CFU-GEMM-derived (*half closed symbols*) colonies are displayed for cultures containing 10% LCM prepared from lymphocytes that were incubated with VLDL (*circles*), LDL (*triangles*) or AcLDL (*squares*). Control plates expressing 100% colony formation contained 21 ± 1 bursts, 48 ± 5 granulocyte-macrophage colonies and 10 ± 1 mixed hematopoietic colonies/6 × 10^4 cells.

AcLDL on release of mitogens for smooth muscle cells. The data indicate that endothelial cells release a quantitatively similar amount of mitogen whether or not incubated with AcLDL.

To corroborate this finding, we next determined the influence of lipid loading on beta-chain PDGF gene expression. Accordingly, cytoplasmic RNA was extracted from human endothelial cells that were incubated under cholesterol-depleted and loaded con-

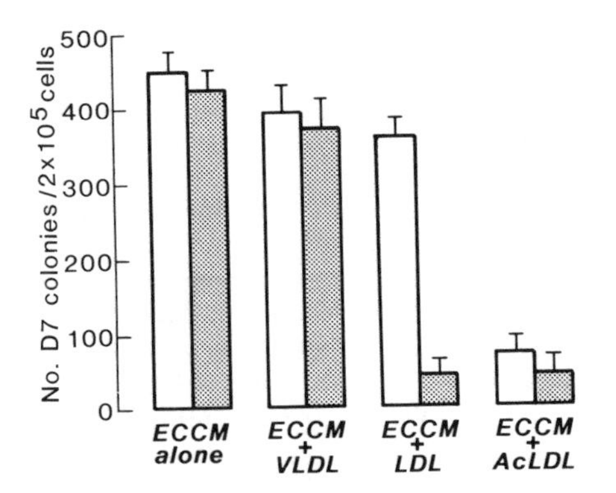

FIGURE 5. Effects of delipidation on growth factors expressed by ECCM. Mean SE CFU-E-derived colonies are displayed for cultures containing 10% supranates (*stippled bars*) or infranates (*open bars*) of ECCM prepared by centrifugation of spent medium at 123,000 × g, 20 h, 10°C after exposure to 7.5% solid KBr. Note that lipid-free infranate of medium from LDL-exposed cells expresses enhancing activity, while that from AcLDL-exposed cells has no stimulatory activity. (From Dainiak *et al.*[12] Reprinted by permission from the *Journal of Clinical Investigation.*)

ditions. RNA transcripts specific for the beta chain were identified by hybridization with a nick-translated 32P-labeled v-sis DNA probe, and relative amounts of transcripts were detected by dot blot analysis.[12] The dot blot in FIGURE 7 compares activity with serial dilutions of RNA extracted from human osteosarcoma cells and fibroblasts as positive and negative controls, respectively. We observed no difference in the relative amounts of specific beta chain PDGF message in endothelial cells prepared in the presence and absence of acetylated LDL in two separate experiments (see FIG. 7). Likewise, similar amounts of immunoprecipitable PDGF were found in the media (data not shown). Furthermore, the erythropoietic factor in ECCM was found to be heat labile and resistant to thiol-containing compounds, suggesting that the activity is biochemically distinct from PDGF.

CONCLUDING REMARKS

It is well known that mammalian cells require cholesterol for membrane synthesis. Lipoproteins of the LDL class function to deliver cholesterol to mammalian cells through the process of receptor-mediated endocytosis. These lipoproteins regulate intracellular cholesterol synthesis through coordinate expression of the genes for the LDL receptor and HMG coA reductase.[17] The LDL receptor has sequence homology with the precursor for

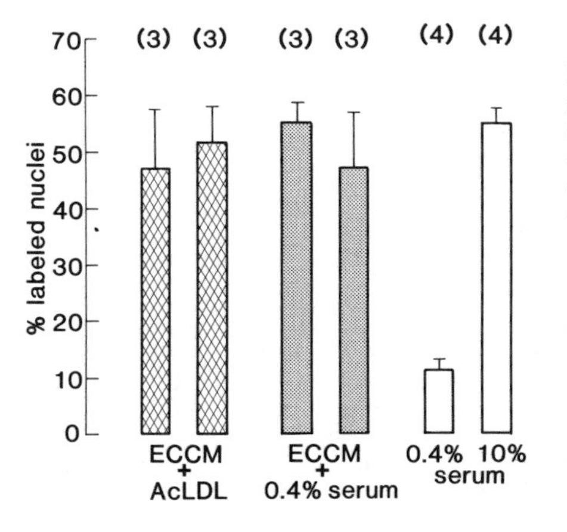

FIGURE 6. Endothelial cells release mitogens for smooth muscle cells. Mean ± SE thymidine labeling of bovine aortic smooth muscle cells is shown for cells incubated with medium conditioned by bovine aortic endothelial cells previously incubated with 0.4% lipoprotein-deficient serum (*stippled bars*) or with 388 μg (*left hatched bar*) or 160 μg (*right hatched bar*) AcLDL/ml. Control cells were incubated with 0.4% nonmitogenic serum or 10% serum (*open bars*). *Numbers in parentheses* refer to number of determinations. Note that endothelial cells release mitogens for smooth muscle cells whether or not incubated with AcLDL.

EGF, although not for EGF itself. While the functional significance of this finding is obscure, it is tempting to speculate that there may be some relation between this sequence homology and the inhibition of LDL-mediated growth factor release. In this communication, we have studied the effects of lipoproteins on growth factor release from endothelial cells and lymphocytes under serum-free conditions. Our results suggest that modified lipoproteins block the release of growth factors from lymphocytes and endothelial cells (see FIG. 8). In addition, the factor(s) released by endothelial cells appears to be distinct from that released by lymphocytes (*e.g.*, burst-promoting activity) as well as distinct from PDGF, a previously described growth factor for erythroid progenitor cells.

In contrast to the effects of AcLDL, unmodified low density lipoproteins may have a direct inhibitory effect on hematopoietic progenitor cells (see FIG. 8). However, since no attempt was made in these studies to purify stem cell populations prior to culture, it is also possible that LDL suppresses the endogenous synthesis and/or release of growth regulators from marrow accessory cells, including monocytes and fibroblasts (reviewed in REF. 2). The latter hypothesis may be analagous to the inhibitory effects that serum LDL are known to have on mammalian cell function such as inhibition of immune responses,[18] suppression of lymphocyte proliferation[19] and inhibition of human marrow cell turnover.[20,21] Since many hematopoietic growth factors are produced at low levels constitutively, and are inducible by agents known to trigger an immunologic response, impaired stem cell proliferation may be due in part to immunosuppression by native LDL. Whether similar effects are observed in cultures containing serum or in cultures containing additives that are heavily contaminated with lipoproteins,[22–24] remains to be determined.

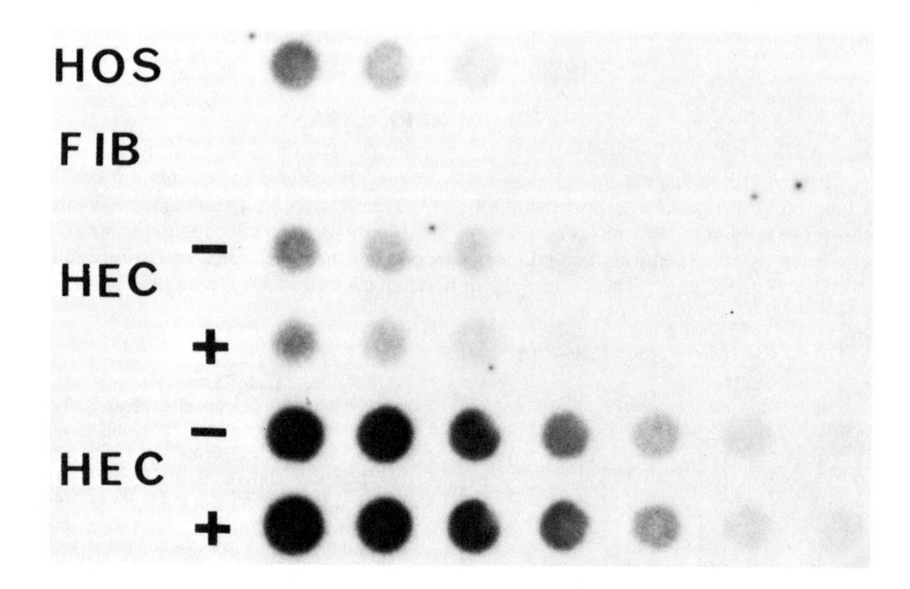

FIGURE 7. Effects of AcLDL on sis levels in endothelial cells. The dot blot compares sis RNA levels in human endothelial cells incubated with (+) or without (−) AcLDL. RNA from two different experiments was serially diluted 1:1. Note that there is no significant difference in the relative amounts of specific beta chain PDGF transcripts. HEC: human endothelial cells; HOS: human osteosarcoma cells; FIB: human dermal fibroblasts. (From Dainiak *et al.*[12] Reprinted by permission from the *Journal of Clinical Investigation.*)

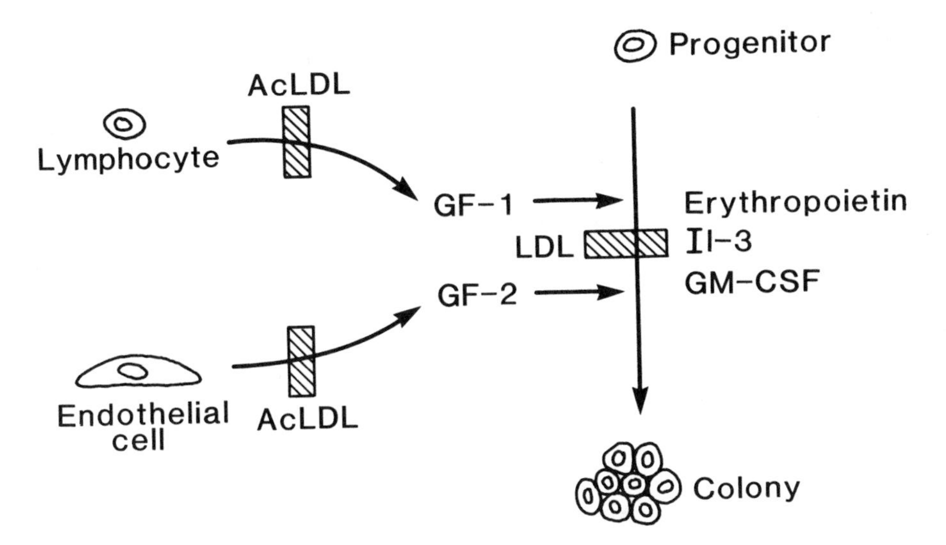

FIGURE 8. Hypothetical model of lipid effects on hematopoietic progenitor cell growth. Whereas LDL may directly inhibit progenitor cell proliferation and/or differentiation under serum-free conditions, AcLDL appears to inhibit the release of hematopoietic growth factors (GF-1 and GF-2) from lymphocytes and endothelial cells.

REFERENCES

1. DAINIAK, N. 1985. *In* Hematopoietic Stem Cell Physiology. E. P. Cronkite, N. Dainiak, R. P. McCaffrey, J. Palek & P. J. Quesenberry, Eds.: 59–76. Alan R. Liss. New York, NY.
2. SIEFF, C. A. 1987. J. Clin. Invest. **79:** 1549–1557.
3. GORDON, M. Y. & E. C. GORDON-SMITH. 1981. Br. J. Haematol. **47:** 163–169.
4. QUESENBERRY, P. J. & M. A. GIMBRONE, JR. 1980. Blood **56:** 1060–1067.
5. ASCENSAO, J. T., G. M. VERCELLOTTI, H. S. JACOB & E. D. ZANJANI. 1984. Blood **63:** 553–558.
6. SIEFF, C. A., S. TSAI & D. V. FALLER. 1987. J. Clin. Invest. **79:** 48–51.
7. KAUSHANSKY, K., N. LIN & J. W. ADAMSON. 1988. J. Clin. Invest. **81:** 92–97.
8. ZSEBO, K. M., V. N. YUSCHENKOFF, S. SCHIFFER, D. CHANG, E. MCCALL, C. A. DINARELLO, M. A. BROWN, B. ALTROCK & G. C. BAGBY, JR. 1988. Blood **71:** 99–103.
9. DAINIAK, N., D. SUTTER & S. KRECZKO. 1986. Blood **68:** 1289–1297.
10. DAINIAK, N., S. KRECZKO, A. COHEN, R. PANNELL & J. LAWLER. 1985. Exp. Hematol. **13:** 1073–1079.
11. DAINIAK, N. & C. M. COHEN. 1982. Blood **60:** 583–594.
12. DAINIAK, N., H. B. WARREN, S. KRECZKO, M. A. RIORDAN, L. FELDMAN, J. LAWLER, A. M. COHEN & P. F. DAVIES. 1988. J. Clin. Invest. **81:** 834–843.
13. FELDMAN, L., C. M. COHEN, M. A. RIORDAN & N. DAINIAK. 1987. Proc. Natl. Acad. Sci. USA **84:** 6775–6779.
14. FOX, P. L. & P. E. DICORLETO. 1986. Proc. Natl. Acad. Sci. USA **83:** 4774–4778.
15. FOX, P. L., G. M. CHISHOLM & P. E. DICORLETO. 1986. J. Cell Biol. **103:** 17a.
16. DAINIAK, N., G. DAVIES, M. KALMANTI, J. LAWLER & V. KULKARNI. 1983. J. Clin. Invest. **71:** 1206–1214.
17. BROWN, M. S. & J. L. GOLDSTEIN. 1986. Science **232:** 34–47.
18. CURTISS, L. K., D. H. DEHEER & T. S. EDINGTON. 1980. Cell. Immunol. **49:** 1–11.
19. WADDELL, C. C., D. TAUNTON & J. J. TWOMEY. 1976. J. Clin. Invst. **58:** 950–954.
20. ZUCKER, S., M. S. MICHAEL, R. M. LYSIK, H. J. GLUCKSMAN, J. REESE & A. RENDIN. 1979. Cell Tissue Kinet. **17:** 393–399.

21. GRANSTROM, M. 1972. Exp. Cell Res. **87:** 426–431.
22. AYE, M. T., J. A. SEGUIN & J. P. MCBURNEY. 1979. J. Cell. Physiol. **99:** 233–241.
23. KONWALINKA, G., C. BREIER, D. GEISSLER, C. PESCHEL, C. J. WIEDERMANN, J. PATSCH & H. BRAUNSTEINER. 1988. Exp. Hematol. **16:** 125–130.
24. DOUAY, L., V. BARBU, C. BAILLOU, A. NAJMAN, N.-C. GORIN, J. POLONOVSKI & G. DUHAMEL. 1983. Exp. Hematol. **11:** 499–505.

The Immune System as Mediator of Virus-Associated Bone Marrow Failure: B19 Parvovirus and Epstein-Barr Virus

NEAL S. YOUNG, BRUCE BARANSKI, AND GARY KURTZMAN

Cell Biology Section
Clinical Hematology Branch
National Heart, Lung, and Blood Institute
Building 10, Room 7C103
Bethesda, Maryland 20892

Virus infection of human bone marrow with resulting hematopoietic failure has been suspected because of clinical clues, such as the occurrence of aplastic anemia after hepatitis or infectious mononucleosis, the frequent occurrence of decreased blood counts during viral illnesses, and epidemics of transient aplastic crisis within families of patients with genetic hemolytic anemias.[1,2] A variety of hematologic abnormalities in acquired immunodeficiency syndrome may be due to marrow infection with the human immunodeficiency retrovirus. Animal models have shown that viruses can cause bone marrow failure. Neutropenia in cats infected with feline panleucopenia virus, a parvovirus, was one of the first observations of virus-transmitted disease.[3] Feline leukemia virus in cats[4] and Friend virus in mice[5] are two retroviruses that cause anemia when injected into animals of the appropriate strain or age.

The ability to study the interaction of some viruses with defined populations of bone marrow cells has increased our understanding of virus-hematopoietic cell interactions. We briefly review here data for two viruses, B19 parvovirus and Epstein-Barr virus (EBV), for which the immune system has been implicated as playing a permissive role (in chronic parvovirus infection) or an effector role (in EBV-associated aplastic anemia).

B19 PARVOVIRUS

Clinical Spectrum of B19 Parvovirus Infection

This virus was discovered serendipitously in the serum of a normal blood donor in 1975 and subsequently linked by seroepidemiologic studies to transient aplastic crisis of chronic hemolytic diseases and to fifth disease in the remainder of the population.[6] The same virus thus causes a hematologic illness in the person with underlying erythroid stress, a rash in normal children, and a polyarthralgia syndrome in affected adults. Normal volunteer studies have shown that the hematologic manifestations of B19 parvovirus infection occur early, coincident with viremia, while the symptoms of fifth disease (rash, joint pains) occur with the development of virus-specific antibodies.[7]

Failure of marrow red blood cell production in the setting of increased peripheral erythrocyte demand is the most dramatic manifestation of acute parvovirus infection.

Acute worsening of anemia has been documented in a wide variety of chronic hemolytic states but also in acute blood loss and iron deficiency. Temporary cessation of erythropoiesis in normal persons also follows infection, with small decrements in hemoglobin that can be detected on serial blood studies, but these blood count changes are not manifested clinically. Both neutropenia and thrombocytopenia are common in parvovirus infection of normal volunteers and in patients with chronic hemolysis (especially when spleen function has been preserved, as in hereditary spherocytosis).

Recently, chronic bone marrow failure has been associated with persistent parvovirus infection in immunosuppressed persons.[8] The first recognized case of chronic parvoviremia occurred in a child with Nezelof's syndrome, a combined immunodeficiency disorder in which some immunoglobulin production is preserved. This child's immunologic disease had been diagnosed when he presented with an unusual pneumonia at the age of six months. One year later, he became abruptly anemic and neutropenic. Bone marrow aspirate smears showed giant pronormoblasts, similar to those described in the bone marrows of patients with transient aplastic crisis. Using molecular hybridization methods, B19 parvovirus was demonstrated in serial serum specimens and in the bone marrow of this patient. Disappearance of the virus from the patient's serum, either spontaneously or as a result of therapy (see below), was associated with improvement in blood counts. Perhaps unsurprisingly, given the underlying immunologic deficit, symptoms and signs of fifth disease have never developed in this patient. The frequent occurrence of anemia and neutropenia reported in Nezelof's syndrome has suggested the possibility that other patients with immunodeficiency may acquire their chronic cytopenias as a result of parovivirus infection.

Two other populations have been identified as at risk for chronic bone marrow failure due to persistent B19 parvovirus infection. We have studied four children with acute lymphocytic leukemia who were receiving maintenance chemotherapy during remission.[9] Anemia developed in all four and was associated with neutropenia in one. Prolonged periods (2–>12 months) of hematologic depression were associated with parvoviremia in all cases and demonstration of bone marrow giant pronormoblasts in 3/3 cases examined. Like the index case, these children had no other symptoms of virus infection. From serologic studies of normal populations, it is clear that school children are the age group at greatest risk of parvovirus infection. B19 parvovirus infection of children with remitted malignant disease who are receiving immunosuppressive chemotherapy must be common and may be a mechanism for the frequently observed phenomenon of "refractoriness" to chemotherapy.

We have also studied two patients who presented with pure red cell aplasia and a positive antibody test for human immunodeficiency virus. In one case, acquired immunodeficiency syndrome (AIDS) manifested as pneumocystis pneumonia with corticosteroid therapy. In the other, an intravenous drug abuser, death rapidly resulted from Staphylococcal sepsis. Both patients were parvoviremic at presentation. However, we have not detected B19 parvovirus in the sera of many other AIDS patients who suffered varying patterns of peripheral blood cytopenias later in their courses. About half the adult population will have circulating neutralizing antibodies to B19 parvovirus.

Mechanism of Marrow Failure in Parvovirus Infection[6]

In vitro, B19 parvovirus is a potent inhibitor of erythroid colony formation, both from CFU-E (late erythroid colony forming cell) and BFU-E (primitive, burst forming cell). Virus can be directly demonstrated in isolated erythroid progenitor cells at the CFU-E stage. B19 parvovirus replicates in human bone marrow cells in suspension culture. In both culture systems, the virus is extraordinarily specific for erythroid cells: CFU-GM derived colony formation is normal even at high virus concentrations, and the virus does not replicate in cultures in the absence of erythropoietin or depleted of erythroid cells.

Although parvoviruses can persist in cells in a latent form, active replication is associated with cell death. B19 parvovirus infection of erythroid cells is accompanied by obvious toxic signs on phase and electron microscopy. Replication may not be required for cytotoxicity, however. The nonstructural protein of the virus is lethal in transfected cells, which are nonpermissive to virus replication.[10] Nonstructural protein expression, in the absence of virus replication, provides a mechanism for the clinical effect of the virus on nonerythroid cells, particularly granulocytic progenitors and megakaryocytes. The role of nonstructural protein in B19 parvovirus infection of permissive erythroid cells is uncertain. An interesting model of action may be the creation of a favorable milieu for parvovirus replication by an effect on the cell's ability to reproduce. Adenoassociated virus, a dependent human parvovirus that ordinarily requires coinfection with another virus for its own replication, can replicate in cells treated with hydroxyurea.[11] Expression of the nonstructural protein gene in erythroid cells may similarly affect some nuclear constituent required for parvovirus replication.

Immune Restriction of Parvovirus Infection

Natural[12] and experimental[7] parvovirus infection is regularly followed by the appearance of anti-B19 parvovirus IgM and IgG antibodies. Serum containing either specific antibodies of either class neutralizes the virus' ability to inhibit erythroid colony formation *in vitro*. A cellular immune response has not been demonstrated *in vitro:* peripheral blood mononuclear cells from persons who have been infected with virus do not proliferate in the presence of B19 parvovirus antigen.

Patients with chronic parvovirus infection have defects in antibody formation against viral antigens. In a few cases, the production of adequate antibodies is quite delayed in comparison to normal, allowing a subacute course. In other patients, antibody is made which can be detected by radioimmunoassay, using plastic plates coated with virus. However, these antibodies have failed to recognize viral antigens on immunoblot, or they bind very weakly to only one of the viral capsid proteins (the minor 83-kd rather than the major 58-kd species). In at least one patient, no antibody response by any criteria has developed against the virus.

Treatment of chronic B19 parvovirus infection with preparations containing anti-B19 parvovirus-specific antibodies has confirmed the modulating role of the immune system on this viral disease. The patient with congenital immunodeficiency was treated with very high doses of a commercial immunoglobulin preparation prescreened for the presence of anti-B19 parvovirus antibodies by Western analysis. Several weeks of therapy resulted in marked reticulocytosis and the appearance of abundant erythroid cells in the bone marrow (anemia did not improve, possibly due to the concurrent transfusion of antibodies to red cell membrane antigens). A patient with chronic parvovirus anemia and acute lymphocytic leukemia in remission has been treated with his father's plasma, which contained high titers of anti-B19 parvovirus IgG. In this case, antibody infusion produced symptomatic fifth disease, with fever, typical skin eruption, and polyarthralgia. The reticulocyte count slightly increased, but more dramatic improvement in erythropoiesis may have been blunted by continuing maintenance cytotoxic chemotherapy. Finally, a patient with AIDS who presented with anemia due to parvovirus infection received commercial immunoglobulin infusions and completely recovered.

EPSTEIN-BARR VIRUS

Hematologic Abnormalities Associated with EBV Infection

Modest neutropenia and thrombocytopenia are common in uncomplicated infectious mononucleosis; whether the mechanism of bone marrow depression is peripheral destruction (as with anti-i antibody hemolysis) or decreased marrow production is unknown.[2] Severe bone marrow failure apparently occurs very rarely after typical infectious mononucleosis, with only about a dozen cases of aplastic anemia and two cases of pure red cell aplasia reported in the English literature. Pancytopenia or more limited forms of bone marrow failure may not be uncommon in patients with chronic or unusual EBV infection, but the bone marrow frequently shows infiltration with lymphocytes or histiocytes (as in hemophagocytic syndrome). Finally, aplastic anemia is reported to be a frequent terminal illness in patients with X-linked lymphoproliferative disease and other similar, genetic syndromes.

The relationship between marrow failure and EBV infection has been strengthened by the recent demonstration of the virus in the bone marrow of some patients with aplastic anemia and pure red cell aplasia.[13] Using immunofluorescence for the virus' nuclear antigen (EBNA) and molecular hybridization (in situ and Southern analysis) for virus-specific DNA sequences, virus has been found in three patients with aplasia studied prospectively; two of them had typical infectious mononucleosis preceding their hematologic disease, the third a viral syndrome that included pharyngitis, cervical lymphadenopathy, and EBV pneumonia. Retrospectively, EBV was identified in the marrow of a single patient with aplastic anemia after screening almost 40 patients' DNA samples. Two unusual cases of pure red cell aplasia have also been associated with EBV infection: atypical relapse (without macrocytosis) in Diamond-Blackfan syndrome after infectious mononucleosis and recurrent erythropoietic failure in a patient with sickle cell disease and chronic EBV infection.

Immune System–Bone Marrow Interactions

EBV infection is associated with activation of many immune system cells, as reflected most obviously in the atypical lymphocytes of the peripheral blood. B cell infection with virus is followed by activation and proliferation of a variety of cells, especially suppressor lymphocytes (CD 8^+) but also helper T cells and natural killer cells.[14] Many of the immunological abnormalities that accompany EBV infection have also been observed in patients with idiopathic aplastic anemia: reversed helper/suppressor lymphocyte ratio, T cell activation, increased circulating soluble interleukin-2 receptor, and abnormal lymphokine production (both gamma and alpha interferons).[15] Some patients with EBV-associated aplasia have recovered following therapy directed against the immune system, such as corticosteroids or antithymocyte globulin.

Recent experiments in our own laboratory have suggested a mechanism of interaction of virus, hematopoietic cells, and cells of the immune system.[16] First and unsurprisingly, T cells activated *in vitro* by exposure to EBV or EBV-infected autologous lymphoid cells can inhibit hematopoiesis in colony culture, employing as target cells bone marrow obtained from the same normal donor. There appear to be both specific and nonspecific components to inhibition of autologous bone marrow progenitors under these conditions. Stimulated T cells produce gamma interferon, an inhibitor of hematopoietic cell proliferation. T cells grown from the blood of a patient with EBV-associated aplastic anemia also produced interferon; these cells have the activated suppressor phenotype. However, EBV-stimulated normal T cells also are more efficient in suppression of autologous marrow progenitors that have been exposed to EBV in comparison to normal bone marrow target cells, consistent with a more specific action on virus-infected cells.

Direct infection of hematopoietic progenitors by EBV was suggested by the finding of EBV in the bone marrow, but not the blood, of patients with aplastic anemia. Using normal bone marrow, EBV infection of hematopoietic progenitors has been supported by the presence of EBV antigen and EBV DNA in the progeny of BFU-E, tested as pooled samples or individually isolated burst colonies. Under our experimental conditions, bone marrow cells are depleted of B lymphocytes by exhaustive panning, and cells bearing B cell antigens are not detectable within the limits of flow cytometry analysis at the completion of the culture period. Only a small proportion of progeny cells are infected by the criteria of *in situ* hybridization and EBNA determination, and only colonies derived from BFU-E, and not CFU-GM, are apparently virus infected. In contrast to parvovirus, EBV is not cytotoxic to colony formation, even in high concentration.

Taken together, the clinical observations and preliminary experiments *in vitro* would indicate that EBV bone marrow failure may be mainly immunologically mediated. EBV would infect but not kill bone marrow cells. Marrow suppression would be effected by the T cell-mediated immune response to EBV infection, with both nonspecific and specific inhibition of hematopoiesis, the former mediated by negative cytokines like gamma-interferon, the latter presumably by cellular recognition of EBV or other antigens on stem cells.

REFERENCES

1. KURTZMAN, G. & N. S. YOUNG. 1988. Viruses and bone marrow failure. Clin. Hematol. In press.
2. BARANSKI, B. & N. S. YOUNG. 1987. Hematologic manifestations of viral illness. Clin. Hematol./Oncol. N. Am. **1:** 167–183.
3. HAMMOND, W. D. & J. F. ENDERS. 1939. A virus disease of cats, principally characterized by aleucocytosis, enteric lesions and the presence of intracellular inclusion bodies. J. Exp. Med. **69:** 327–351.
4. SCOTT, F. W. 1987. Panleukopenia. *In* Diseases of the Cat. J. Holzworth, Ed.: 182–193. Saunders. Philadelphia, PA.
5. HANKINS, W. D. & J. KAMINCHIK. 1984. Modification of erythropoiesis and hormone sensitivity by RNA tumor viruses. *In* Aplastic Anemia: Stem Cell Biology and Advances in Treatment. N. S. Young, A. Levine & R. K. Humphries, Eds.: 141–152. Alan R. Liss. New York, NY.
6. YOUNG, N.S. 1988. Hematologic and hematopoietic consequences of B19 parvovirus infection. Semin. Hematol. **25:** 159–172.
7. ANDERSON M. J., P. G. HIGGINS, L. R. DAVIS, J. S. WILLMAN, S. E. JONES, I. M. KIDD, J. R. PATTISON & D. A. J. TYRRELL. 1985. Experimental parvoviral infection in humans. J. Infect. Dis. **152:** 257–265.
8. KURTZMAN, G. J., K. OZAWA, G. R. HANSON, B. COHEN, R. OSEAS & N. S. YOUNG. 1987. Chronic bone marrow failure due to persistent B19 parvovirus infection. N. Engl. J. Med. **317:** 287–294.
9. VAN HORN, D. K., P. P. MORTIMER, N. YOUNG & G. R. HANSON. 1986. Human parvovirus associated red cell aplasia in the absence of underlying hemolytic anemia. Am. J. Ped. Hematol./Oncol. **8:** 235–238.
10. OZAWA, K., J. AYUB, S. KAJIGAYA, T. SHIMADA & N. YOUNG. 1988. The gene encoding the nonstructural protein of B19 (human) parvovirus may be lethal in transfected cells. J. Virol. **61:** 2884–2889.
11. YAKOBSON, B. T. KOCH & E. WINOCUR. 1987. Replication of adeno-associated virus in synchronized cells without the addition of a helper virus. J. Virol. **61:** 972–981.
12. CHORBA, T. L., P. COCCIA, R. C. HOLMAN, P. TATTERSALL, L. J. ANDERSON, J. SUDMAN, N. S. YOUNG, E. KURCZYNSKI, U. M. SAARINEN, D. N. LAWRENCE, J. M. JASON & B.

Evatt. 1986. Parvovirus B19—the cause of epidemic aplastic crisis and erythema infectiosum (fifth disease). J. Infect. Dis. **154:** 383–393.

13. Baranski, B., G. Armstrong, J. T. Truman, G. R. Quinnan, S. E. Strauss & N. S. Young. 1988. Epstein-Barr virus in the bone marrow of patients with aplastic anemia. Ann. Int. Med. **109:** 695–704.
14. Tosato, G. 1987. The Epstein-Barr virus and the immune system. Adv. Cancer Res. **49:** 75–125.
15. Baranski, B. & N. S. Young. 1988. Autoimmune aspects of aplastic anemia. In Vivo **2:** 91–94.
16. Baranski, B., J. Moore, G. Armstrong, I. Magrath & N. Young. 1987. Epstein-Barr virus associated aplastic anemia: Demonstration of virus in clinical bone marrow samples and in hematopoietic progenitor cells infected in vitro. Clin. Res. **35:** 419A.

Genetic Relatedness of the Human Immunodeficiency Viruses Type 1 and 2 (HIV-1, HIV-2) and the Simian Immunodeficiency Virus (SIV)

G. FRANCHINI AND M. L. BOSCH

Laboratory of Tumor Cell Biology
National Cancer Institute
National Institutes of Health
Bethesda, Maryland 20892

INTRODUCTION

Since 1981, when the first cases of acquired immunodeficiency syndrome (AIDS) were described,[1] a rapid progress in scientific research has led to the discovery and characterization of human retroviruses associated with the disease.[2–5] HIV-1 and HIV-2 have been isolated from patients with AIDS,[2–5] and SIV has been obtained from macaques rhesus with a syndrome very similar to human AIDS.[6,7]

Interestingly, the spread of the two closely related human viruses differs markedly. Epidemiological studies indicate that HIV-1 infection is found worldwide,[8] while HIV-2 appears to be contained mainly in West Africa. Also in Africa, HIV-1 and HIV-2 have different distributions of sero-prevalence as is summarized in FIG. 1. Similarly, antibodies reactive to SIV, originally isolated from captive macaques, are present in monkey species which are spread all over Africa.[9] This indicates the existence of viruses immunologically related to SIV in these animals. The relationship of these viruses with each other may be relevant in the understanding of their origin and host range. In this paper, we shall try to summarize the data available to date in this regard.

A common feature of all these viruses is their biological behavior *in vitro*. HIV-1, HIV-2 and SIV infect only OKT4+cells and they apparently use the same receptor.[10,11] The *in vivo* host range for HIV-1 appears to be limited to the chimpanzee and human while HIV-2 infects also macaque rhesus (Zagury *et al.*, personal communication). The genomes of these retroviruses have been molecularly cloned and characterized,[12–17] and the basic genetic organization of HIV-1, HIV-2 and SIV is very similar (FIG. 2).

The overall genomic organization of HIV-1 is *LTR-gag-pol-sor-R-env-LTR*. Two other crucial genes *tat* and *trs* are extensively overlapping with the envelope gene. HIV-2 and SIV also have a genomic organization similar to HIV-1, but an extra gene (X) is present between *sor* and *R* genes (FIG. 2). The function of the nonstructural genes of HIV-1 has been the subject of extensive studies for the last three years.

The *sor* gene product is a 23-Kd protein[18,19] which is located in the cytoplasm of the infected cells (G. Franchini, unpublished results). Studies on *sor* deletion mutants indicated that the gene plays an important role in regulating virus infectivity.[20–22] The 3′-*orf*, located at the 3′-*end* of the genome is encoding for a 27-Kd protein[23] which is phosphorylated and apparently binds GTP.[24] In infected cells the 3′-*orf* is localized in the cytosol and membrane fractions.[25] The 3′-*orf* gene product appears to be dispensable for viral replication *in vitro* and its absence enhances virus production in infected cells.[26]

The *R* open reading frame has been shown to encode for a protein product which is immunogenic in HIV-1 infected individuals,[27] but the native protein in infected cells has not been identified yet.

More data are available on the biological activity of the *tat* and *trs* genes (these two viral proteins are 14Kd and 18 Kd, respectively).[28–31] *Tat* acts in trans and modulates the expression of all viral genes. *Tat* presence is essential for virus replication,[31] and its activity seems to be exerted at least at two main levels: 1) at transcriptional level increasing the amount of viral mRNAs probably inhibiting early termination of initiated transcripts, (*tat* would, therefore, be an antiterminator;[32] 2) at posttranscriptional level increasing the expression of viral proteins by as yet unknown mechanisms.[33–36]

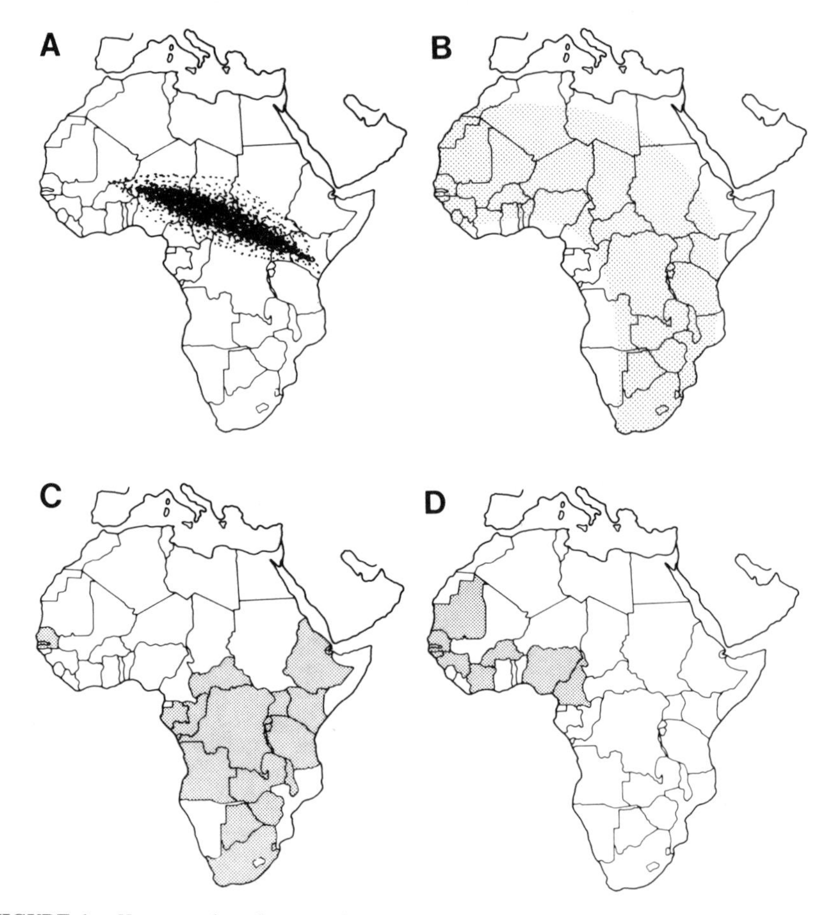

FIGURE 1. Human and nonhuman primate retroviruses in Africa. Distribution **(A)** of HTLV-1 seropositive humans; **(B)** of SIV seropositive guenons; **(C)** of HIV-1 seropositive humans; and **(D)** of HIV-2 seropositive humans.

Trs (also designated *art*) influences the balancing of the various viral mRNAs; its absence favors the expression of regulatory genes (*sor*, 3′-*orf*, *tat*) rather than structural genes (envelope and *gag* proteins).[28,36] It is not clearly defined yet whether *trs* acts

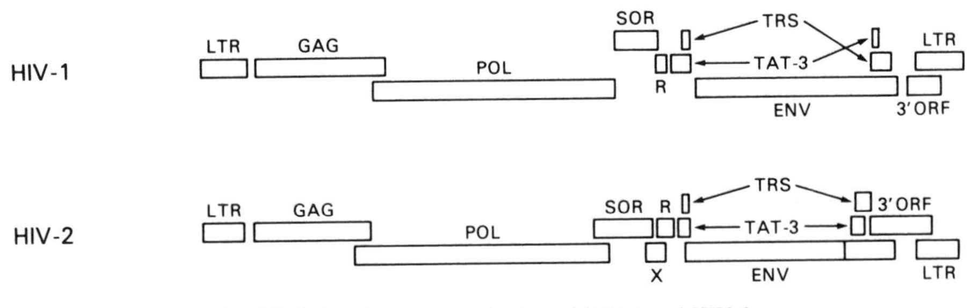

FIGURE 2. Genetic organization of HIV-1 and HIV-2.

directly on the splicing events that regulate viral expression or influence the transport of the processed viral mRNAs into the cytoplasm. Interestingly, *tat* could be envisioned as a gene that always promotes its own expression as well as the expression of all other viral genes, while *trs* expression indirectly would lead to a reduction of *tat* expression and, therefore, viral expression. In summary, the HIV-genome is equipped with two genes that reduce viral replication (*trs* and 3′-*orf*) and their expression may be the key to latency.

A major difference in the genome of HIV-1 and HIV-2/SIV is the presence of an extra open reading frame (designated X-*orf*) in HIV-2 that does not have an obvious counterpart in HIV-1 (FIG. 2). Recently, the X-*orf* has been shown to encode for a 16-Kd protein, which is located in the cytoplasm of HIV-2 infected cells and is associated with mature virions.[37] SIV also has been shown to encode for the same protein which has a slightly lower molecular weight (p14).[38] At present, no information is available on the function of this HIV-2/SIV-specific gene but its amino acid sequence, rich in proline, suggests that it may be a nucleic acid-binding protein.

The overall sequence of the structural and regulatory genes of HIV-1, HIV-2 and SIV is conserved. We analyzed the amino acid identity among the different genes of two independent isolates of HIV-2 (HIV-2_{NIH-Z}, HIV-2_{ROD}), SIV and one strain of HIV-1 (HTLV-IIIB).[12–17] The highest degree of amino acid conservation among the HIV-1, HIV-2 and SIV studied[12–17] is in the *gag* and *pol* genes (TABLE 1). The envelope glycoproteins, both the extracellular portion (ECP) and the transmembrane portion (TMP), of these viruses are less conserved. This finding is consistent with the high degree of variability in the envelope genes found among the different members of the HIV-1 and HIV-2 families.[12–17] The overall conservation of the regulatory genes *sor, tat* and *trs* is comparable (TABLE 1).

Taken together, this comparative analysis indicates that members of the HIV-2 family and SIV are equally related (average 43%) to HIV-1 (FIG. 3A). Furthermore, the degree of variability among the HIV-2 isolates and SIV (ranging between 20 and 30%) is comparable to that of some HIV-1 strains among themselves (FIG. 3B). The two HIV-2

TABLE 1. Amino Acid Identity among the Different Genes of Two Isolates of HIV-2, SIV, and One Strain of HIV-1[a]

			env							
HIV-2_{NIH-Z}	*gag*	*pol*	TOTAL	ECP	TMP	*sor*	*tat*	TRS/ART	X	R
HIV-2_{ROD}	92	91	80	78	84	86	85	86	78	69
SIV	82	75	70	68	72	64	63	59	—	—
HIV-1 (HTLV-IIIB)	52	54	35	32	39	28	29	34	—	37

[a]The numbers express the percentage of amino acid identity.

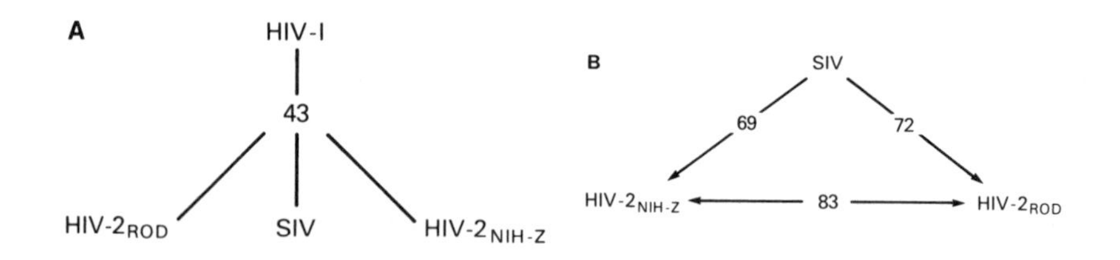

FIGURE 3. Amino acid homology **(A)** of HIV-1, SIV, $HIV^{-2}_{NIH\text{-}Z}$ and $HIV\text{-}2_{ROD}$; and **(B)** of human and nonhuman primate African retroviruses.

isolates are more closely related to each other than to SIV, which is compatible with their origin from different species.

DISCUSSION

In this paper we have described the structural and functional homologies between the different members of the AIDS-associated retrovirus family HIV-1, HIV-2 and SIV. The overall degree of genetic relatedness among these viruses suggests their origin from a common ancestor. In the course of evolution they diverged not only in the nucleotide sequence of the different structural genes, but also in the structure as exemplified by the fact that both HIV-2 and SIV contain the X-*orf,* which is absent from HIV-1.

The greater degree of amino acid identity between HIV-2 and SIV suggests that they diverged more recently from each other than they did from HIV-1. This is summarized in the phylogenetic tree (FIG. 4); the members of the HIV-1 and HIV-2 families are equidistant from each other. The fact that overall SIV differs more from the individual HIV-2

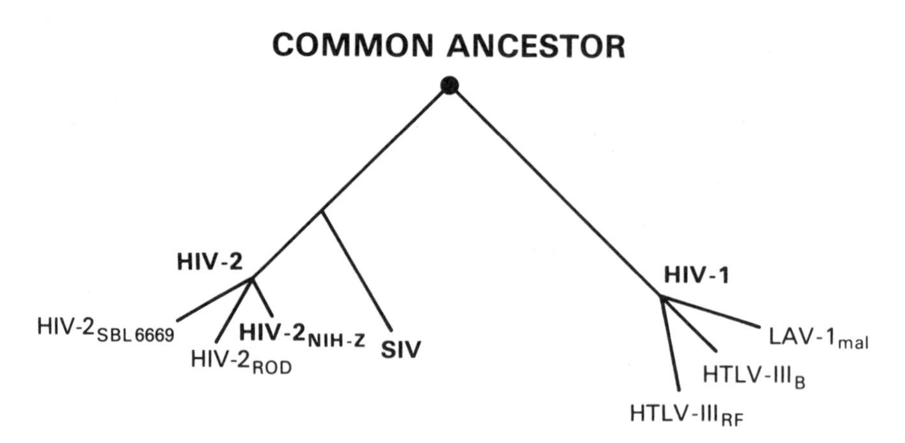

FIGURE 4. Phylogenetic tree of HIV-1, HIV-2, and SIV.

isolates than they do from each other suggests that SIV branched off at a point in evolution before the HIV-2-strain differences arose.

Although their overall biological behavior, both *in vivo* and *in vitro*, appears to be similar, some differences between *in vivo* and *in vitro* are worth noting. Despite the fact that all HIV-2 isolates to date were derived from patients with AIDS,[4,5] HIV-2 infection is associated with a lower morbidity rate as indicated by a seroepidemiological study on Senegalese prostitutes.[39] *In vitro* HIV-1 and HIV-2 seem to have a slightly different host range and display markedly different cytopathic effects on different cell lines (DeRossi *et al.*, unpublished data). It is tempting to speculate that the differences in the biological behavior between HIV-1 and HIV-2 may be due to the fact that HIV-2 expresses the X-*orf*, which is absent from HIV-1. Another biological difference between HIV-1 and HIV-2/SIV is the expression of a truncated enveloped transmembrane protein (gp41 versus gp32–34)[4,5] in HIV-2/SIV. The premature termination of the transmembrane protein appears to be related to the presence of a termination codon in some of the HIV-2 and SIV genomes. Recent studies indicate that despite the presence of the termination codon responsible for this truncation, the envelope region after the termination codon can still be expressed in SIV-infected monkeys.[40] The function of both the X-*orf* and the c-terminus of the small envelope are still unclear and merit further studies since they may hold part of the key to the functional differences between these related human and nonhuman primate retroviruses.

REFERENCES

1. GOTTLIEB, M. S., R. SCHROFF, H. M. SCHAUHER, J. D. WEISMAN, P. T. FAN, R. A. WOLF & A. SAXON. 1981. Pneumocystis carinii pneumonia and mucosal candidasis in previously healthy homosexual men. Evidence of a new acquired cellular immunodeficiency. N. Engl. J. Med. **305:** 1425–1431.
2. BARRE-SINOUSSI, F., J. C. CHERMANN, F. REY, M. T. NUGEYRE, S. CHAMARET, J. GRUEST, C. DAUGUET, C. AXLER-BLIN, F. VEZINET-BRUN, C. ROUZIOUX, W. ROZENBAUM & L. MONTAGNIER. 1983. Isolation of a T-lymphotropic retrovirus from a patient at risk for acquired immune deficiency syndrome (AIDS). Science **220:** 868–871.
3. GALLO, R. C., S. Z. SALAHUDDIN, M. POPOVIC, G. M. SHEARER, M. KAPLAN, B. F. HAYNES, T. J. PALHER, R. REDFIELD, J. OLESKE, B. SAFAI, G. WHITE, P. FOSTER & P. D. MARKHAM. 1984. Frequent detection and isolation of cytopathic retroviruses (HTLV-III) from patients with AIDS and at risk for AIDS. Science **224:** 500–503.
4. CLAVEL, F., D. GUETARD, F. BRUN-VERINET, S. CHAMARET, M. A. REY, M. O. SANTOS-FERREIRA, A. G. LAURENT, C. DAUGUET, C. KATLAINA, C. ROUZIOUX, D. KLATZMANN, J. L. CHAMPALIMAND & L. MONTAGNIER. 1986. Isolation of a new human retrovirus from West African patients with AIDS. Science **233:** 343–346.
5. ALBERT, J., U. BREDBERG, F. CHIODI, B. BOTTIGER, E. M. FENYO, E. NORRBY & G. BIBERFELD. 1987. A new pathogenic human retrovirus of West African origin (SBL-6669) and its relationship to HTLV-IV, LAV-II and HTLV-IIIB. AIDS Res. Hum. Retroviruses **3:** 1–10.
6. DANIEL, M. D., N. L. LETVIN, N. W. KING, M. KANNAGI, P. K. SEHGAL, R. D. HUNT, P. J. KANKI, M. ESSEX & R. C. DESROSIERS. 1985. Isolation of T-cell tropic HTLV-III-like retrovirus from macaques. Science **228:** 1201–1204.
7. MURPHEY-CORB, M., L. N. MARTIN, S. R. S. RAUGAN, G. B. BASKIN, B. J. GORMUS, R. H. WOLF, W. A. ANDES, M. WEST & R. C. MONTELARO. 1986. Isolation of an HTLV-III-related retrovirus from macaques with simian AIDS and its possible origin in asymptomatic mangabeys. Nature **321:** 435–437.
8. PIOT, P., F. A. PLUMMER, F. S. MHALU, J. L. LAMBORAY, J. CHIN & J. M. MANN. 1988. AIDS, an international perspective. Science **239:** 573–579.
9. KANKI, P. J., P. KURTH, W. BECKER, G. DREESMAN, M. F. MCLANE & M. ESSEX. 1985. Antibodies to simian T lymphotropic retrovirus type III in African green monkeys and recognition of STLV-III viral proteins by AIDS and related sera. Lancet **i:** 1330–1332.

10. Kannagi, M., J. M. Yetz & N. L. Lelwin. 1986. *In vitro* growth characteristics of simian T-lymphotropic virus type III. Proc. Natl. Acad. Sci. USA **82:** 7053–7057.
11. Maddon, P. J., A. G. Dalgleish, J. S. McDougal, P. R. Clapham, R. A. Weiss & R. Axel. 1986. The T4 gene encodes the AIDS virus receptor and is expressed in the immune system and the brain. Cell **47:** 333–348.
12. Franchini, G., R. C. Gallo, C. Gurgo, H.-G. Guo, E. Collalti, K. A. Fargnoli, L. F. Hall, F. Wong-Staal & M. S. Reitz, Jr. 1987. Sequence of simian immunodeficiency virus and its relationship to the human immunodeficiency viruses. Nature **328:** 539–543.
13. Chakrabarti, L., M. Guyader, M. Alizon, M. D. Daniel, R. C. Desrosiers, P. Tiollais & P. Sonigo. 1987. Sequences of simian immunodeficiency virus from macaque and its relationship to other human and simian retrovirus. Nature **328:** 543–547.
14. Sanchez-Pescador, R., M. D. Power, P. J. Barr, K. S. Steimer, M. M. Stempien, S. L. Brown-Shimer, W. W. Gee, A. Renard, A. Randolph, J. A. Levy, D. Dina & P. A. Luciw. 1985. Nucleotide sequence and expression of an AIDS associated retrovirus (ARV-2). Science **227:** 484–492.
15. Muesing, M. A., P. H. Smith, C. D. Cabradilla, C. V. Benton, L. A. Lashy & D. J. Capon. 1985. Nucleic acid structure and expression of the human AIDS/lymphadenopathy retrovirus. Nature **313:** 450–458.
16. Guyader, M., M. Emerman, P. Sonigo, F. Clavel, L. Montagnier & M. Alizon. 1987. Genome organization and transactivation of the human immunodeficiency virus type 2. Nature **326:** 662–669.
17. Zagury, J. F., G. Franchini, M. S. Reitz, Jr., E. Collalti, B. Starcich, L. Hall, K. A. Fargnoli, L. Jagodzinski, H.-G. Guo, F. Laure, D. Zagury, S. K. Arya, S. F. Josephs, F. Wong-Staal & R. C. Gallo. 1988. The genetic variability between HIV-2 isolates is comparable to the variability among HIV-1. Proc. Natl. Acad. Sci. USA **85:** 5941–5945.
18. Lee, T. H., J. E. Coligan, J. S. Allen, M. F. McLane, J. E. Groopman & M. Essex. 1986. A new HTLV-III/LAV protein encoded by a gene found in cytopathic retroviruses. Science **231:** 1546–1549.
19. Kan, N. C., G. Franchini, F. Wong-Staal, G. C. DuBois, W. G. Robey, J. A. Lautenberger & T. S. Papas. 1986. Identification of HTLV-III/LAV *sor* gene product and detection of antibodies in human sera. Science **231:** 1553–1557.
20. Fischer, A. G., B. Ensoli, L. Ivanoff, M. Chamberlain, S. Petteway, L. Ratner, R. C. Gallo & F. Wong-Staal. 1987. The *sor* gene of HIV-1 is required for efficient virus transmission *in vitro*. Science **237:** 888–893.
21. Sodroski, J., W. C. Goh, C. Rosen, A. Tartar, D. Portetelle, A. Burny & W. Haseltine. 1986. Replicative and cytopathic potential of HTLV-III/LAV with *sor* gene deletions. Science **231:** 1549–1553.
22. Strebel, K., D. Daugherty, K. Clouse, D. Cohen, T. Folks & M. A. Martin. 1987. The HIV A′ (*sor*) gene product is essential for virus infectivity. Nature **328:** 728–730.
23. Franchini, G., M. Robert-Guroff, F. Wong-Staal, J. Ghrayeb, I. Nato, T. W. Chang & N. T. Chang. 1986. Expression of the protein encoded by the 3′ open reading frame of HTLV-III in bacteria. Demonstration of its immunoreactivity with human sera. Proc. Natl. Acad. Sci. USA **83:** 5282–5285.
24. Guy, B., M. P. Kieny, Y. Riviere, C. L. Peuch, K. Dott, M. Girard, L. Montagnier & J.-P. Lecocq. 1987. HIV F/3′ *orf* encodes a phosphorylated GTP-binding protein resembling an oncogene product. Nature **330:** 266–269.
25. Franchini, G., M. Robert-Guroff, J. Ghrayeb, N. T. Chang & F. Wong-Staal. 1986. Cytoplasmic localization of the HTLV-III 3′ *orf* protein in cultured T cells. Virology **155:** 593–599.
26. Fisher, A. G., L. Ratner, H. Mitsuya, L. M. Marselle, M. E. Harper, S. Broder, R. C. Gallo & F. Wong-Staal. 1986. Infectious mutants of HTLV-III with changes in the 3′ region and markedly reduced cytopathic effects. Science **233:** 655–659.
27. Wong-Staal, F., P. K. Chanda & J. Ghrayeb. 1987. Human immunodeficiency virus type III: The eighth gene. AIDS Res. Hum. Retroviruses **3:** 33–40.
28. Knight, D. M., F. A. Flomerfelt & J. Ghrayeb. 1987. Expression of the *art* / *trs* protein of HIV and study of its role in viral envelope synthesis. Science **236:** 837–840.

29. GOH, W. C., C. ROSEN, J. SODROSKI, D. HO & W. D. HASELTINE. 1986. Identification of protein encoded by the trans-activator gene *tat*III of human T-cell lymphotropic retrovirus type III. J. Virol. **59:** 181–184.
30. SODROSKI, J. G., R. PATARCA, C. A. ROSEN, F. WONG-STAAL & W. D. HASELTINE. 1985. Location of the trans-activation region on the genome of human T-cell lymphotropic virus type III. Science **229:** 74–77.
31. FISHER, A. G., M. B. FEINBERG, S. F. JOSEPHS, M. E. HARPER, L. M. MARSELLE, G. REYES, M. A. GONDA, A. ALDOVINI, C. DEBOUK, R. C. GALLO & F. WONG-STAAL. 1986. The trans-activator gene of HTLV-III is essential for virus replication. Nature **320:** 367–371.
32. KAO, S.-Y., A. F. CALMAN, P. A. LUCIN & B. M. PETERTIN. 1987. Antitermination of transcription within the long terminal repeat of HIV-1 by *tat* gene product. Nature **330:** 489–493.
33. CULLEN, B. R. 1986. Trans-activation of human immunodeficiency virus occurs via a bio-model mechanism. Cell **46:** 973–982.
34. WRIGHT, C. M., B. K. FELBER, H. PASHALIS & G. N. PAVLAKIS. 1986. Expression and characterization of the trans-activator of HTLV-III/LAV virus. Science **234:** 988–992.
35. ROSEN, C. A., J. G. SODROSKI, W. C. GOH, A. I. DAYTON, J. LIPPE & W. A. HASELTINE. 1986. Post-transcriptional regulation accounts for the trans-activation of the human T-lymphotropic virus type III. Nature **314:** 555–559.
36. FEINBERG, M. B., R. F. JARRETT, A. ALDOVINI, R. C. GALLO & F. WONG-STAAL. 1986. HTLV-III expression and production involve complex regulation at the levels of splicing and translation of viral RNA. Cell **46:** 807–817.
37. FRANCHINI, G., J. R. RUSCHE, T. J. O'KEEFFE & F. WONG-STAAL. 1988. The human immunodeficiency virus type 2 (HIV-2) contains a novel gene encoding a 16 Kd protein associated with mature virions. Aids Research and Human Retroviruses **4:** 243–250.
38. HENDERSON, L. E., R. C. SOWDER, T. D. COPELAND, R. E. BENVERISTE & S. OROSZLAN. 1988. A newly identified unique protein in simian and human type 2 immunodeficiency viruses. Science **241:** 199–201.
39. KANKI, P. J., S. M'BOUP, D. RICARD, F. BARIN, F. DENIS, C. BOYE, L. SANGARE, K. TRAVERS, M. ALBAUM, R. MARLINK, J. L. ROMET-LEMME & M. ESSEX. 1987. Human T-lymphotropic virus type 4 and the human immunodeficiency virus in West Africa. Science **236:** 827–831.
40. FRANCHINI, G., P. J. KANKI, M. L. BOSCH, K. FARGNOLI & F. WONG-STAAL. 1988. The Simian immunodeficiency virus envelope open reading frame located after the termination codon is expressed in vivo in infected animals. AIDS Research and Human Retroviruses **4:** 251–258.

Sustained Hypertransfusion and Induction of a Transplantable Myeloid Leukemia in RLV-A-Infected BALB/c Mice[a]

GARY PAUL LEONARDI, JOSEPH LOBUE, MICHAEL MANTHOS, DONALD ORLIC,[b] AND JYOTIRMAY MITRA

Department of Biology
New York University
New York, New York 10003
and
[b]*Department of Anatomy*
New York Medical College
Valhalla, New York 10595

INTRODUCTION

In 1972, Fredrickson *et al.*[1] reported the occurrence of a transplantable monomyelocytic leukemia in a female BALB/c mouse inoculated with a high dilution passage of Rauscher Leukemia Virus (RLV). The transplantable leukemic cells were found to produce a virus which when inoculated into BALB/c mice induces a protracted erythroid dyscrasia characterized by hepatosplenomegaly, erythroblastosis, erythroblastemia, thrombocytopenia and a severe microcytic, hypochromic anemia.[2–5] There is no evidence of myeloid or lymphoid leukemia or thymic lymphoma. Time to death varies between 16 to 20 weeks postinoculation of virus.[6] Because of the severe terminal anemia which is induced the virus has been designated RLV-A.[7]

RLV-A is a replication competent, SFFV-negative and erythropoietin-dependent virus.[8] The genome appears consistent with that of the typical chronic MuLV retroviruses,[9] and though the *pol* and *gag* gene products of RLV-A are immunologically identical to RLV, the *env* gene Gp70 is not.[10] The usual impact of RLV-A viral infection is derangement of erythropoiesis resulting in what appears to be a sideroachrestic anemia.

The present investigations were conducted to determine the effects of suppression of erythropoietin production with virtual irradication of erythropoiesis (the apparent ''target'' lineage) and induction of a hematopoietic microenvironment favoring myelopoiesis on RLV-A infection and expression. Inhibition of erythropoiesis was done by sustained hypertransfusion. This procedure has long been used to modify hemopoiesis[11] since it results in a drastic reduction in both recognizable red cell elements and erythropoietin-dependent progenitors (*e.g.*, BFU-E, CFU-E) but increases erythropoietin-independent progenitors (*e.g.*, CFU-S, CFU-GM).[12]

The use of hypertransfusion to alter erythroleukemic pathogenesis has previously been described[13] in mice inoculated with RLV. Long-term hypertransfusion resulted in in-

[a]The authors acknowledge the generous financial support of Dr. and Mrs. P.-C. Chan.

creased survival time and inhibition of the erythroblastic phase; however, the basic RLV pathogenesis was seen in 5 of the 7 animals surviving past 40 days of hypertransfusion. The other 2 animals developed either a severe erythroblastic or granulocytic response. Attempts to establish leukemogenicity by transplantation of leukocytes was unsuccessful. In studies by Dunn *et al.*[13] virus was inoculated 5 days after the initiation of hypertransfusion; thus an altered microenvironment may not yet have been established prior to infection. This could be the reason why the ''typical'' RLV response failed to be modified.

Recently Brookoff and Weiss[14] have induced specific bone marrow stromal cell changes by sustained hypertransfusion in mice. The sequence of events accompanying the marrow shift from reduction in erythropoiesis to enhanced granulopoiesis included the loss of medullary macrophages and an increase in adipocyte development and reticular adventitial cell numbers. Their suggestion that the marrow stromal changes induced a shift in the microenvironment to curtail erythropoiesis was strengthened when 2-week hypertransfused mice were unable to mount an erythropoietic response when challenged by hemolytic anemia or phlebotomy until a unique stromal cell element (''dark stromal cell'') appeared.[14] In the present study, this technique of sustained hypertransfusion was initiated in animals prior to inoculation of RLV-A. How the pathogenesis of this virally-induced erythroblastic disease was altered in animals, that: (1) have presumably had erythropoietin and susceptible erythroid elements practically eliminated and (2) possess an hematopoietic microenvironment which may be changed to preferentially support granulopoiesis, is herein reported.

METHODS

Animals

Male BALB/c mice from a breeding colony maintained at New York University (NYU) were used in this study. Both female mice and purchased retired breeders (Charles River, Wilmington, MA) were used as blood donors for hypertransfusion due to the shortage of available male mice in the NYU colony. All mice were housed 4 per cage and maintained on a diet of Purina lab chow and water *ad libitum*.

RLV-A Plasma Preparation and Inoculation

RLV-A viremic plasma was obtained from mice bearing a transplantable monomyelocytic leukemia (MML).[1] Briefly, blood from terminal MML mice was collected using heparinized syringes via cardiocentesis. The blood was centrifuged at 2,000 × g for 10 minutes at 2°C. The supernatant was then collected and recentrifuged a second time at the above conditions. The supernatant was again collected and recentrifuged a third time at 10,000 × g for 15 minutes. Following this, the supernatant was collected in plastic vials and stored in liquid nitrogen. For RLV-A infection, mice were inoculated intraperitoneally with 0.2 ml of rapidly thawed viremic plasma. Care was taken that for each experiment the viral potency controls and hypertransfused-viral groups received plasma from the same preparation lot.

Hypertransfusion Studies

This study was divided into 2 experiments: short-term and long-term hypertransfusions. In both studies, 5- to 6-week-old animals were inoculated with 1 ml of 75%-

packed, saline-washed erythrocytes, administered intraperitoneally every 7 days for 42 days. Hypertransfusion in mice eliminates erythropoiesis and modifies bone marrow stromal elements so as to favor granulopoiesis.[14] Following the initial 42-day hypertransfusion, groups of at least 4 animals were given RLV-A viremic plasma. In the short-term hypertransfusion experiment, no further hypertransfusion of these animals was done. In the long-term experiment, however, hypertransfusion every 7 days continued throughout the experiment. In both experiments, nonhypertransfused littermates were inoculated with viremic plasma to serve as viral potency controls and normal untreated animal groups were also established. In the long-term experiment, a group of littermates hypertransfused throughout the experimental period but not inoculated with virus was established.

In each experiment, at least 4 animals per group were used and over 1,000 donor animals were required throughout the experiment to maintain the hypertransfused state. The preparation of blood for hypertransfusion followed a standard protocol.[14] Blood was collected from etherized donor animals via cardiocentesis using heparinized syringes. This blood was pooled into centrifuge tubes cooled in an ice bath. The blood was then centrifuged at 1,600 × g for 5 minutes at 2°C in a refrigerated centrifuge. The supernatant and buffy coat was discarded and the pellet resuspended in sterile, physiologic saline. This procedure was repeated three times in order to wash the pellet. After the final centrifugation and resuspension, saline was not added but the packed red cell volume was determined using the Strumia *et al.* micro-micro capillary technique[15] (Sherwood Medical Industries, St. Louis, MO). Packed red cell volumes in the range of 72–77% were used for hypertransfusion.

Hematologic Parameters

Throughout the experiment, packed red cell volumes were determined as indicated earlier. Peripheral blood reticulocyte counts were made according to the technique of Brecher.[16] Peripheral blood nucleated cell counts were made using a model "ZM" Coulter counter according to prescribed techniques (Coulter Diagnostics, Hialeah, FL). Peripheral blood differential counts were also done.

At termination bone marrow and splenic smears were made using a fine sable hair brush dipped into 3 to 1 mixture of fetal bovine serum and distilled water. Slides were dried and fixed in absolute ethanol. Benzidine-hematoxylin-stained slides were used to determine erythroblast percentages.[17]

Microscopy

Various tissues from experimental and control animals were prepared for light and transmission electron microscopy. Specimens were initially immersed in a paraformaldehyde-glutaraldehyde fixative buffered in 0.2 M PIPES at room temperature for 1 hour.[18] Following this, specimens were placed in a modified fixative containing half the original paraformaldehyde and glutaraldehyde concentrations for a period of 6–48 hours under refrigeration. At this point, light microscopy samples were dehydrated and placed in a glycol-methacrylate resin according to outlined procedures (Polysciences, Inc., Warrington, PA). Sections (1–2 μ thick) were cut with a JB4 microtome (Polysciences, Inc.) and stained with hematoxylin and eosin.

Electron microscopy samples were postfixed in 2% osmium tetroxide buffered in 0.2 M PIPES for 1 hour on ice. Specimens were dehydrated in ethanol and embedded in epoxy resin (Polysciences, Inc.). Thin sections (60–90 mu) were cut on a Reichert OMU2 ultramicrotome (C. Reichert A.G., Vienna, Austria) and placed on copper grids. Grid staining with uranyl acetate and lead citrate were done to increase image contrast. Sections were examined on a Phillips 300 TEM (Phillips Electronics, Mahwah, NJ). Peripheral blood elements were prepared for electron microscopy according to previously described methods.[19]

Transplantability Studies

In the long-term hypertransfusion RLV-A experiment in which mice became leukemic, blood and splenic suspensions from terminal animals were transplanted into normal recipients. Whole blood from RLV-A viral potency control terminal animals was also transplanted to normal recipients. This was done to determine whether leukemic cells were present.

For long-term hypertransfusion RLV-A leukemic mice a small amount of blood (0.2–0.4 ml) was collected from terminal animals via cardiocentesis just prior to termination. Four recipient animals received 0.1 to 0.2 ml of this blood via the infraorbital sinus while under light ether anesthesia. A portion of the spleen was rapidly excised from terminal donors and placed into a sterile petri dish containing fetal bovine serum (GIBCO, Grand Island, NY). The spleen was pressed through nylon mesh and then taken up and expelled through a syringe several times in order to make a single cell suspension. Four recipient mice were given 0.2 ml of this spleen cell suspension intraperitoneally. Hematologic parameters were monitored as described above.

Terminal RLV-A animals exhibit increases in leukocytic elements in the peripheral blood.[20] To ascertain whether these cells were neoplastic, 0.2 ml of whole blood was obtained from terminal animals and inoculated into 4 recipients according to the procedures described above.

Myeloid Leukemia Plasma Passage

To examine whether animals bearing the transplantable granulocytic leukemia possessed the RLV-A virus, plasma passage into recipient mice was done. This plasma was prepared and 4 recipient animals were inoculated according to the procedures outlined above for RLV-A plasma preparation and inoculation.

RESULTS

Preliminary Hypertransfusion Study

A group of 8 animals was hypertransfused weekly to ascertain if the findings of Brookoff and Weiss[14] could be confirmed in BALB/c mice. On days 7, 14, 21 and 42 a pair of mice were sacrificed and prepared for microscopic examination. Hematologic and morphologic observations in these mice established that sustained hypertransfusion in BALB/c mice essentially produced the bone marrow changes found in Swiss mice. Briefly, these changes included elimination of erythropoiesis, transitory lipid accumulation in adventitial cells concomitant with increased granulopoiesis and marked reduction in marrow macrophages.[14]

Hematologic and Morphologic Findings

Throughout the experimental period, mean packed red cell volume (PRCV) was maintained in a range from 45.0 ± 1.0 to 52.1 ± 1.0 in the normal control group. Mean reticulocyte percentages in this group ranged from 1.5 ± 0.04 to 2.5 ± 0.2 and mean nucleated cell counts varied between 5,300 ± 279 and 11,300 ± 784 cells per mm^3.

The RLV-A virus potency control group showed a reduction in PRCV from 49.8 ± 0.5 to 34.8 ± 1.0 following virus inoculation (FIG. 1). After this initial decrease, PRCVs showed little change until approximately 20 weeks postinoculation. Animals then began succumbing to the disease and a rapid decline in PRCV to less than 20% with concomitant erythroblastemia (FIG. 2) but no compensatory reticulocytosis occurred. Accompanying these peripheral changes were hepatosplenomegaly with marked erythroblastosis in these organs (FIGS. 3 and 4). Bone marrow involvement was minimal although TEM examination showed virus particles present in megakaryoctyes (FIG. 5) and other blood cells. Bone marrow and splenic smears showed significant increases in benzidine-positive cells over control samples (TABLE 1). This is a typical disease course which has previously been described in detail.[2,3,6]

In the short-term hypertransfusion group, cessation of hypertransfusion and viral inoculation at 42 days resulted in the development of the usual RLV-A erythroblastosis after a slightly increased latent period (FIG. 1; TABLE 1). A "typical" RLV-A pathologic picture was observed in histological examination of organs. Thus, inoculation of RLV-A in mice in which erythropoiesis had been virtually eliminated and which possessed blood-forming organs presumably more conducive to granulopoiesis had no appreciable effect on the erythroblastic action of RLV-A if hypertransfusions were stopped.

In the long-term hypertransfusion control group, 4 of the 6 animals had mean PRCVs of 62.8 ± 2.1 to 73.3 ± 1.0% (FIG. 1). Mean reticulocyte counts never rose above 0.05 ± 0.03% in these four animals (E3, FIG. 6). The other 2 animals in this group manifested leukemoid reactions during the experimental period which were associated with an unexpected drop in PRCV to 35% with moderate reticulocytosis in one of the animals (C2, FIG. 6). This leukemoid reaction was characterized by a massive increase in peripheral nucleated cell count with a left shift (168,000 cells per mm^3). A less severe leukemoid reaction occurred in the other animal (PRCV drop to 48% with mild reticulocytosis and leukocytosis) 25 weeks after sustained hypertransfusion. In both animals, continued hypertransfusion was effective in elevating the PRCV and correcting the reticulocytosis and leukocytosis.

Upon termination, histologic examination of the 4 hypertransfusion control animals exemplified in panel E3, FIGURE 6, showed considerable phagocytosis in the liver (FIGS. 7 and 8) and spleen (FIG. 9) although splenic architecture remained intact. Electron microscopic examination of the bone marrow showed the occurrence of phagocytic activity in adventitial and endothelial cells (FIG. 10).

As for all virus-infected groups, the PRCV of long-term hypertransfusion RLV-A animals dropped slightly following viral inoculation on day 42 (73.8 ± 1.1% to 67.3 ± 0.8%; FIG. 1). However, no animals in this group ever showed the "typical" RLV-A pathology. Instead, in 5 of 6 cases (one animal died early of unknown causes) a myeloid leukemia developed which was characterized by hepatomegaly, with ectopic myelopoiesis (FIGS. 11 and 12), splenomegaly with greatly increased myelopoiesis and loss of splenic architecture (FIG. 13), infiltration of the kidney with leukemic cells (FIGS. 14 and 15), and a peripheral blood picture characterized by elevated nucleated cell counts (FIG. 16). These peripheral blood cells were used for passage of this leukemia (see below). Thus the continued suppression of erythropoiesis and the sustained induction of a presumably granulopoietic microenvironment before and after RLV-A viral inoculation led to the development of a transplantable granulocytic leukemia in these mice.

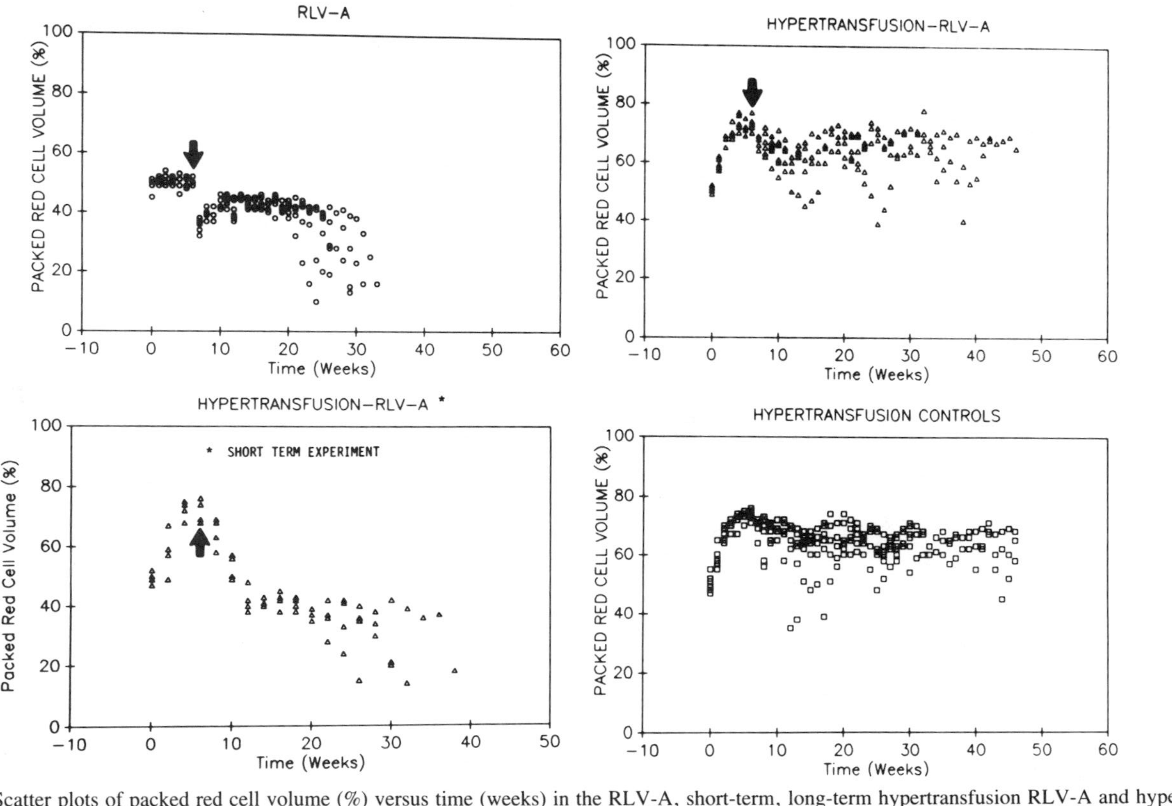

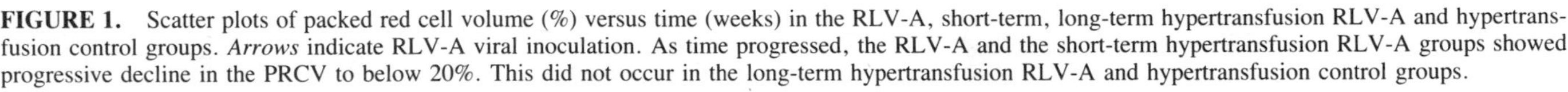
FIGURE 1. Scatter plots of packed red cell volume (%) versus time (weeks) in the RLV-A, short-term, long-term hypertransfusion RLV-A and hypertransfusion control groups. *Arrows* indicate RLV-A viral inoculation. As time progressed, the RLV-A and the short-term hypertransfusion RLV-A groups showed progressive decline in the PRCV to below 20%. This did not occur in the long-term hypertransfusion RLV-A and hypertransfusion control groups.

Passage and Transplantability Studies

Successful passage of the leukocytic cells found in the long-term hypertransfusion RLV-A group resulted in the establishment of a transplantable monomyelocytic type of leukemia (FIG. 17; TABLE 2). Both splenic cell suspensions and peripheral blood passage resulted in leukemia in recipients in approximately 10–14 days postinoculation. To establish that neoplastic cells had been transplanted and that the observed leukemia was not due to the inoculation of a leukemia-inducing virus, karyotype analysis was done on leukemic cells from a recipient female which had been inoculated with donor male cells. The occurrence of the "Y" chromosome in the terminal female blood samples established transplantability.[21] To date, some 11 successful cell passages have been done.

Preliminary histopathologic examination of these animals has indicated that the

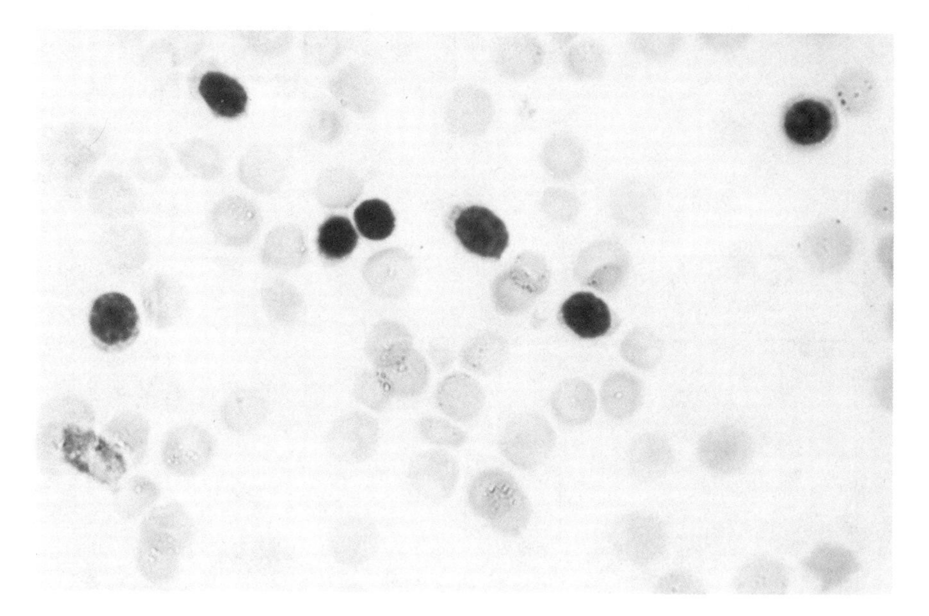

FIGURE 2. Light micrograph of terminal RLV-A peripheral blood. Erythroblasts become the major nucleated cell present.

spleen, liver and bone marrow are the major organs affected in the leukemia. The spleen shows considerable enlargement with the loss of white pulp. It appears to be composed primarily of immature granulocytic and monocytic cells (FIG. 18). Cellular infiltration is also seen in the liver, especially around portal regions (FIG. 19). Hepatocytes generally contained what appeared to be numerous lipid inclusions and seemed to be degenerating, although electron microscopic examination indicated that the sinusoidal endothelium was generally intact (FIG. 20). There was considerable bone marrow involvement with the parenchyma generally being comprised of immature granulocytic and monocytic elements. Electron microscopic examination of the peripheral blood indicated the presence of virus particles in platelets which varied considerably in size and shape (FIG. 21). Numerous granulocytic and monocytic cells in various stages of maturation were observed, some of which contained what appeared to be intracellular lipid-like substances.

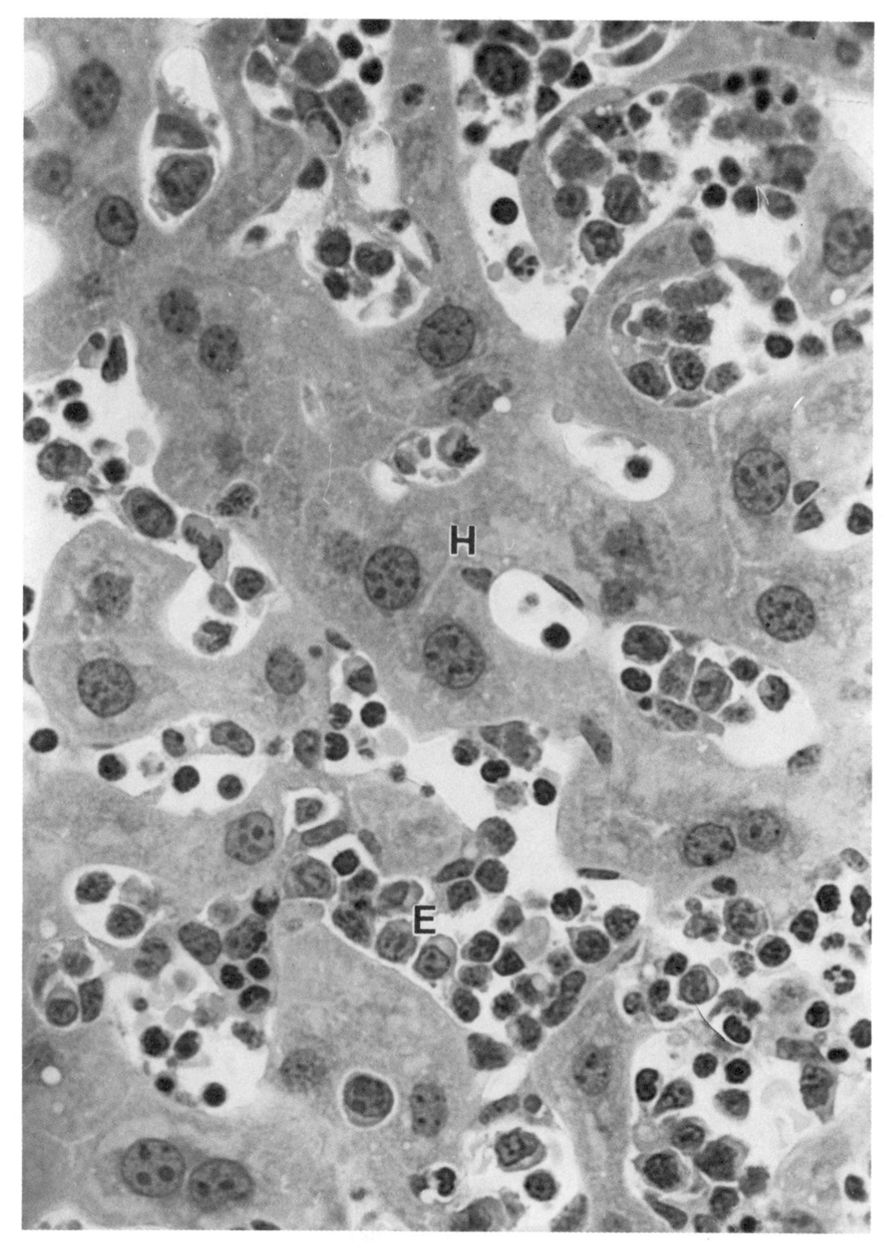

FIGURE 3. Light micrograph of RLV-A liver. Numerous erythroblasts (E) were seen in the sinusoids, generally associated with Kupffer cells. The hepatocytes (H) generally appeared histologically normal except for those in close proximity to portal regions (not shown) where massive erythroid involvement was found. These portal regions showed what appeared to be lipid accumulation and degenerative changes in the hepatocyte cytoplasm.

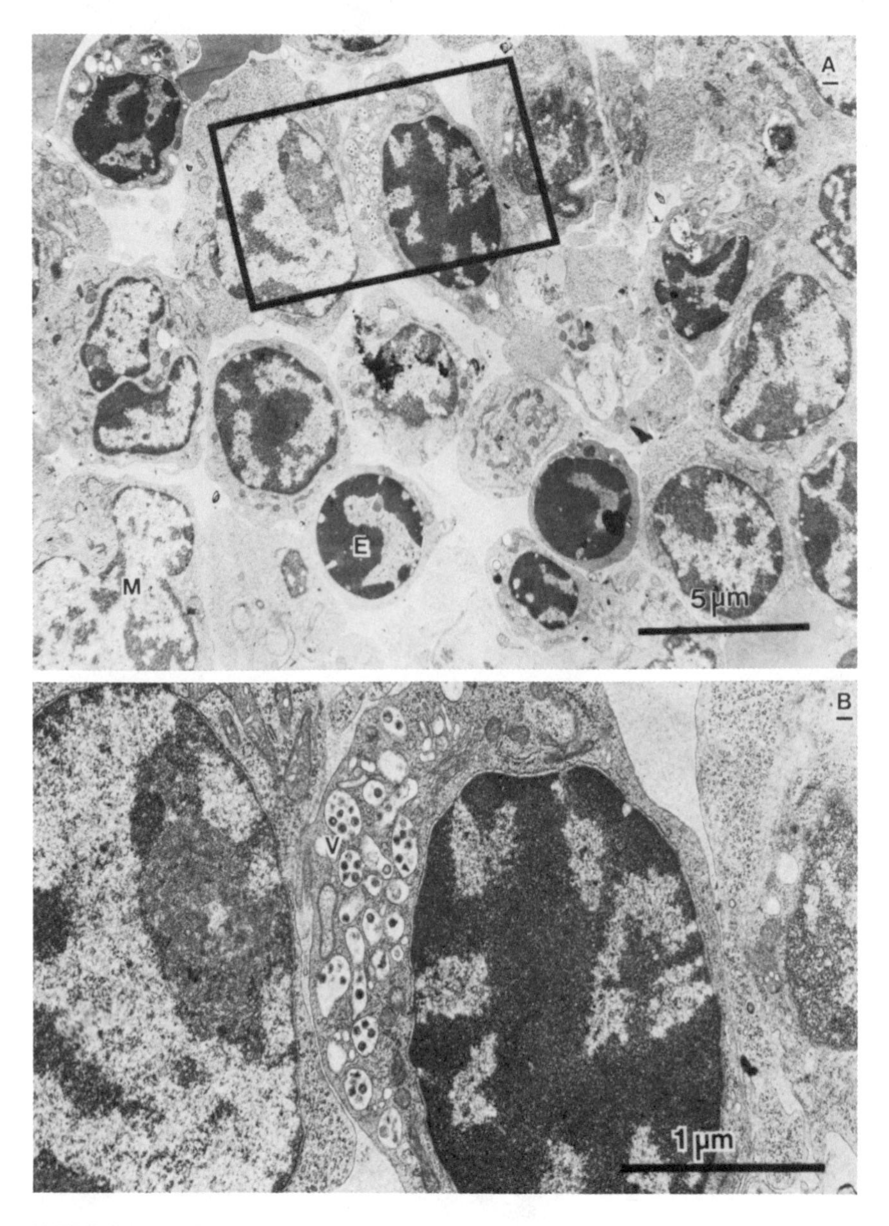

FIGURE 4. **(A)** Electron micrograph of terminal RLV-A spleen containing numerous erythroblasts (E) and monocytic cells (M). One erythroblast (*box*) was found to contain virus particles. **(B)** A high-magnification electron micrograph of the virus-containing erythroblast seen above. Note the presence of virus particles (V).

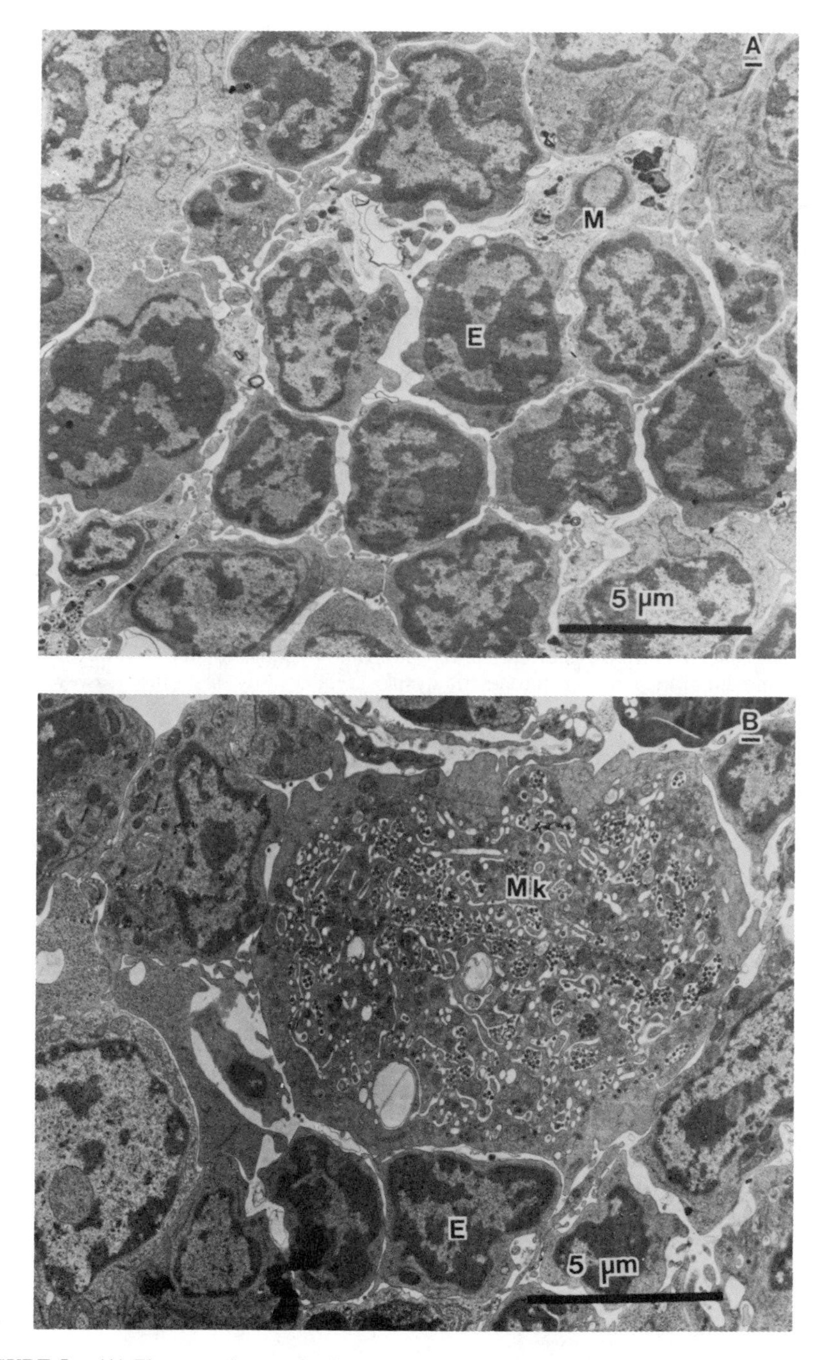

FIGURE 5. **(A)** Electron micrograph of terminal RLV-A bone marrow. Present are a number of erythroblasts (E) associated with a portion of a macrophage (M). The marrow is not appreciably altered in RLV-A disease. **(B)** Electron micrograph of RLV-A bone marrow showing an erythroid element (E) and a virus-containing megakaryocyte (Mk).

TABLE 1. Splenic and Bone Marrow Smear Counts of Benzidine Positive and Negative Cells[a]

Group[b]	Spleen		Bone Marrow	
	Benz +	Benz −	Benz +	Benz −
Normal controls (6)	5.50 ± 1.5	94.5 ± 1.5	10.75 ± 1.02	89.25 ± 1.02
RLV-A injected (4)	29.98 ± 1.96	70.02 ± 1.96	7.35 ± 2.17	92.65 ± 2.17
Hypertransfusion controls (6)	.06 ± .04	99.94 ± .04	0.1 ± .04	99.9 ± .04
Hypertransfusion RLV-A (4) Long experiment	.03 ± .03	99.97 ± .03	0.1 ± .06	99.9 ± .06
Hypertransfusion RLV-A (3) Short experiment	21.4 ± 9.25	78.6 ± 9.25	8.53 ± 2.8	91.5 ± 2.8

[a]Smear counts are represented as percents. Values represent means ± 1 SEM.
[b]The number in parentheses after each group shows the number of mice studied per group.

To ascertain whether the RLV-A virus was present in animals bearing this transplantable leukemia, plasma from these leukemic mice was given to recipients. The "typical" RLV-A pathogenesis was observed. However, attempts to establish a transplantable leukemia from terminal RLV-A peripheral blood which contains high numbers of leukocytic cells did not succeed. Instead, the "typical" RLV-A erythroblastic response was also observed in these recipient mice.

DISCUSSION

Sustained hypertransfusion for 42 days followed by inoculation of RLV-A resulted in development of the usual erythropoietic dysplasia induced by this virus. However, when hypertransfusion was continued postinoculation (long-term experiment) this typical erythroblastic disease did not develop. Instead, these virally infected mice eventually developed a transplantable myeloid leukemia. One explanation of these findings might be that in the short-term experiment, although the microenvironment was granulopoietic at the time of RLV-A inoculation, erythropoiesis was allowed to resume, replenishing this category of viral "target" element. Infection of erythroid elements resulted in erythro-

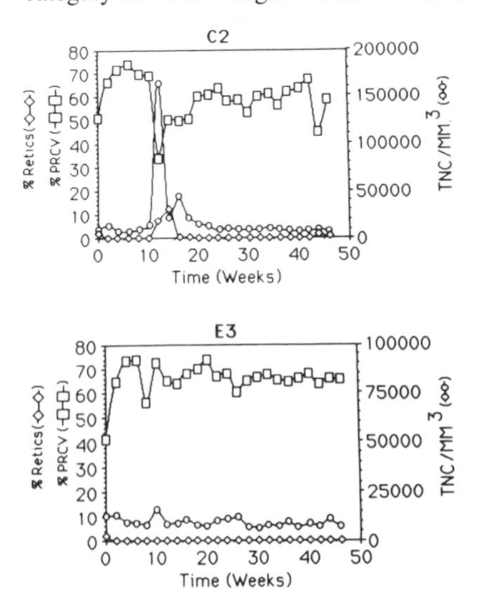

FIGURE 6. Reticulocytes (%), PRCV (%) and peripheral blood total nucleated cell count (TNC/mm^3) in two hypertransfusion control animals. One animal **(C2)** showed an unexpected sharp decline in PRCV below control values. This was associated with large increases in reticulocytes and nucleated cell counts (leukemoid reaction). Continued hypertransfusion restored the hematocrit, and eliminated reticulocytosis and leukocytosis. The other animal **(E3)** is representative of the four members of this group which maintained elevated PRCV and showed no leukocytosis or reticulocytosis throughout the experiment.

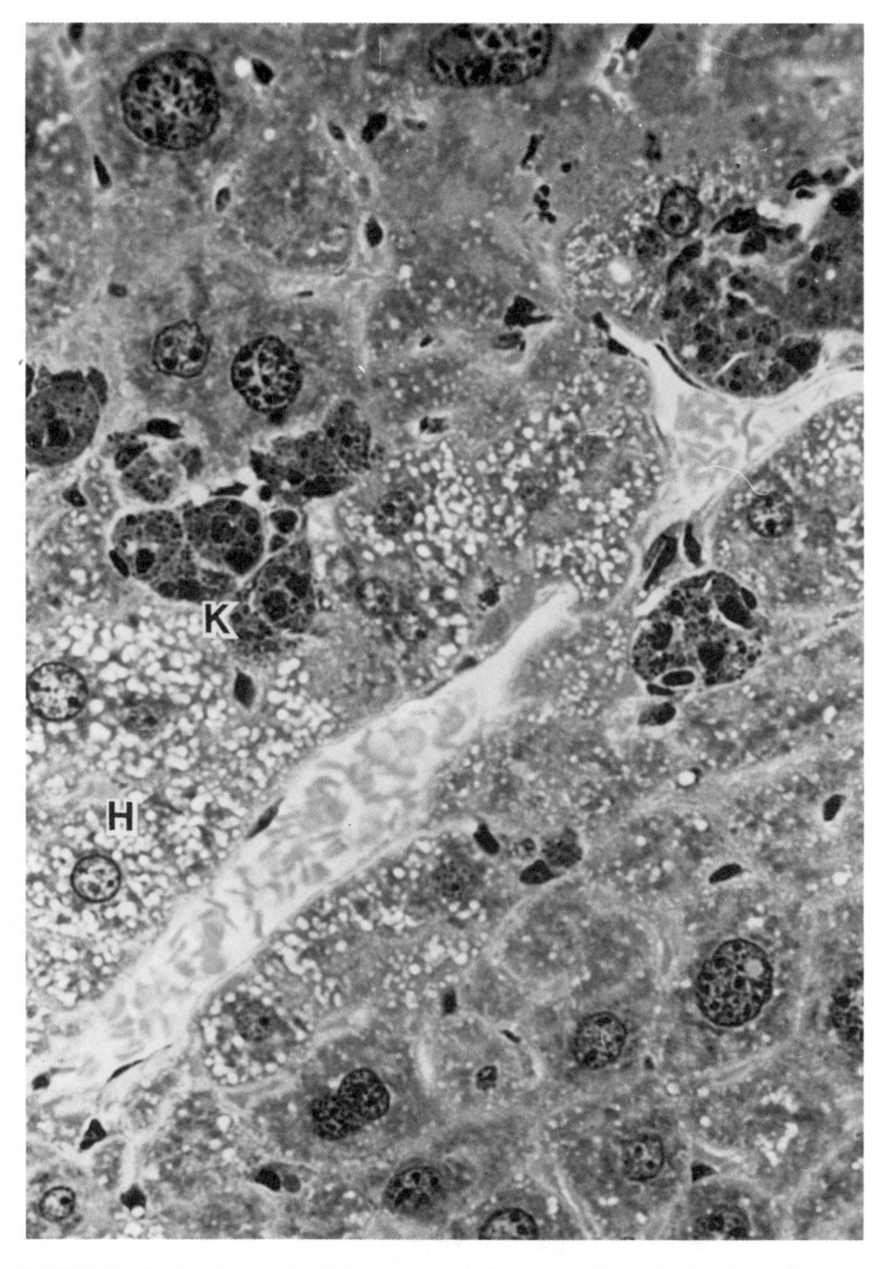

FIGURE 7. Light micrograph of the hypertransfusion control liver showing large phagocytic Kupffer cells (K) present around a blood vessel. An increase in phagocytic activity was seen in sinusoids and portal regions. Hepatocytes (H) in this region contained vacuoles and appeared to be degenerating.

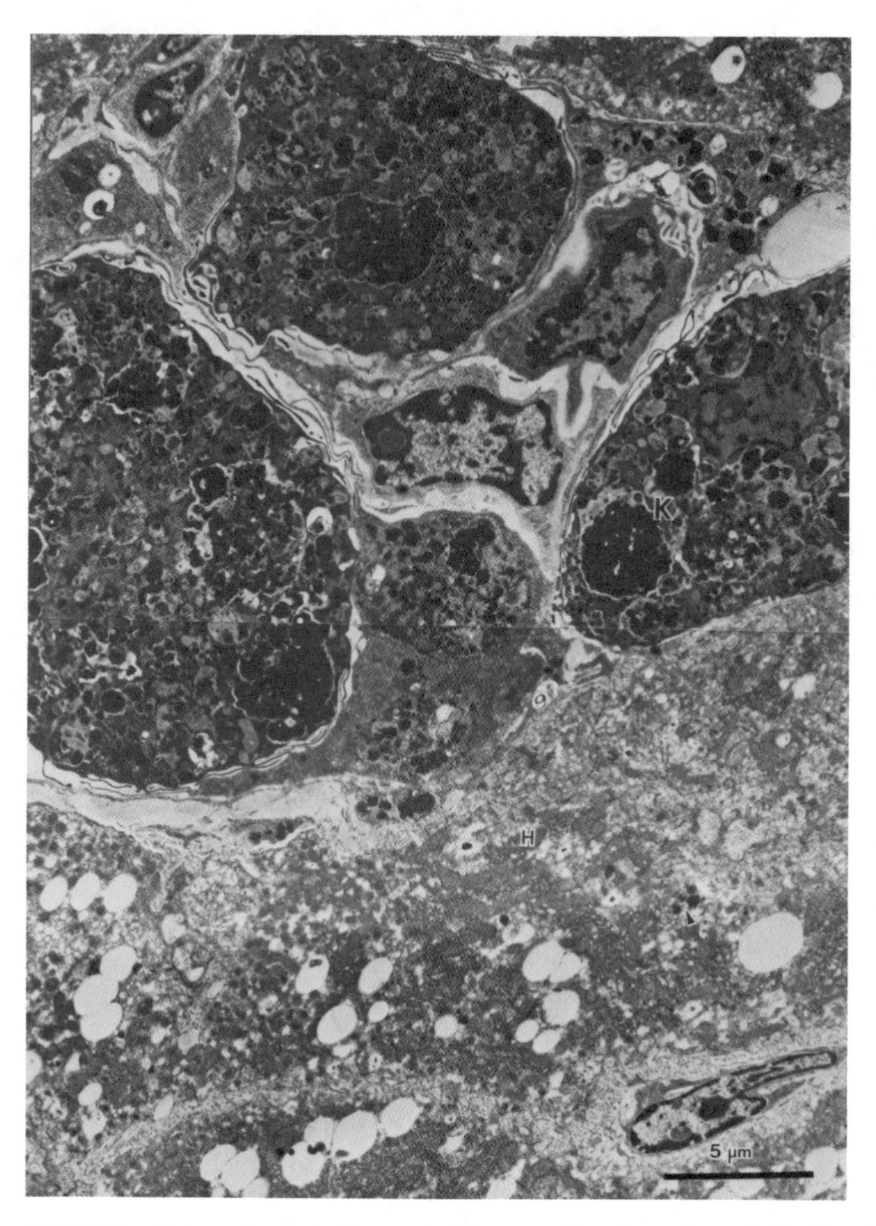

FIGURE 8. Electron micrograph of hypertransfusion control liver. Large phagocytic Kupffer cells (K) are seen in the sinusoid lumen. Hepatocytes (H) that are in close proximity to Kupffer cells show vacuolization.

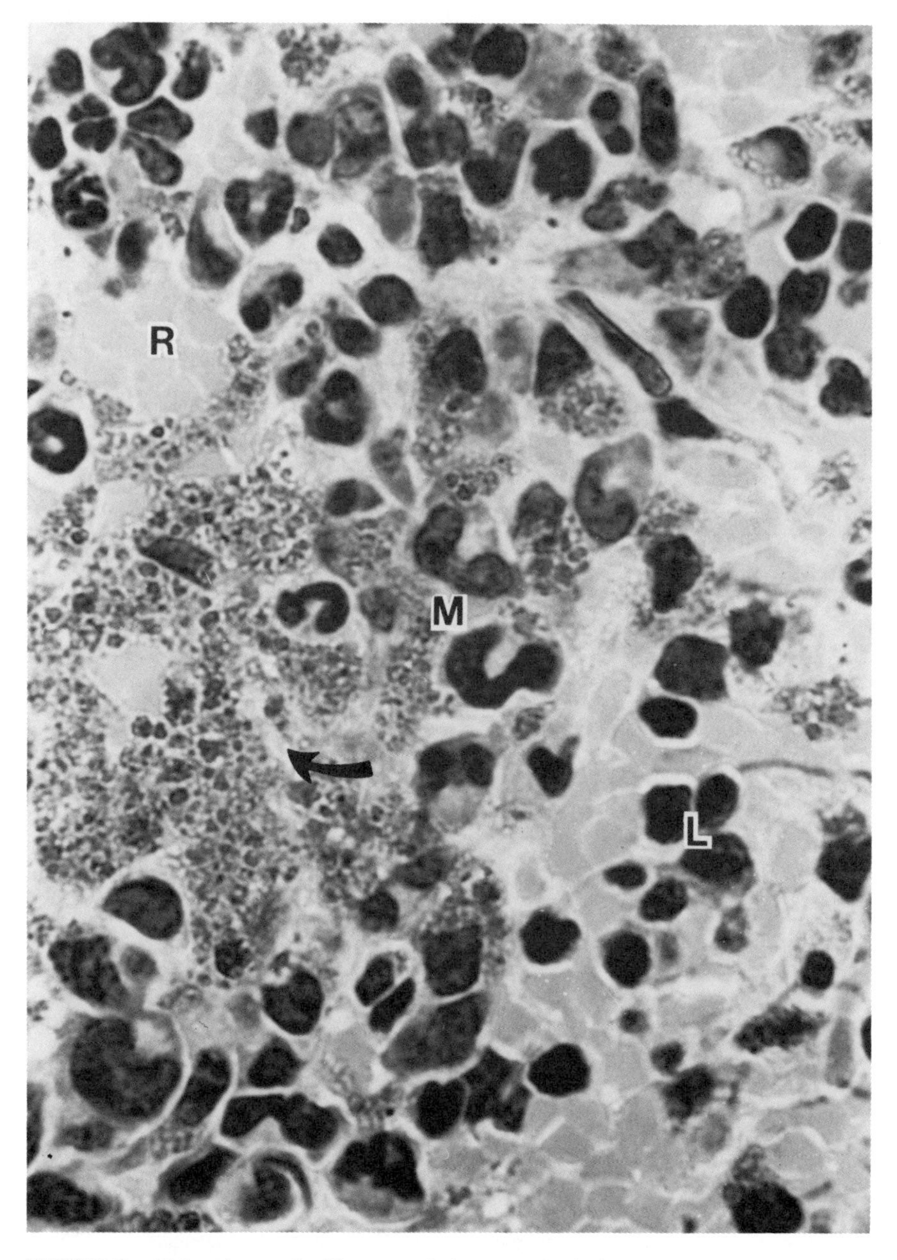

FIGURE 9. Light micrograph of hypertransfusion control splenic red pulp. Considerable phagocytic activity (*arrow*) was seen in this region. Red cells (R), monocytes (M) and lymphocytes (L) are among cellular elements present. Although there was marked phagocytic activity, normal splenic red and white pulp morphology was maintained.

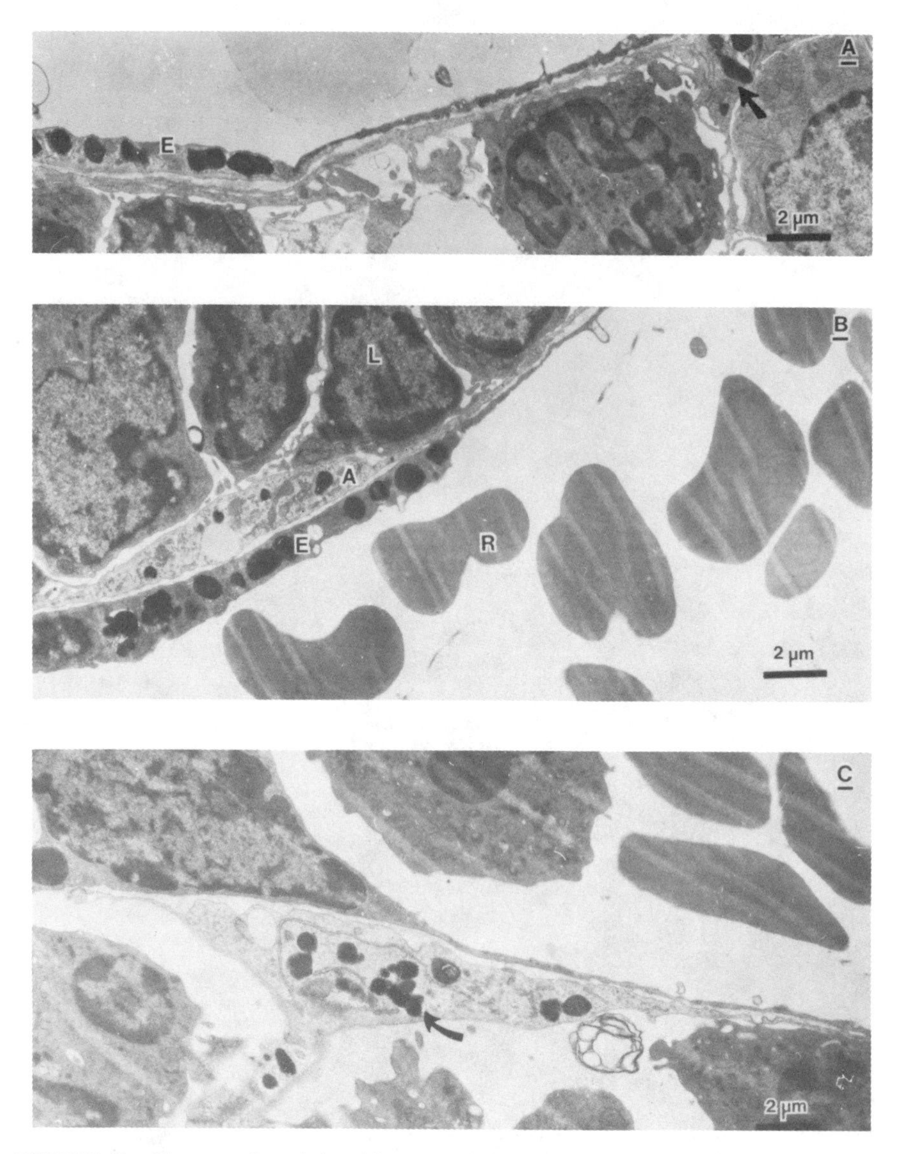

FIGURE 10. Electron micrographs of hypertransfusion control bone marrow sinus regions. **(A)** Material is seen in endothelium (E) and in adventitial cells (*arrow*). **(B)** Bone marrow sinus region containing erythrocytes (R). The endothelium (E) and an adventitial cell (A) contain debris. Bone marrow parenchymal elements include lymphocytes (L). **(C)** Adventitial cell containing debris (*arrow*).

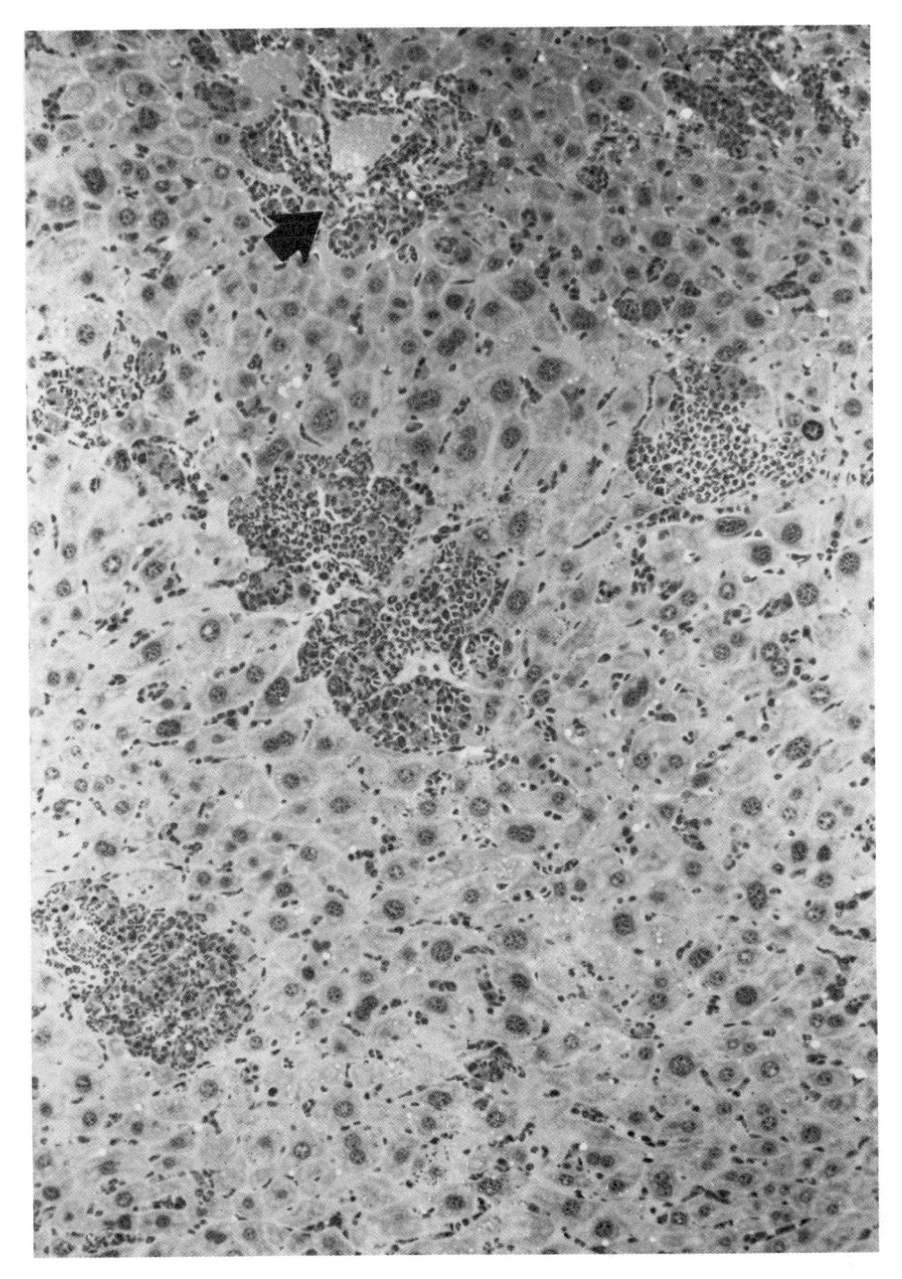

FIGURE 11. Light micrograph of long-term hypertransfusion RLV-A liver. Note the numerous leukocytic elements present around portal areas (*arrow*). These were both mature and immature granulocytic and monocytic elements.

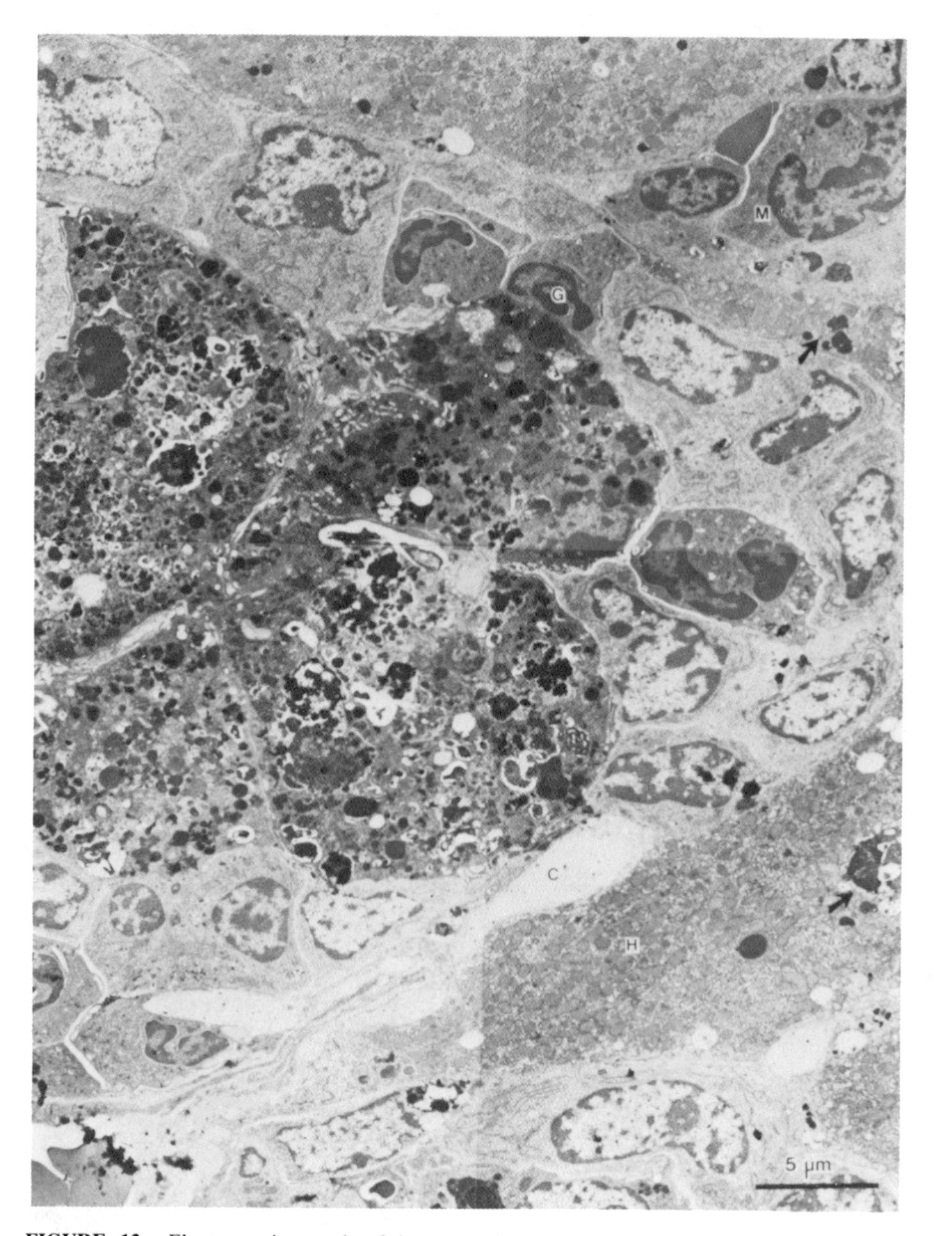

FIGURE 12. Electron micrograph of long-term hypertransfusion RLV-A liver. Phagocytic Kupffer cells and numerous leukocytic granulocytic (G) and monocytic (M) elements are occupying the sinus. The adjacent hepatocytes (H) contain intracytoplasmic debris (*arrows*). An area containing collagen fibers (C) is evident.

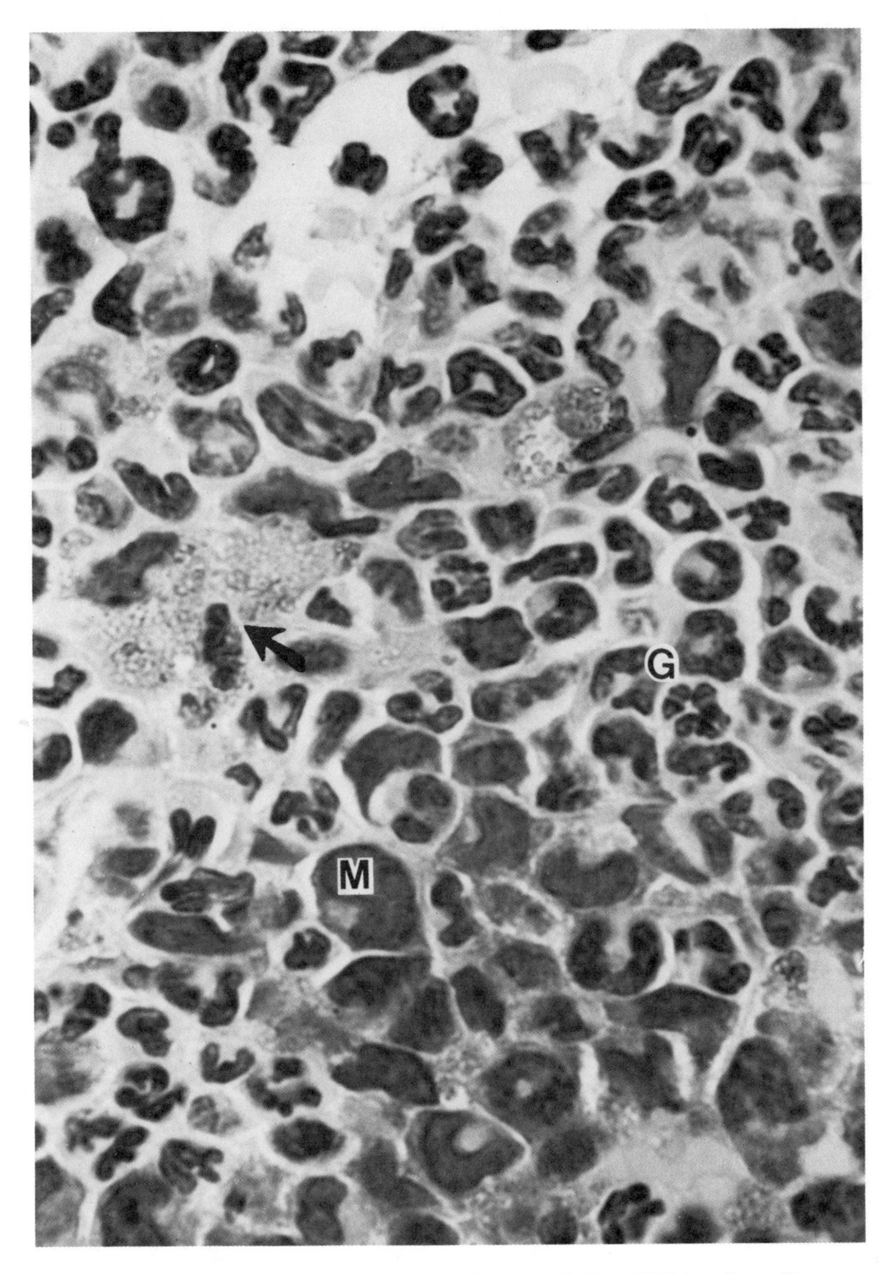

FIGURE 13. Light micrograph of long-term hypertransfusion RLV-A spleen. Present are monocytic (M) and granulocytic (G) elements which have obliterated the splenic lymphoid follicles. Phagocytic activity (*arrow*) was found throughout the splenic red pulp.

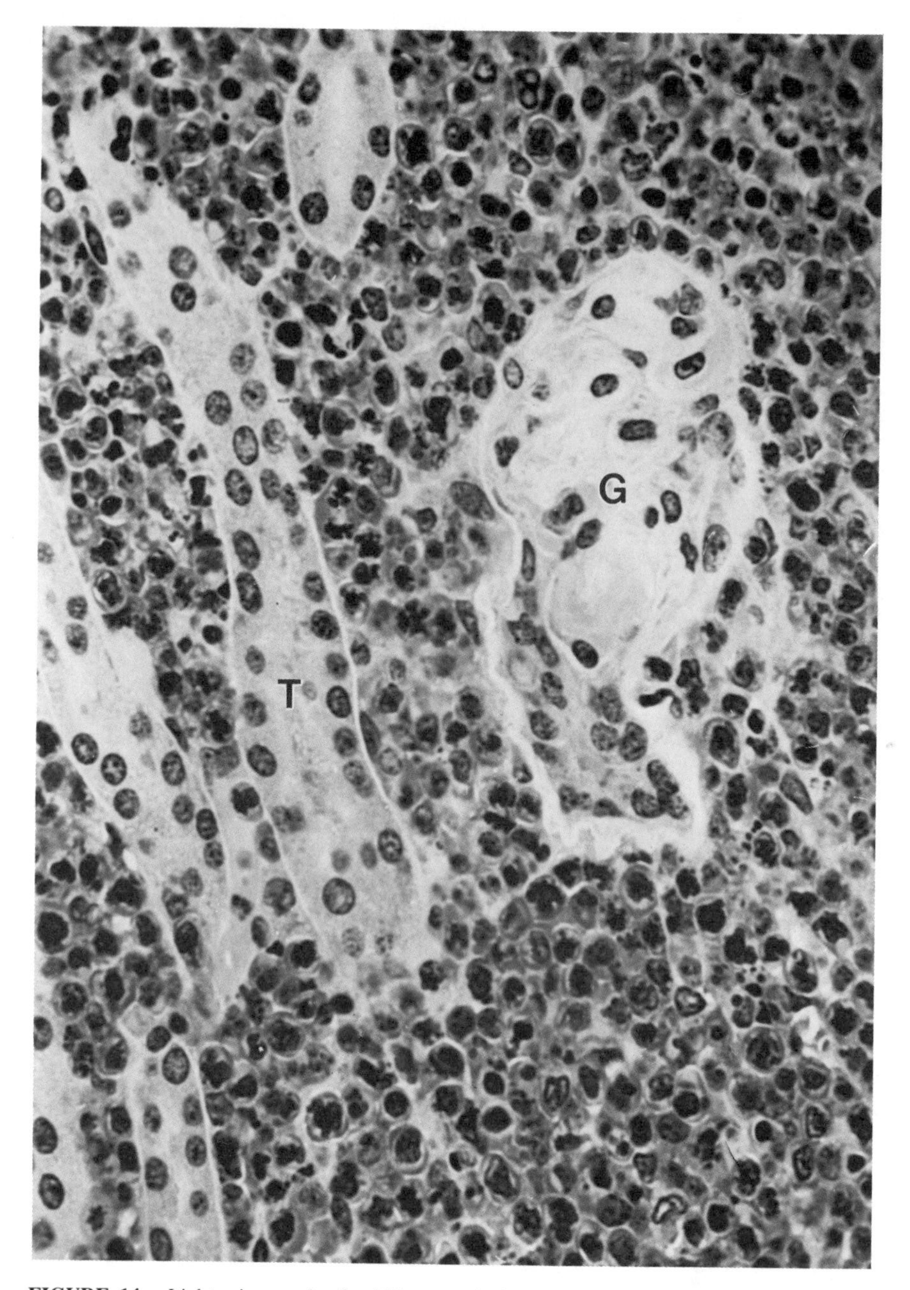

FIGURE 14. Light micrograph of a kidney cortical region from a long-term hypertransfusion RLV-A animal. Tubules (T) and a glomerulus (G) are surrounded by massive leukemic infiltration. Although the kidney in this animal was increased 4 times normal size, the medullary region showed little morphologic change.

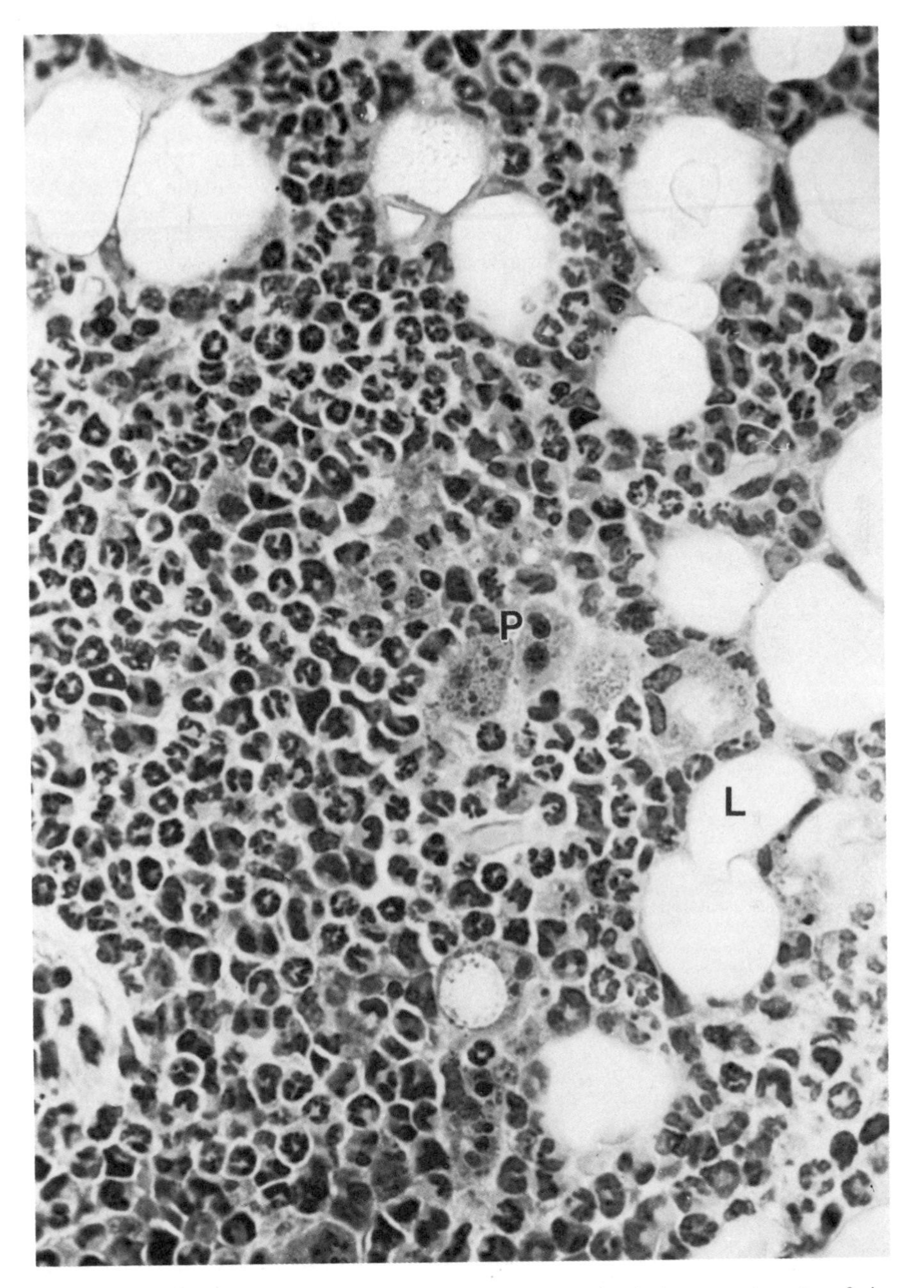

FIGURE 15. Light micrograph of the subcapsular kidney region in long-term hypertransfusion RLV-A. Note the numerous granulocytic and monocytic elements occurring among adipocytes (L). Phagocytic activity (P) was also found.

blastic disease and death in a time too short to allow any other effect of RLV-A infection to exert itself. It is known[22] that injection of low doses of R-MuLV allows a recovery from the initial erythroblastic phase and the subsequent development of lymphatic leukemia. In the long-term experiments since hypertransfusion spared the mice by preventing erythroblastic disease, the opportunity for the expression of the leukemogenicity of RLV-A (an apparent variant R-MuLV) could be realized. Experiments reported by LoBue *et al.*[23] and Gallicchio[24] argue against this as the complete answer. They found that when RLV-A-infected mice were treated with exogenous or endogenous erythropoietin (induced by phenylhydrazine) at the time of onset of moderate anemia (PRCV approximately 30%) the anemia could be ameliorated and survival time increased to as much as 600 days, with no evidence of myeloid leukemia. In several instances, complete remission of the erythropoietic dysplasia was effected; organomegaly and erythroblastosis were resolved, anemia was overcome, and the blood picture and stem cell numbers were returned to normal. Gallicchio[24] and LoBue *et al.*[25] also found that treatment of RLV-A anemic mice with

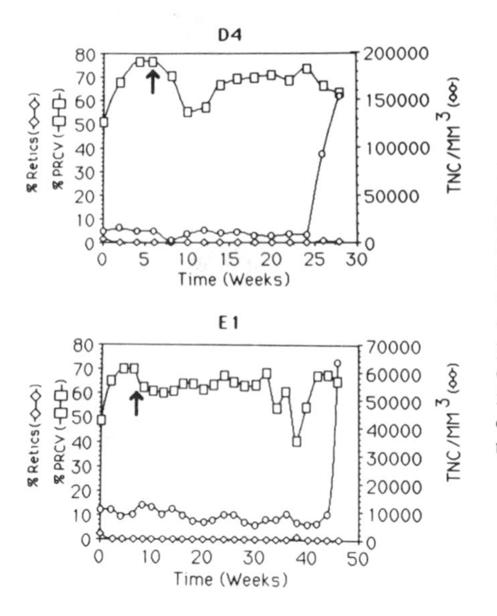

FIGURE 16. Reticulocytes (%), PRCV (%) and total nucleated cell counts (TNC/mm^3) versus time (weeks) for 2 animals in the long-term hypertransfusion RLV-A group. One animal **(D4)** never showed a drop in PRCV below normal control values. In the other animal **(E1)**, a drop in PRCV below control values was accompanied by a slight reticulocyte response. In all cases the leukemia was manifested by a large increase in peripheral blood nucleated cell counts. *Arrows* indicate RLV-A virus inoculation.

antierythropoietin (anti-EP) exacerbated the disease with animals becoming anemic earlier and dying sooner than virus potency controls. Histopathologic examination of RLV-A-anti-EP-treated mice showed leukocytic involvement suggestive of myeloid leukemia. These findings seem to indicate that induction of active erythropoiesis with correction of anemia can be life sparing and even "curative" in RLV-A-infected mice. However, suppression of erythropoiesis with or without the maintenance of PRCV appears to encourage (via alterations in microenvironment?) development of myeloid leukemia.

Delwel *et al.*[26] have made an excellent case for the proposition that R-MuLV may not be "specific for the transformation of definite hemopoietic target cells" and that other features, such as proviral integration and recombination with endogenous viruses, are responsible for the particular leukemia which is induced postinfection. It seems unlikely that proviral integration events, uninfluenced by other factors, were responsible for the leukemogenic expression of RLV-A in the present studies. This would have required identical proviral insertions in five different mice all producing onset of a phenotypically

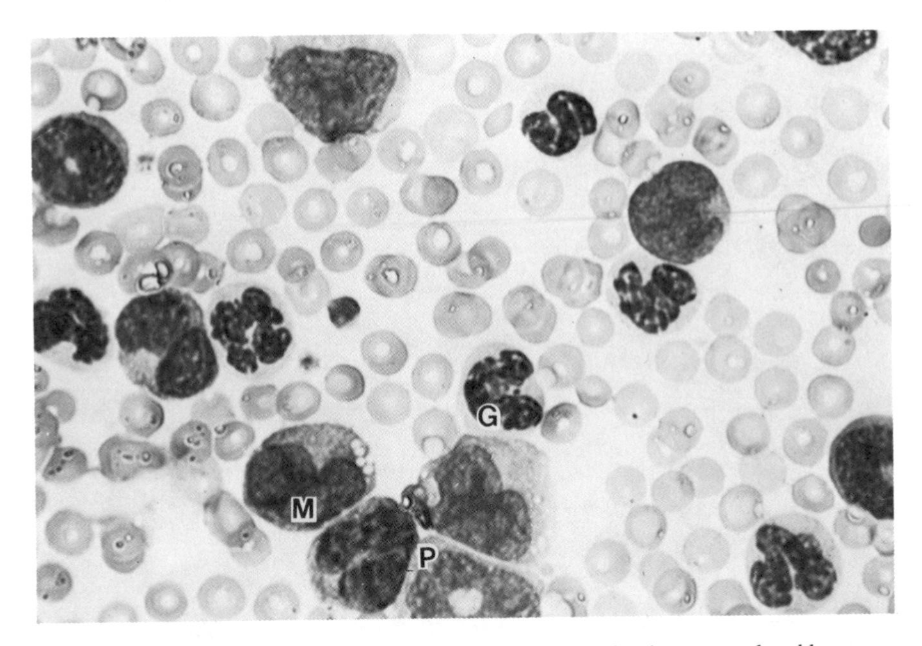

FIGURE 17. Light micrograph of peripheral blood from mice bearing a transplantable mono-myelocytic leukemia. Present are numerous granulocytes (G), promyelocytes (P) and monocytic elements (M).

identical leukemia. Alternatively, five independent, identical recombinations with endogenous virus would be needed to explain the RLV-A leukemogenesis observed. It seems unlikely that recombination occurred since passage of viremic plasma from myeloid leukemic mice induced RLV-A erythroblastic disease.

Neither proviral insertion alone, recombination events nor suppression of the erythroblastic effects of RLV-A are sufficient to explain the development of myeloid leukemia in the long-term hypertransfused RLV-A-infected mice. However, it seems plausible to suggest that creation and maintenance of a granulopoietic microenvironment, through sustained hypertransfusion, may have had an important role in the expression of leukemogenicity of RLV-A. There is some literature to support this contention. For example, Greenberger *et al.*[27] utilizing a modified Dexter long-term marrow culture system infected

TABLE 2. Hematologic Data Obtained from 11 Animals Bearing a Transplantable Myeloid Leukemia

Item	Value[a]
Packed red cell volume (%)	31.4 ± 0.6
Blood reticulocytes (%)	3.8 ± 1.1
Blood nucleated cell count (cells per mm^3)	172,440 ± 8,780
Blood nucleated cell count (%):	
Erythroblasts	0.3 ± 0.1
Lymphocytes	12.6 ± 0.8
Granulocytic	73.5 ± 1.0
Monocytic	13.6 ± 1.5

[a]Values were derived from terminal stage animals (10–14 days postinoculation). All values represent means ± 1 SEM.

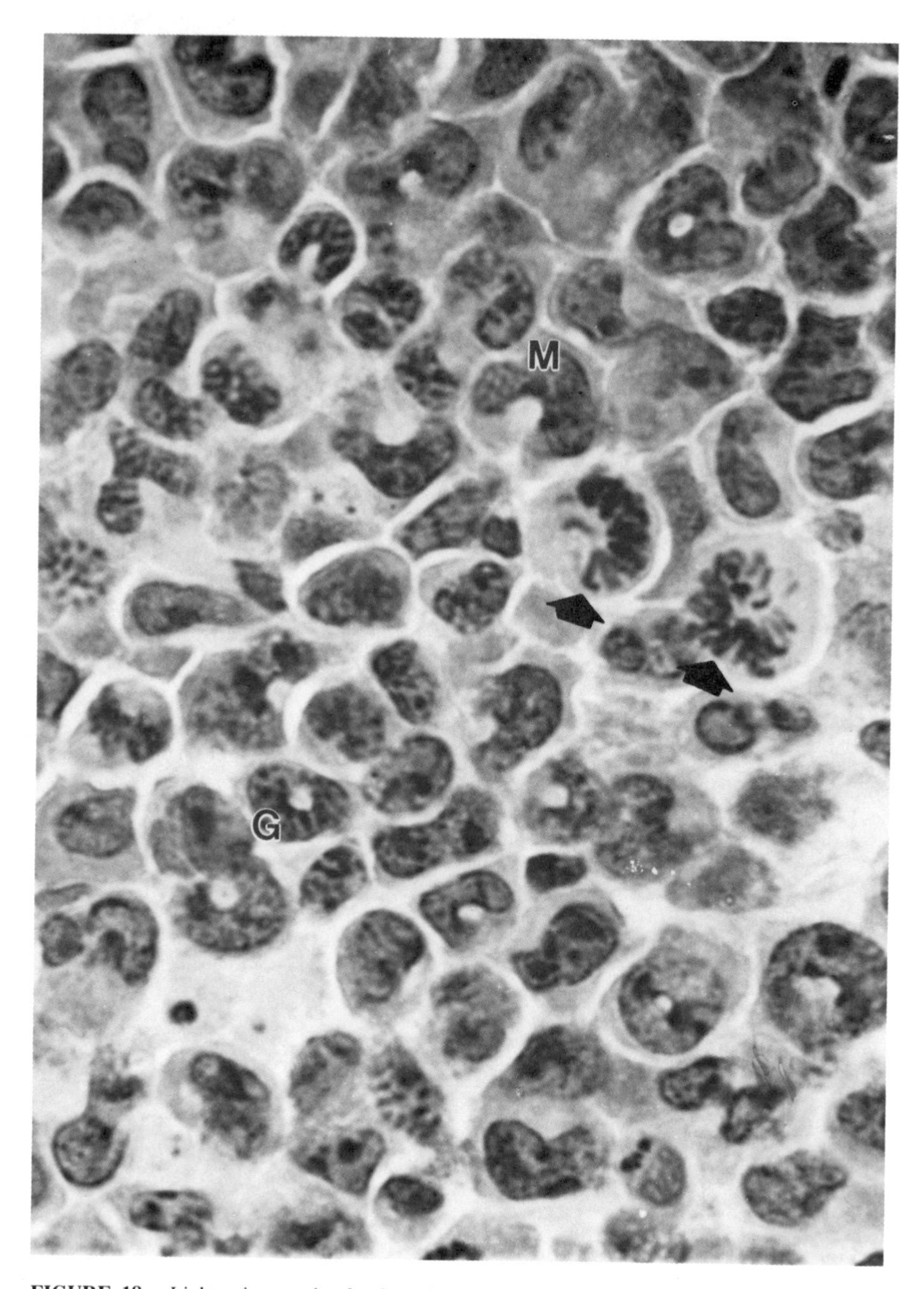

FIGURE 18. Light micrograph of spleen from mice bearing a transplantable monomyelocytic leukemia. Granulocytic (G) and monocytic (M) elements are seen. The splenic white pulp has been eliminated. Mitotic cells (*arrows*) are seen.

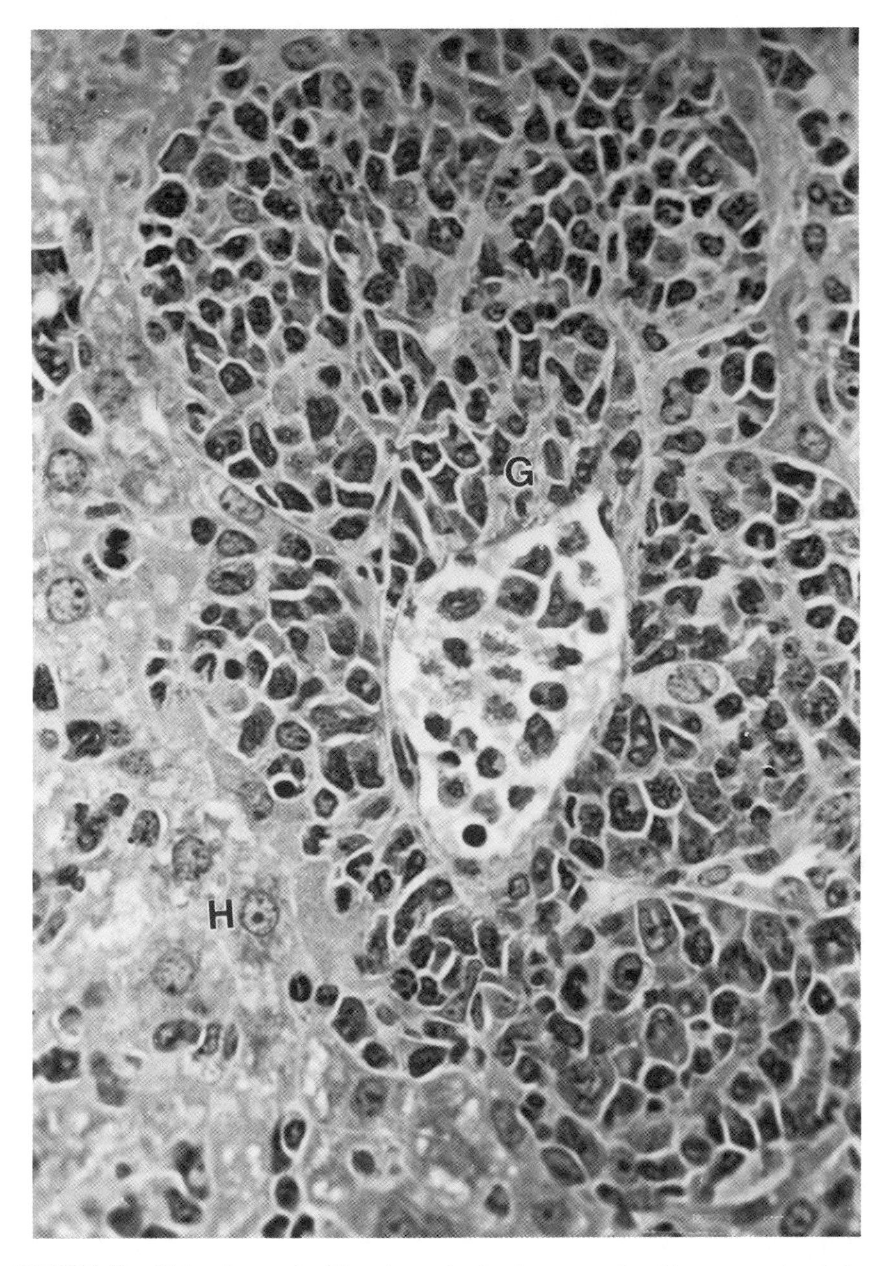

FIGURE 19. Light micrograph of liver from mice bearing a transplantable monomyelocytic leukemia. A cuff of myeloid cells (G) is seen around a blood vessel. The hepatocytes (H) generally show vacuolization, especially in cells bordering portal regions.

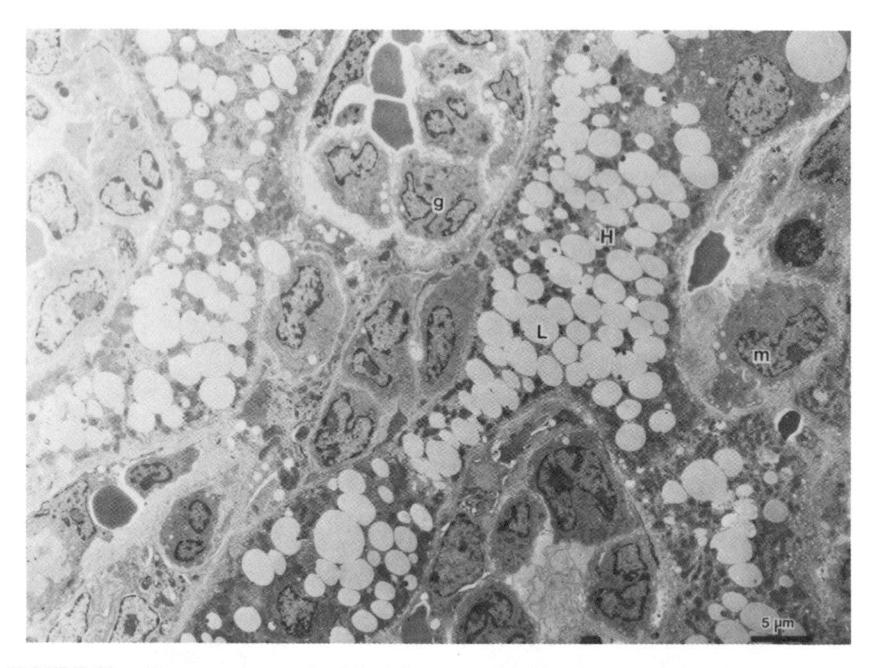

FIGURE 20. Electron micrograph of liver from an animal bearing a transplantable monomyelocytic leukemia. Numerous myeloid (g) and monocytic (m) elements are seen occupying liver sinusoids. Hepatocytes (H) appear to be undergoing lipid degeneration (L).

NIH/Swiss mouse marrow cells with either Friend (FLV-A) or Abelson (A-MuLV) virus and induced transplantable promyelocytic leukemic cell lines. The virus produced by these cell lines, however, did not cause granulocytic leukemia *in vivo* but rather produced diseases identical to those produced by the parent viruses. The Dexter culture system produces a decidedly granulocytic microenvironment, apparently owing to the presence of large numbers of stromal elements producing myeloid growth factors.[28] Thus, Greenberger *et al.*[27] found that ". . . malignant, phenotypically granulocytic cell lines were induced *in vitro* in a microenvironment that greatly favored granulocytic differentiation by two viruses that normally cause nongranulocytic leukemias *in vivo*. . . ." They suggest that it is the virally infected CFU-S, not CFU-C or myeloid cells, subsequently differentiating in a granulocytic hematopoietic microenvironment that results in the development of the leukemia. When Rauscher Virus (R-MuLV) was used to infect cells *in vitro*, leukemic cell lines could not be established. However, there was an increase in immature leukocytes resembling preleukemic dysmyelopoietic granulocytes. Failure to induce leukemic lines could have been due to viral replication rates or specific culture requirements for R-MuLV-infected cells that were lacking in the Dexter culture system.

SUMMARY

Infection of BALB/c mice with the RLV-A virus typically results in an erythropoietic dysplasia characterized by hepatosplenomegaly, erythroblastosis, erythroblastemia and severe anemia without reticulocytosis. Mice hypertransfused weekly with 75%-packed red cells for 42 days prior to RLV-A infection and viral potency controls manifested this typical RLV-A response. Mice that were hypertransfused prior to and following RLV-A infection never developed the "typical" RLV-A pathogenesis. Instead, a transplantable myeloid leukemia was established. Although the reason for altered pathogenesis remains uncertain, it seems plausible that continued hypertransfusion, presumably after establishment of an altered granulopoietic microenvironment, resulted in a completely different viral expression and development of the transplantable myeloid leukemia.

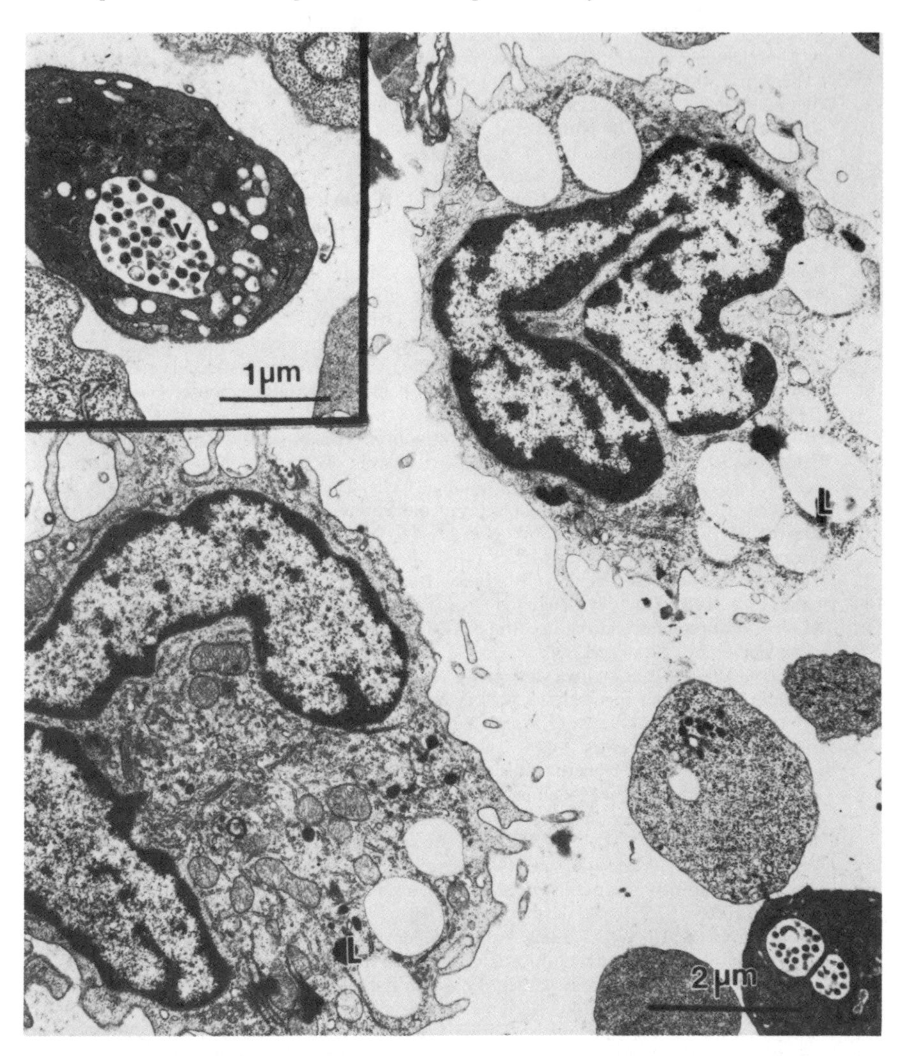

FIGURE 21. Electron micrograph of peripheral blood from an animal bearing a transplantable monomyelocytic leukemia. Myeloid cells which appear to contain lipid (L) are among leukocytic elements found in this leukemia. The *inset* shows a higher magnification of a virus-containing platelet (V).

ACKNOWLEDGMENTS

The authors express their gratitude to Ms. Ann Marie Snow for her technical assistance.

REFERENCES

1. FREDRICKSON, T. N., J. LOBUE, P. ALEXANDER, JR., E. SCHULTZ & A. S. GORDON. 1972. A transplantable leukemia from mice inoculated with Rauscher leukemia virus. J. Natl. Cancer Inst. **48:** 1597–1605.
2. LOBUE, J., P. A. ALEXANDER JR., T. N. FREDRICKSON, E. F. SCHULTZ & A. S. GORDON. 1972. Erythrokinetics in normal and diseased states. Virally induced erythroblastosis: A model system. *In* Regulation of Erythropoiesis. A. S. GORDON, M. CONDORELLI & C. PESCHLE, Eds.: 89–101. Il. Ponte. Milan.
3. BERGSON, A., J. LOBUE, A. S. GORDON & T. N. FREDRICKSON. 1978. Effect of phenylhydrazine pretreatment on splenectomized Rauscher leukemia virus infected mice. J. Surg. Oncol.**10:** 239–243.
4. MORSE, B., D. GIULIANI, T. FREDRICKSON & J. LOBUE. 1978. Erythrokinetics and ferrokinetics of a virally induced erythroblastosis. Blood **51:** 623–632.
5. WEITZ-HAMBURGER, A., T. N. FREDRICKSON, J. LOBUE, W. D. HARDY, JR., P. FERDINAND & A. S. GORDON. 1973. Inhibition of erythroleukemia in mice by induction of hemolytic anemia prior to infection with Rauscher leukemia virus. Cancer Res. **33:** 104–111.
6. LOBUE, J. 1984. Viral-associated alterations in hemopoiesis in the mouse. Blood Cells **10:** 211–222.
7. LOBUE, J., A. S. GORDON, A. WEITZ-HAMBURGER, P. FERDINAND, J. F. CAMISCOLI, T. N. FREDRICKSON & W. D. HARVEY, JR. 1974. Erythroid differentiation in murine erythroleukemia. *In* Control of Proliferation in Animal Cells. B. CLARKSON & R. BASERGA, Eds.: 863–884. Cold Spring Harbor Laboratory, Cold Spring Harbor, NY.
8. ARMAND, R. & R. STEEVES. 1980. How many types of erythroleukemia are induced by retroviruses? Nature **230:** 615–616.
9. PIMENTEL, E. 1986. Oncogenes. CRC Press. Boca Raton, FL.
10. BLAKELY, A. 1985. Rauscher leukemia virus (variant A): The isolation and characterization of RLV-A viral proteins and the occurrence of P^{30} during RLV-A disease. Ph.D. thesis, New York University, New York, NY.
11. JACOBSON, L. O., E. GOLDWASSER & C. W. GURNEY. 1960. Transfusion-induced polycythemia as a model for studying factors influencing erythropoiesis. *In* CIBA Foundation Symposium on Hemopoiesis. G. E. W. WOLSTENHOLME & M. O'CONNOR, Eds.: 423–451. Little, Brown. Boston, MA.
12. MONETTE, F. C. 1985. Hypertransfusion-experimental results: Effect on erythropoietic progenitors. *In* Mathematical Modeling of Cell Proliferation: Stem Cell Regulation in Hemopoiesis. Vol. II. Erythropoietic Suppression, Combined Stresses, Drug Effects. H. E. WICHMANN & M. LOEFFLER, Eds.: 4–18. CRC Press. Boca Raton, FL.
13. DUNN, T. B., R. A. MALMGREN, P. G. CARNEY & A. W. GREEN. 1966. Propylthiouracil and transfusion modifications of the effects of the Rauscher virus in BALB/c mice. J. Natl. Cancer Inst. **36:** 1003–1025.
14. BROOKOFF, D. & L. WEISS. 1982. Adipocyte development and the loss of erythropoietic capacity in the bone marrow of mice after sustained hypertransfusion. Blood **60:** 1337–1344.
15. STRUMIA, M., A. SAMPLE & E. HART. 1954. An improved microhematocrit method. Am. J. Clin. Pathol. **24:** 1016–1024.
16. BRECHER, G. 1949. New methylene blue as a reticulocyte stain. Am. J. Pathol **19:** 895–896.
17. LOBUE, J., B. S. DORNFEST, A. S. GORDON, J. HURST & H. QUASTLER. 1963. Marrow distribution in rat femurs determined by cell enumeration and Fe^{59} labeling. Proc. Soc. Exp. Biol. Med. **112:** 1058–1062.
18. KARNOVSKY, M. J. 1965. A formaldehyde-glutaraldehyde fixative of high osmolality for use in electron microscopy. J. Cell Biol. **27:** 137–138.

19. ANDERSON, D. R. 1965. A method of preparing leukocytes for electron microscopy. J. Ultrastruct. Res. **13:** 263–268.
20. BROXMEYER, H., L. KOLTUN, J. LOBUE, T. N. FREDRICKSON & A. S. GORDON. 1975. Granulopoiesis in "preleukemic" mice with anemia induced by Rauscher leukemia virus, variant a. J. Natl. Cancer Inst. **55:** 1123–1127.
21. RODRIGUEZ, E. Unpublished data.
22. DE BOTH, N. J., M. VERMEY, E. VAN'T HULL, E. KLOOTWIJK-VAN-DIJKE, L. J. L. D. VAN GRIENSVEN, J. N. M. MOL & T. J. STOOF. 1978. A new erythroid cell line induced by Rauscher leukemia virus. Nature **272:** 626–628.
23. LOBUE, J., T. N. FREDRICKSON, V. GALLICCHIO, A. RONQUILLO & A. S. GORDON. 1975. Endogenous and exogenous erythropoietin in the treatment of the fatal anemia of RLV disease. *In* Erythropoiesis. K. NAKAO, J. W. FISHER & F. TAKAKU, Eds.: 455–461. University of Tokyo Press. Tokyo.
24. GALLICCHIO, V. S. 1976. The in vivo stimulation of erythroid differentiation in Rauscher leukemia virus (RLV-A) induced erythroleukemia. Ph.D. thesis, New York University, New York, NY.
25. LOBUE, J., T. N. FREDRICKSON, V. S. GALLICCHIO, B. S. MORSE, F. EDINGER, M. ROY, A. BERGSON, L. A. KOLTON & A. S. GORDON. 1976. Erythropoietin and the pathophysiology of murine erythroleukemia. *In* Topics in Hematology. S. SENO, F. TAKAKU & S. IRINO, Eds.: 792–796. Excerpta Medica. Amsterdam.
26. DELWEL, H. R., P. J. M. LEENEW, D. BERENDS, N. J. DE BOTH & W. VAN EWIJK. 1987. The expression of differentiation antigens by Rauscher virus-induced erythroid, lymphoid and myeloid cell lines. Leuk. Res. **11:** 25–30.
27. GREENBERGER, J. S., P. A. DAVISSON, P. J. GANS & W. C. MOLONEY. 1979. In vitro induction of continous acute promyelocytic leukemia cell lines by Friend or Abelson murine leukemia virus. Blood **53** (5): 987–1001.
28. DEXTER, T. M. & N. G. TESTA. 1976. Differentiation and proliferation of hemopoietic cells in culture. Methods Cell Biol.**14:** 387–405.

Stromal Regulation of Hematopoiesis

PETER J. QUESENBERRY, IAN K. MCNIECE,
H. ELIZABETH MCGRATH, DANIEL S. TEMELES,
GWEN B. BABER, AND DONNA H. DEACON

Division of Hematology/Oncology
Department of Internal Medicine
University of Virginia School of Medicine
P.O. Box 502
Charlottesville, Virginia 22908

INTRODUCTION

Hematopoiesis *in vivo* is regulated both by the generation from peripheral tissues of feedback signals, *i.e.*, long-range regulation, and by the modulation of cell production by the interaction between stromal cells in the bone marrow and target hematopoietic stem/progenitor cells. The latter may be due to direct cell contact or more likely to the transfer of regulatory molecules from cell to cell. The introduction by Dexter and colleagues of long-term *in vitro* culture systems, dependent upon the formation of adherent "stromal" layers, has provided a model system to approach the role of stroma and various stromal cell types in modulating hemopoiesis.[1–3] In the Dexter murine hemopoietic system, growth occurs when there is a formation of an adequate stromal layer consisting of a variety of cell types. The growth in general consists of the production of granulocytes, megakaryocytes, and monocyte/macrophages along with the adherent cell types.[4] In addition, a wide range of hematopoietic stem cells are maintained including colony forming unit spleen day 9 (CFU-S d 9), colony forming unit diffusion chamber (CFU-D), and colony forming unit granulocyte-macrophage (CFU-GM) stimulated by a number of regulators, colony forming unit megakaryocyte (CFU-Meg) and the high proliferative potential colony forming cell (HPP-CFC)[5–8] originally described by Bradley and Hodgson.[9] Initially it was felt that this sytem did not produce conventional regulators or colony stimulating factors for myelopoiesis.[3,10,11] However, as we will outline below, this was not the case, as this system proved to be a rich source of a variety of growth factors. These were difficult to detect in conditioned media (cm), in part, because of the avid and rapid binding, utilization and/or internalization of these factors by hemopoietic cells within this system.[12,13] Our initial studies sought to define the cellular components of the adherent layer critical for maintaining long-term hemopoiesis.

Stromal Cell Layer

A number of approaches were taken to attempt to isolate stromal cells from hemopoietic cells. Standard short-term adherence or density separations were unsuccessful and it was found that extensive washing of the cultures did not separate the hemopoietic progenitors from the stromal cells.[14] However, irradiation of the stromal layers with 950–1000 rads effectively abolished ongoing hemopoiesis preserving a hemopoietic stromal layer capable of supporting the ongoing production of granulocytes, megakaryocytes and macrophages and the support of CFU-S and CFU-GM.[12,14,15] When stromal layers were irradiated at 3–4 weeks of culture there were two predominant cell types, a nonspecific esterase, acid phosphatase positive cell which actively phagocytosed and had the phenotype of a macrophage, and an alkaline phosphatase positive cell which accumulated fat in the presence of steroids and which was negative for multiple other markers including factor VIII, nonspecific esterase and acid phosphatase. This latter large alkaline phosphatase positive preadipocyte had at least some of the characteristics of the adventitial reticular cell of the bone marrow.[16] In an alternative model, mice were irradiated with 1000R and their marrow explanted 24 hours later into Dexter cultures. Under these conditions explants at high cell levels (two tibias and two femurs) resulted in a confluent monolayer capable of supporting ongoing hemopoiesis, while explantation at lower cell levels gave rise to clones of adherent cells. Analysis of these colonies showed that the majority consisted of mixtures of cells with the macrophage phenotype and the phenotype of a large alkaline phosphatase positive preadipocyte. A smaller percentage of colonies were pure preadipocyte-type cells, while rare colonies were pure macrophage. Intercolony macrophages were scattered throughout these cultures. When these cultures were refed with nonadherent cells from long-term Dexter cultures, incapable themselves of forming adherent layers, active hemopoiesis occurred upon colonies which contained the preadi-

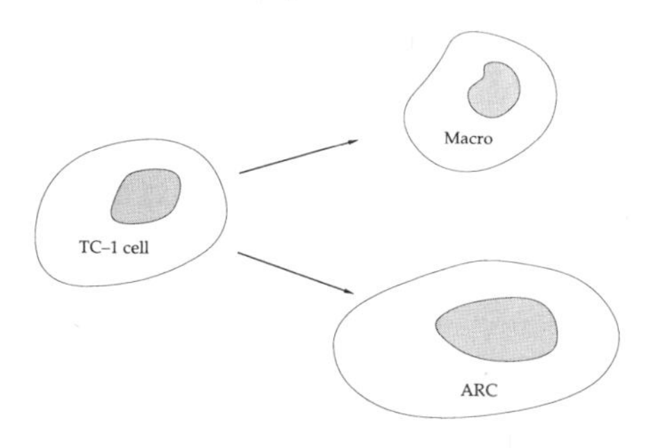

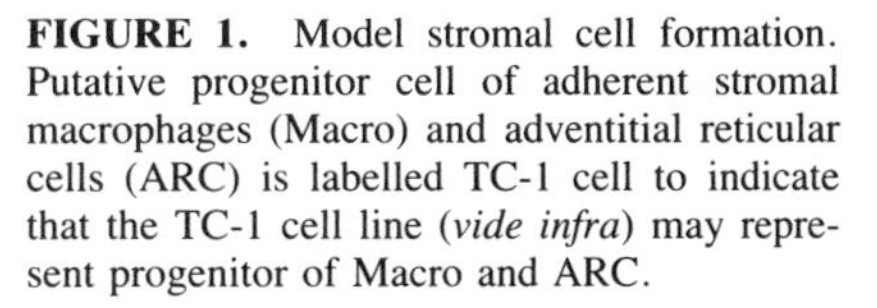

FIGURE 1. Model stromal cell formation. Putative progenitor cell of adherent stromal macrophages (Macro) and adventitial reticular cells (ARC) is labelled TC-1 cell to indicate that the TC-1 cell line (*vide infra*) may represent progenitor of Macro and ARC.

pocyte alkaline phosphatase positive cell, but not upon pure macrophage colonies suggesting potential important cell/cell regulatory influences. Further analysis of these models, studying explant marrow from irradiated mice, at limiting dilution levels, showed that a majority of adherent colonies formed (at levels at which no intercolony cells were apparent) were mixed in phenotype containing macrophagelike and preadipocytelike cells.[17] These data suggested that there may be common progenitor cells for the adherent macrophage and the alkaline phosphatase positive preadipocyte. These data further suggest that either there are unique lineage macrophages or that the preadipocytic fibroblast-type cells may be an inherent component of the hemopoietic lineages. This concept is presented in FIGURE 1, the progenitor cell being labeled TC-1 to indicate that an isolated cell line from long-term marrow culture may in fact represent the common progenitor of these two cell types (*vide infra*).

Modulation of Hemopoiesis in Long-Term Marrow Cultures

Lithium chloride when administered to humans stimulates production of granulocytes and probably platelets.[18–21] Accordingly, we studied its effects in the Dexter long-term murine marrow cultures. We found that lithium stimulated production of granulocytes, megakaryocytes and adherent macrophages over long periods of time in a dose-related

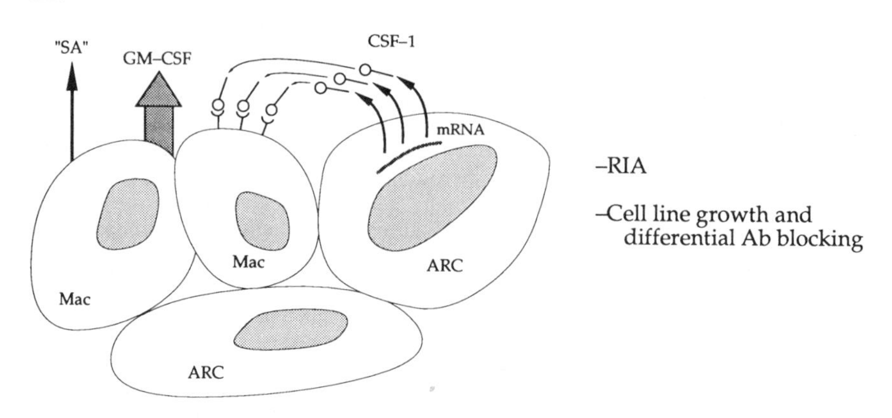

FIGURE 2. Irradiation effect on stromal growth factor production. RT = irradiation; SA = synergistic activity.

fashion in the Dexter culture system.[5,14] In addition virtually all of the hemopoietic stem cells defined by growth in *in vitro* culture or in *in vivo* models were also stimulated by lithium in murine Dexter cultures. These included CFU-S day 9, CFU-GM, HPP-CFC, CFU-D, and CFU-Meg.[5–7] In addition, interleukin-3 (IL-3) responsive stem cells *in vitro* were also increased in long-term cultures exposed to lithium. Higher doses of lithium appeared to selectively stimulate the high proliferative potential colony forming cells along with macrophage and eosinophil production.[8] In this latter situation, granulocyte production and CFU-GM numbers were decreased. Lithium was also found to stimulate CFU-D colony formation *in vivo* apparently by indirect means.[22] In further studies with this model, stromal layers from Dexter cultures preexposed to lithium or no lithium were irradiated and their capacity to support active hemopoiesis was then assessed.[14] Irradiated lithium exposed or nonexposed stroma were refed with nonadherent cells from long-term marrow cultures which were incapable of forming an adherent stromal layer. Under these conditions, the lithium preexposed irradiated stroma stimulated increased differentiated cell production, CFU-GM and CFU-S as compared to control irradiated stroma. These data indicated that lithium acted indirectly via on effect on an adherent radioresistant stromal cell.

Growth Factor Production by Stromal Layers

Based upon the above cited studies, it was speculated that interleukin-3,[23] a multifactor stimulator, might be the active stimulator in Dexter culture. This was an attractive hypothesis because of the known effect of interleukin-3 in being able to induce cell lines from Dexter culture.[24–25] The IL-3 responsive cell line, FDC-Pl,[26] was tested in agar overlay cultures addressing the question of whether or not the stromal layers produced a stimulator for this cell type. It was found that the stromal layers did in fact produce a stimulator for FDC-Pl cell growth and that irradiation of stromal layers increased levels of this stimulator[27] (FIG. 2). Because lectin stimulation of T lymphocytes induces high levels of IL-3,[26] the lectin pokeweed mitogen was added to irradiated stroma to assess its effect on levels of stimulators for FDC-Pl cells. This manipulation caused the induction of high levels of stimulator for FDC-Pl cells which could now be detected with ease in supernatants from the cultures. However, in further studies assessing the effect of anti-interleukin-3 antibody it was found that the stimulator present in conditioned media or produced in agar overlay culture models was not blocked by this antibody to interleukin-3 but in fact was totally blocked by anti-GM-CSF antibody.[27] When supernatants from irradiated lectin-stimulated Dexter stromal cells were tested against whole murine bone marrow they were found to stimulate macrophage, granulocyte, granulocyte-macrophage, megakaryocyte and mixed colony formation. Anti-GM-CSF and anti-CSF-l antibodies each partially blocked the stimulation.[27] These data altogether suggest that the stroma were sources of CSF-l and GM-CSF and that irradiation or lectin exposure selectively increased GM-CSF levels (FIGS. 2 and 3). Evaluation of supernatants in these cultures for CSF-l levels by radioimmunoassay indicated detectable levels of CSF-l in each of these situations but no increases with the various manipulations. Further preliminary studies, however, assessing messenger RNA production for CSF-l in these lectin-stimulated irradiated cultures indicate the apparent production of increased amounts of messenger RNA stimulated by lectin suggesting that CSF-l may in fact be increased under these conditions but not detected in supernatants of these cultures. This latter circumstance may be explained by the rapid binding and utilization of CSF-l by macrophages in the adherent layer.

Lithium-Induced Growth Factors in Dexter Culture

Similar studies were carried out with nonirradiated or irradiated stromal layers exposed to various levels of lithium. Under these conditions utilizing selective cell line stimulation and antibody blocking, it was clear that no IL-3 was present and that lithium was inducing increased levels of GM-CSF (FIG. 4). CSF-1 was also detected but again not increased by lithium stimulation. These data suggest that the driving terminal differentiating hormones in murine Dexter culture may be GM-CSF and CSF-1.

Cell Line Isolation from Long-Term Marrow Culture

In order to further investigate growth factor production by cell types in Dexter culture, adherent cells from long-term culture were isolated by a differential trypsinization and sequential passage technique.[28] Conditions utilized for these experiments involved the omission of hydrocortisone and the substitution of fetal calf serum for horse serum. These conditions begin to approach those for the Whitlock-Witte long-term lymphoid cultures[29] and may in fact explain the type of stromal cells isolated in these experiments. The

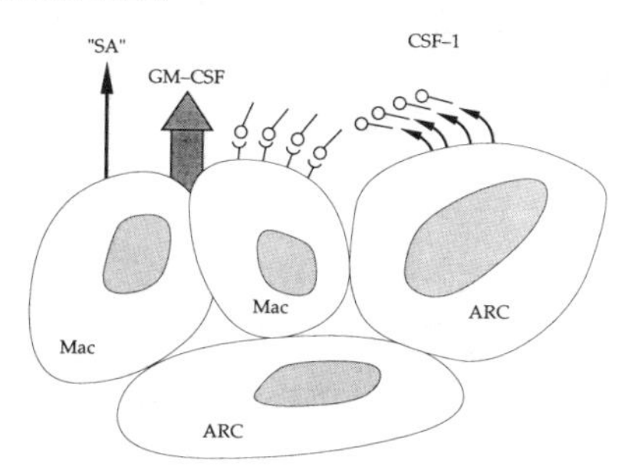

FIGURE 3. Lectin induction of stromal growth factors. SA = synergistic activity.

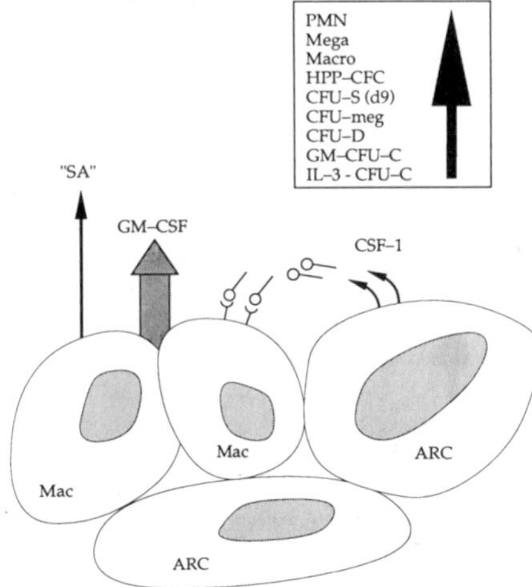

FIGURE 4. Effect of lithium on stromal growth factor production.

prototype cell line which we have extensively studied is the TC-1 line isolated under the above conditions. This line has now been found to produce a variety of growth factors including CSF-1 (RIA, mRNA and antibody inhibition), GM-CSF (cell line stimulation and antibody blocking), G-CSF (selected cell line stimulation), a BCGF-like activity (bioassay), and an apparently unique bioactivity active on early B cell and hemopoietic cell lineages. This latter activity has been termed hemopoietic lymphopoietic growth factor-1 (HLGF-1). It is of particular interest because it is capable of synergizing with CSF-1 to give macrophage high proliferative potential type colonies or synergizing with IL-3 or GM-CSF to give giant colonies of mixed lineages.[30] In addition, in short-term liquid cultures it induces the generation of pre-B cells from marrow cell populations depleted of B and pre-B cells prior to culture.[30] This activity now has been partially characterized biochemically and found to be a relatively large molecular weight (over 100,000 daltons), glycoprotein, relatively resistant to proteases, and heat stable. The six-step purification scheme including anion exchange, Conconavalin A and S-300 chromatography, followed by chromatography on mono Q and Superose 12 columns by FPLC, has resulted in preparations giving 1–2 bands on SDS gel chromatography. When the TC-1 cell line is analyzed for the presence of messenger RNA for different growth factors, message for CSF-1, and probably GM-CSF have been detected. Biochemical and functional studies suggest that the HLGF-1 is separate from these and other activities identified in TC-1 conditioned media. However, it should be noted that preliminary results indicate that G-CSF may in fact have pre-B inducing potential although molecular weight determinations would indicate that the activity HLGF-1 is separate from G-CSF.

Multifactor Stimulation of Hemopoiesis

A cautionary note in interpreting the above results is the recent observation in our laboratory that multiple growth factor combinations result in the generation of HPP-CFC-type murine colonies. We have found that GM-CSF + IL-3, G + GM-CSF, CSF-1 + GM-CSF, and G + IL-3 all result in HPP-CFC-like colonies.[31,32] Thus the potential exists for multiple growth factor interactions producing dramatic synergistic effects. However, the separative studies with HLGF-1 continue to indicate that this is a separate factor and that the observed biologic phenomenon is not based upon the multiple interactions of different growth factors. The fact that multiple growth factor combinations can result in a macrophage-type HPP-CFC suggests that there may be a common final pathway when several regulatory signals impact upon early hemopoietic progenitor cells. The cell populations may be separate progenitor classes or alternatively may be one progenitor class which when maximally stimulated channels towards macrophage differentiation. One of several possibilities is the endogenous production of CSF-1 by the targert cells themselves or of the receptor for CSF-1.

Model of In Vitro *Growth Regulation*

A modification of the Dexter murine myeloid marrow culture termed the Whitlock-Witte culture[29] result in the long-term production of pre-B and B cells. These cultures are different from the Dexter cultures in that the temperature is 37°C rather than 33°C, hydrocortisone is omitted, 2-mercaptoethanol is added and selected lots of fetal calf serum are substituted for horse serum. Under these conditions lymphoid growth ensues. Of particular interest are the observations of Johnson and Dorshkind[33] that if Dexter cultures are switched to Whitlock-Witte conditions, myelopoiesis stops and after a lag, B lymphopoiesis progresses. Kincade and colleagues[34] have suggested that the basic stromal cells in each system are similar indicating that the alterations in culture conditions might

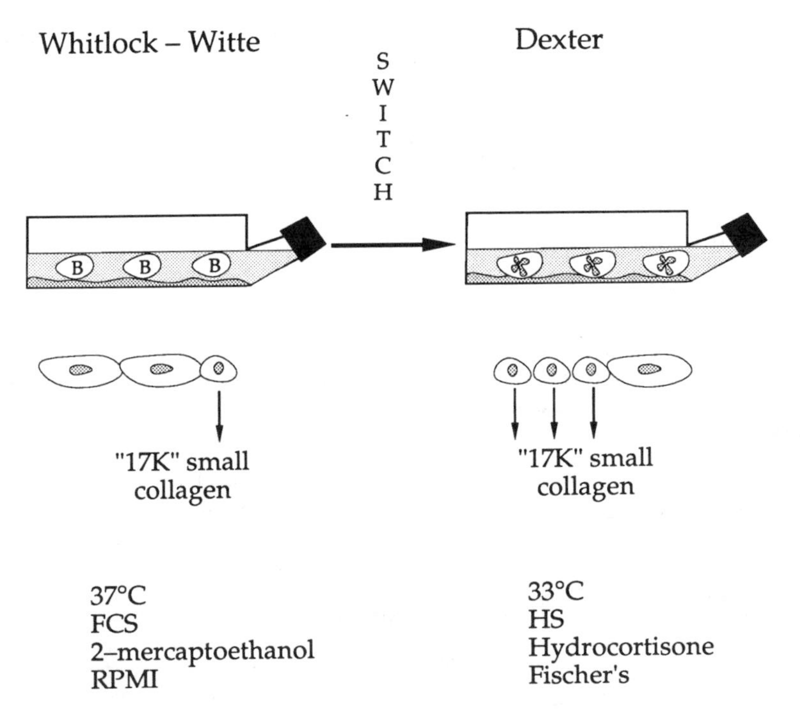

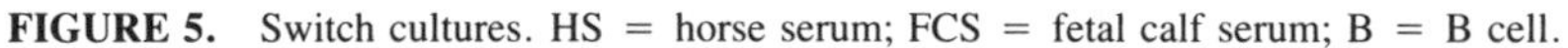

FIGURE 5. Switch cultures. HS = horse serum; FCS = fetal calf serum; B = B cell.

be modulating growth factor production and shutting on or off myeloid or B lymphoid growth factors dependent upon local conditions. However, observations in our own laboratory indicate that there may be a relative predominance of the large preadipocyte epithelioid cells under Whitlock-Witte conditions with a relative predominance of macrophage-type cells under Dexter conditions. Thus it is possible that selective alterations in the ratio of different stromal cell types may underly alterations in growth factor production and changes in the final differentiated phenotype of the cultures. Along this line it is of interest that we have found a small molecular weight collagenase sensitive protein produced in Dexter cultures[35] and apparently from adherent macrophages. Lesser amounts are produced in Whitlock-Witte cultures. The role of this small molecular weight "collagen" in modulating growth in these two different systems, however, remains to be determined.

The data on alterations in cell support when culture conditions are switched from Dexter to Whitlock-Witte (FIG. 5), and the observations on factor production in Dexter culture and in the TC-1 cell line suggest an overall model for the regulation of *in vitro* and

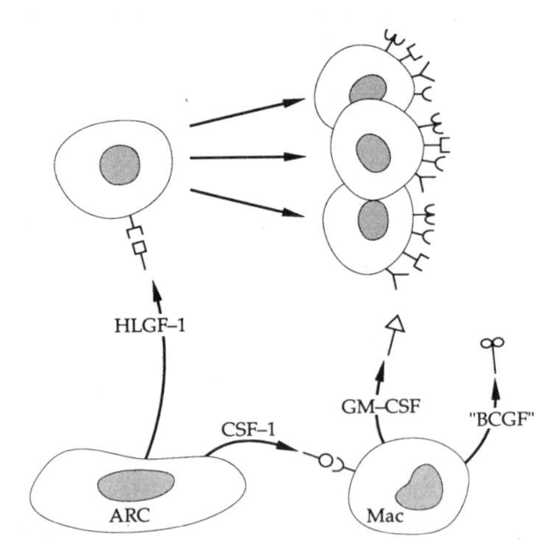

FIGURE 6. Model stromal cell regulation of Dexter culture.

possibly *in vivo* myelopoiesis. This model supposes that the preadipocyte alkaline phosphatase positive fibroblast-type cell produces CSF-1 which supports marrow macrophages. Further, the macrophage may be the source of GM-CSF and also a BCGF-like or B cell differentiating factor. Also different culture conditions may favor the production of either GM-CSF or BCGF activity and this may determine whether the final lineages seen in culture relate to the B cell lineage or myeloid lineages. In addition, one of the two cell types presumably produces the HLGF-1-like molecule which may be a bioactivity acting in a permissive fashion to support the survival of early stem cells which can then express multiple receptors for different growth factors. Thus regulation of this system may involve the elaboration of molecules supporting the survival of early primitive hematopoietic stem cells which express receptors for multiple different growth factors. The final differentiation pathways may then be determined by the selective production of relatively lineage-specific growth factors (FIG. 6). This selective production *in vitro* may be dependent upon the specific culture conditions used or alternatively *in vivo* may be dependent upon local environmental features such as pulsatile flow, oxygen gradients, or pH.

REFERENCES

1. DEXTER, T. M., T. D. ALLEN & L. G. LAJTHA. 1977. Conditions controlling the proliferation of haemopoietic stem cells in vitro. J. Cell. Physiol. **91:** 335–344.
2. DEXTER, T. M. & L. G. LAJTHA. 1974. Proliferation of haemopoietic stem cells in vitro. Br. J. Haematol. **28:** 525–530.
3. DEXTER, T. M. 1979. Cell interactions in vitro. *In* Clinical Haematology. L. LAJTHA, Ed. Vol. **8:** 453–468. W. B. Saunders. Philadelphia, PA.
4. QUESENBERRY, P. J. 1987. Stromal cells in long-term bone marrow cultures. *In* Handbook of the Hemopoietic Microenvironment. M. TAVASSOLI, Ed. Humana Press. Clifton, NJ. In press.
5. LEVITT, L. & P. J. QUESENBERRY. 1980. The effect of lithium on murine hematopoiesis in a liquid culture system. N. Engl. J. Med. **302:** 713–719.
6. DOUKAS, M. A., E. O. NISKANEN & P. J. QUESENBERRY. 1986. The effect of lithium on stem cell and stromal cell proliferation in vitro. Exp. Hematol. **14:** 215–221.
7. MCGRATH, H. E., C.-M. LIANG, T. A. ALBERICO & P. J. QUESENBERRY. 1987. The effect of lithium on growth factor production in long-term bone marrow cultures. Blood **70:** 1136–1142.
8. WADE, P. M., F. M. STEWART, M. A. DOUKAS & P. J. QUESENBERRY. 1982. The effect of lithium chloride (LiCl) on murine Dexter culture high proliferative potential colony forming cells (HPP-CFC) and stromal colony stimulating activity (CSA) production (abstract). Blood **60:** 348A.
9. BRADLEY, T. R. & G. S. HODGSON. 1979. Detection of primitive macrophage progenitor cells in mouse bone marrow. Blood **54(6):** 1446–1450.
10. WILLIAMS, N., H. JACKSON, A. P. SHERIDAN, M. J. MURPHY, A. ELSTE & M. A. S. MOORE. 1978. Regulation of megakaryopoiesis in long-term murine bone marrow cultures. Blood **51:** 245–255.
11. DEXTER, T. M. & R. K. SHADDUCK. 1980. The regulation of hematopoiesis in long term bone marrow cultures. I. Role of L-cell CSF. J. Cell. Physiol. **102:** 279–280.
12. GUALTIERI, R. J., R. K. SHADDUCK & P. J. QUESENBERRY. 1984. Hematopoietic regulatory factors produced in long term murine bone marrow cultures and the effect on in vitro irradiation. Blood **64:** 516–525.
13. HEARD, J. M., S. FICHELSON & B. VARET. 1982. Role of colony stimulating activity in murine long term bone marrow cultures: Evidence for its production and consumption by adherent cells. Blood **59:** 761–767.
14. QUESENBERRY, P. J., M. A. COPPOLA, R. J. GUALTIERI, P. J. WADE, Z. X. SONG, M. A. DOUKAS, C. E. SHIDELER, D. G. BAKER & H. E. MCGRATH. 1984. Lithium stimulation of murine hematopoiesis in a liquid culture system: An effect mediated by radioresistant stromal cells. Blood **63:** 121–127.
15. SONG, Z. X. & P. J. QUESENBERRY. 1984. Radioresistant murine marrow stromal cells: A morphologic and functional characterization. Exp. Hematol. **12:** 523–533.
16. LICHTMAN, M. A. 1981. The ultrastructure of the hemopoietic environment of the marrow: A review. Exp. Hematol. **9:** 391–410.
17. QUESENBERRY, P. J. & H. E. MCGRATH. Unpublished data.
18. SHOPSIN, B., R. FRIEDMAN & S. GERSHON. 1971. Lithium and leukocytosis. Clin. Pharmacol. Ther. **12:** 923.
19. BILLIE, P. E., M. K. JENSEN, J. P. K. JENSEN & J. C. POULSEN. 1975. Studies on the hematologic and cytogenic effect of lithium. Acta Med. Scand. **198:** 281.
20. JOYCE, R. A. & P. A. CHERVENICK. 1981. Selective effects of lithium on hematopoiesis (abstract). Exp. Hematol. **9:** 160.
21. HAMMOND, W. P. & D. C. DALE. 1982. Cyclic hematopoiesis: Effects of lithium on colony forming cells and colony stimulating activity in grey collie dogs. Blood **59:** 179.
22. DOUKAS, M. A., E. O. NISKANEN & P. J. QUESENBERRY. 1985. Lithium stimulation of granulopoiesis in diffusion chambers—a model of a humoral, indirect stimulation of stem cell proliferation. Blood **65:** 163–168.
23. IHLE, J. N., J. KELLER, L. HENDERSON, F. KLEIN & W. PALASZYNSKI. 1982. Procedures for the purification of interleukin 3 to homogeneity. J. Immunol. **129:** 2431.
24. GREENBERGER, J. S., M. A. SAKAKEENY, A. KRENSKY, S. BURAKOFF, D. REID, T. NOVAK &

S. BOGGS. 1984. Experimental evidence for the potential usefulness of permanent IL-3-dependent multipotential hematopoietic progenitor cell lines in transfusion granulocyte support therapy: Need for expansion of early passage cells. *In* Long Term Bone Marrow Culture. D. G. WRIGHT & J. S. GREENBERGER, Eds. Vol. **18:** 255–269. Alan R. Liss. New York, NY.

25. IHLE, J. N., J. KELLER, J. S. GREENBERGER, L. HENDERSON, R. A. YETTER & H. C. MORSE III. 1982. Phenotypic characteristics of cell lines requiring interleukin-3 for growth. J. Immunol. **129:** 1377.
26. IHLE, J. N., J. KELLER, S. OROSZLAN, L. D. HENDERSON, T. D. COPELAND, F. FITCH, M. B. PRYSTEWSKY, E. GOLDWASSER, J. W. SCHRODER, E. PALASZYNSKI, M. DY & B. LEBEL. 1983. Biologic properties of homogeneous IL-3. J. Immunol. **131:** 282.
27. ALBERICO, T. A., J. N. IHLE, C.-M. LIANG, H. E. MCGRATH & P. J. QUESENBERRY. 1987. Stromal growth factor production in irradiated lectin exposed long term murine bone marrow cultures. Blood **69:** 1120–1127.
28. SONG, Z. X., R. SHADDUCK, D. INNES, A. WAHEED & P. J. QUESENBERRY. 1985. Hematopoietic factor production by a cell line (TC-1) derived from adherent murine marrow cells. Blood **66:** 273–281.
29. WHITLOCK, C. A. & O. N. WITTE. 1982. Long term culture of B lymphocytes and their precursors from murine bone marrow. Proc. Natl. Acad. Sci. USA **79:** 3608–3612.
30. QUESENBERRY, P., Z. SONG, H. MCGRATH, I. MCNIECE, R. SHADDUCK, A. WAHEED, G. BABER, E. KLEEMAN & D. KAISER. 1987. Multilineage synergistic activity produced by a murine adherent marrow cell line. Blood **69:** 827–835.
31. MCNIECE, I. K., F. STEWART, D. DEACON & P. J. QUESENBERRY. 1988. Synergistic interactions between hematopoietic factors as detected by in vitro murine bone marrow colony formation. Exp. Hematol. **16:** 383.
32. MCNIECE, I. K., B. ROBINSON & P. QUESENBERRY. 1987. Stimulation of murine high proliferative potential colony forming cells by the combination of GM-CSF plus CSF-l. Blood. In press.
33. JOHNSON, A. & K. DORSHKIND. 1986. Stromal cells in myeloid and lymphoid long term bone marrow cultures can support multiple hemopoietic lineages and modulate their production of hemopoietic growth factors. Blood. In press.
34. KINCADE, P. W., P. L. WITTE & K. S. LANDRETH. 1986. Stromal cell and factor dependent B lymphopoiesis in culture. *In* Current Topics in Microbiology and Immunology. R. GISLER & C. PAIGE, Eds. Springer-Verlag. In press.
35. WATERHOUSE, E. J., P. J. QUESENBERRY & G. BALIAN. 1986. Collagen synthesis by murine bone marrow cell culture. J. Cell. Physiol. **127:** 397–402.

Hematopoiesis on Nylon Mesh Templates

Comparative Long-Term Bone Marrow Culture and the Influence of Stromal Support Cells[a]

BRIAN A. NAUGHTON[b] AND GAIL K. NAUGHTON[c]

[b]*Medical Laboratory Sciences*
Hunter College School of Health Sciences
425 East 25th Street
New York, New York 10010
[c]*Department of Pediatric Hematology*
Mount Sinai School of Medicine
One Gustave L. Levy Place
New York, New York 10029

INTRODUCTION

Feeder layers consisting of fibroblasts or other connective tissue elements have been employed since 1955 to enhance the growth of cultured cells, presumably by conditioning the substrate with trophic factors and other cell products.[1] This method was first applied to the culture of bone marrow cells by Dexter and co-workers[2] who demonstrated that hematopoietic stem cell renewal was sustained in cultures inoculated onto a preestablished support layer of medullary stromal cells. Earlier attempts to culture hematopoietic cells in the absence of this support layer promoted only the terminal differentiation of the stem cells in the initial inoculum with no evidence of self-renewal.[3] Several different types of cells have been identified in the stromal compartment of the bone marrow. These include macrophages which support erythropoiesis[4] and influence granulopoiesis,[5] reticular cells which have been implicated in the synthesis of extracellular matrix components,[6] endothelial cells which are associated with granulopoiesis and have been reported to elaborate colony stimulating factors[7] and interstitial (type III) collagen,[8] several subsets of fibroblasts which synthesize other collagens and essential components of the marrow microenvironment,[9] and adipocytes which are associated with mature granulocytic precursors in long-term bone marrow cultures (LTBMC)[10] and the perisinal region of marrow *in vivo*.[11] Stromal cells appear to regulate hematopoiesis by synthesizing colony stimulating factors (CSF) and other trophic substances[12] and form the hematopoietic microenvironment. Qualitative differences in stroma capable of supporting myelopoiesis and erythropoiesis have been reported.[13] Other studies indicate that stromal cell regeneration precedes hematopoietic recovery after local bone marrow injury[14] and following bone marrow transplantation (BMT).[15] The influence of stromal cell monolayers at different stages of confluence on suspended nylon screen LTMBC[16] is described in the present paper. Also included is data indicating that an immortal cell line of human fetal fibroblasts can support the initial seeding and long-term growth of human and nonhuman primate but not rat bone marrow cells.

[a]Supported by Marrow Tech, Inc. Grants 7-70984 and 6-6638 to the Research Foundation of the City University of New York.

MATERIALS AND METHODS

Bone Marrow Samples

Human. Bone marrow was aspirated from multiple sites on the posterior iliac crest of hematologically normal adult volunteers after informed consent was obtained. Specimens were collected into heparinized tubes and suspended in 8 ml of RPMI 1640 medium which was conditioned with 10% fetal bovine serum (FBS) and 5–10% horse serum (HS) and supplemented with hydrocortisone, fungizone, and streptomycin. The cell clumps were disaggregated and divided into aliquots of 5×10^6 nucleated cells.

Nonhuman Primate. Intact cynomologous macaque monkey femurs were purchased from the Charles River Primate Center (Port Washington, NY). The epiphyseal ends of the femurs were separated from the bone shaft under sterile conditions. The red marrow was removed, suspended in medium, and aliquoted as above.

Rat. Adult male Long-Evans rats (225–400 gm) were anesthetized with ether and after removal of their femurs, were exsanguinated from the abdominal aorta using heparinized syringes. The femurs were split and the marrow contents were scraped into a sterile petri dish containing 3 ml of Fischer's medium conditioned with 10% FBS and 10% HS and supplemented with hydrocortisone, fungizone, heparin, and antibiotics.[16] Aliquots of 5–7 $\times 10^6$ cells were prepared.

Support Matrices

Nylon filtration screen (#3-210/36, Tetko, Inc., Elmsford, NY) was used as a template to support all LTBMC. This consisted of fibers which were 90 μm in diameter and assembled into a square weave pattern with sieve openings of 210 μm.

Comparative Long-Term Cultures

Comparative long-term cultures of human, cynomologous macaque, and Long-Evans rat bone marrow cells were established as follows.

Preparation of the Screen and Inoculation of Support Cells

Human. 8-mm × 45-mm pieces of screen were soaked in 0.1 M acetic acid for 30 min and treated with 10 mM polylysine suspension for 1 hr to enhance attachment of support cells. These were placed in a sterile petri dish and inoculated with either 5×10^6 human bone marrow cells or with equal numbers of human fetal fibroblasts (#GM 1380, Coriell Institute, Camden, NJ). Human fetal fibroblasts were grown to confluence in monolayers using RPMI 1640 medium conditioned with 10% FBS, 5–10% HS, and supplemented with hydrocortisone, fungizone, and streptomycin (35°C/5% CO_2/>90% humidity). These cells were lifted using collagenase (10 μg/ml for 15 min) and transferred onto the screen. After 1–2 hr of incubation at 5% CO_2 the screens were placed in a Corning 25-cm^2 culture flask and floated with an additional 5 ml of medium. Screens inoculated with marrow stromal cells were transferred in a similar manner.

Nonhuman Primate. Two matrices were employed for LTBMC of monkey cells: human fetal fibroblasts (as described previously) and nylon mesh that was inoculated with 5 × 10^6 femoral marrow cells from a cynomologous macaque. Culture conditions and screen pretreatment protocols were identical to those used for the human cultures.

Rat. 8-mm × 45-mm pieces of nylon screen were soaked in 0.1 M acetic acid for 30 min and coated with solubilized type IV mouse collagen (GIBCO Labs, Grand Island, NY) for 1–2 hr. The screen was inoculated with 5–7 × 10^6 Long-Evans rat femoral marrow cells and after 1–2 hr of incubation in 5% CO_2 at 33°C, the mesh was transferred to a 25-cm^2 culture flask. 5 ml of medium was added to float the screen.

Adherence and subsequent growth of the stromal elements was monitored using inverted phase contrast microscopy and scanning electron microscopy (SEM).

Inoculation of Hematopoietic Cells

When approximately 70% of the mesh openings were bridged with support cells (10–14 d for rat stroma, 7–13 d for human or monkey stroma, and 4–7 d for fetal fibroblasts), the screens were transferred to sterile petri dishes and inoculated with 5 × 10^6 human or monkey nucleated bone marrow cells or 2–5 × 10^6 rat femoral marrow cells, respectively. After 2 hr of incubation in 5% CO_2 the screens were gently floated in 25-cm^2 Corning flasks to which 5 ml of medium was added.

Methods of Evaluation

Total cell counts and cytospin preparations were made using spent medium removed when the cultures were fed (every 5 d). Cell counts were performed using the hemacytometer method. Cytospins were stained with Wright's-Giemsa and differential counts were performed on random fields. Cell counts of the adherent zone were done at different intervals of LTBMC by treating the screen with a 1:1 mixture of collagenase and trypsin (10 μg/ml) and mildly ultrasonicating.

The CFU-C content of the adherent zone of rat LTBMC was determined using a modification of the method of Bradley and Metcalf.[3] Briefly, 4 × 10^4 cells were plated and incubated at 37°C in 7–7.5% CO_2. Pokeweed mitogen rat spleen cell conditioned medium was utilized as a source of colony stimulating activity (CSA) for rat CFU-C which were counted after 14 d in culture. Human CFU-C were determined by aliquoting 10^5 nucleated cells/ml in Iscove's modified Dulbecco's medium supplemented with 20% FBS and plating over a layer of 10^6 peripheral blood leukocytes in 0.5% agar.[17] Colonies were scored on days 7 and 14 after plating (37°C/7% CO_2). The number of BFU-E in the adherent zone of human LTBMC were determined at various intervals using 0.8% methylcellulose in Iscove's medium containing 30% FBS, 1% bovine serum albumin, 10^{-4} M mercaptoethanol, 2.5–5 I.U./ml of partially purified human urinary erythropoietin,[30] and 4.5% of phytohemagglutinin-stimulated human leukocyte conditioned medium.[18] Hemoglobinized colonies were scored after 12–14 d of culture.

Cytofluorographic analysis of the cellular content of the adherent zones of human and monkey LTBMC was performed using the EPICS system (Coulter Electronics, Hialeah, FL). Cells were separated from the nylon screen at various intervals after the inoculation of hematopoietic cells using collagenase and trypsin followed by extensive washing. Then cells were incubated for 45–60 min in Hank's balanced salt solution without Ca^{++} or Mg^{++}. These were reacted with the following monoclonal antibodies which were conjugated to fluorescein isothiocyanate (FITC): Mo-1, T 3, B 1, Plt-1, and MY-9 (Coulter Immunology, Hialeah, FL). MsIgM-FITC-treated cells were used as controls. Sorting windows were chosen on the basis of fluorescence and light scatter histograms. A 0.255 window was appropriately gated and the cellular profiles were determined.

Cultures were sacrificed at various intervals following the first (support cells) and second (hematopoietic cells) inoculation for electron microscopic study. Briefly, nylon screens were cut into approximately 4 equal parts and were fixed in 3% gluteraldehyde-phosphate buffer solution, washed, dehydrated in acetone, and placed in a Denton critical point dryer. In some instances, the stromal layer was physically disrupted to permit the visualization of the underlying cell growth.[16] Specimens were coated with 60% gold and 40% palladium and studied with an Amray SEM.

Studies on the Effect of Confluent Stromal Cell Monolayers on the Growth of Nylon Screen Cultures

Femoral marrow cells from Long-Evans rats were poured through a packed Fenwal wool column as described by Boswell and co-workers.[19] Briefly, 10^7–10^8 femoral marrow cells were placed in 4 ml of medium and poured over a nylon wool column which was preincubated (37°C/45 min) in medium. After an additional 45 min of incubation the nonadherent cells were drained and the adherent cells were removed by extensive washing and elution with EDTA-Versene solution (1:5000 in saline; GIBCO). Approximately 5×10^6 cells were inoculated in parallel into 25-cm^2 flasks and grown to 50% and 100% confluence. Preestablished nylon screen LTBMC which were standardized with respect to time following the second inoculation (2 wk), were inserted into each flask. Growth on the nylon screen LTBMC and the monolayer was observed microscopically. Cell counts and cytospins of the nonadherent zone were performed every 5 d. Differential counts of cytospin preparations of the enzyme dissociated adherent cells were performed at 7, 15, and 28 d after insertion of the nylon screen LTBMC.

RESULTS

The growth pattern of human and cynomologous macaque cells in nylon screen LTBMC was similar to that for rat bone marrow, which was described in detail elsewhere.[16] Briefly, support cells (either marrow-derived or fetal human fibroblasts) grew linearly along and enveloped each nylon strand before starting to span the mesh openings (FIG. 1). Hematopoietic (and stromal) cells of the second inoculum seed in the natural interstices formed by the stromal cell processes which are present in at least 70% of the openings in the 3.6-cm^2 mesh. Hematopoietic cells did not bind directly to the nylon but adhered to only those areas where support cells were attached (FIG. 2). Colonization was evident in all cultures by 3–6 d after the second inoculation (FIG. 3). The 210-μm sieve provided sufficient area for the expression of erythroid, myeloid and other colonies (FIG. 4). Hematopoiesis was observed on the outer surfaces of the nylon screen LTBMC but was most extensive in the interstices of the developing support cells (FIG. 5). Analysis of cytospin slides prepared after each feeding revealed the presence of late stage precursors of the erythroid, myeloid, and lymphoid lineages in the human and monkey cultures (TABLE 1). These persisted for the term of culture of each species tested (39 wk for the rat,[16] 12.5 wk for the primates) although the relative percentages of the cell types varied. Macrophages/monocytes/fibroblasts released into the nonadherent zone of the human cultures increased with time, mainly at the expense of the myeloid cells (TABLE 1). Analysis of the adherent zone of human and cynomologous LTBMC revealed that the relative percentage of stromal to hematopoietic cell numbers increased with time in culture (TABLE 2). Cellular proliferation achieved a steady state condition after several wks in culture; similar numbers of cells were found in the adherent and nonadherent zones when the LTBMC were examined on a weekly basis (FIG. 6). The numbers of cells in the nonadherent zone for the first 1–2 wk of culture were somewhat misleading. In our experience, many of the cells which appear in the medium in the early stages of culture were formerly loosely attached to the matrix. These become detached easily causing an artificially high cell count. Likewise, because of the relatively low seeding efficiency only 5×10^5 to 10^6 cells initially adhere to the mesh even though the inoculation volume was 5×10^6 cells. This obscures the 2–3-fold cellular proliferation which occurs during the first wk of culture. As hematopoietic colonization proceeded, the relative percentage of stromal elements dropped (TABLE 2). Stromal cell growth at later periods of the LTBMC

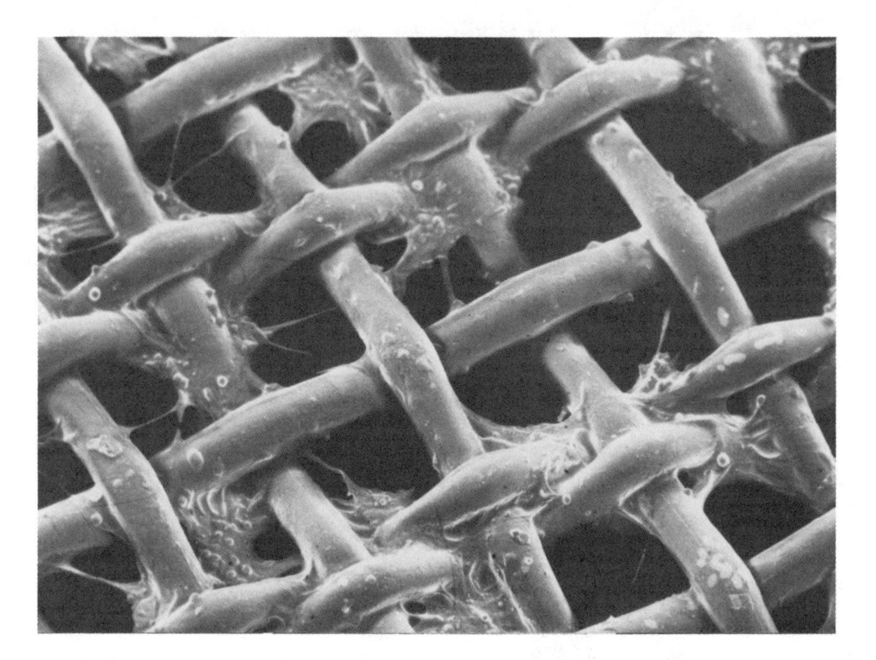

FIGURE 1. SEM of nylon screen 3 d following inoculation with human fetal fibroblasts showing the degree of subconfluence necessary to support hematopoietic cell proliferation. 210×.

occurs at the expense of hematopoiesis. Cytofluorographic analysis of adherent zones of the human cultures at 2, 7, and 10.5 wk confirmed the presence of early (MY-9) and late (Mo-1) myeloid cells, B (B-1) and T (T-3) lymphocytes, megakaryocytes/platelets (Plt-1), and monocytes/macrophages (Mo-1) (TABLE 3). Substantial numbers of CFU-C were recovered from the adherent zone of the rat[16] and human LTBMC relative to those present in the initial inoculum (FIG. 7). Preliminary findings indicate that BFU-E persisted in the human LTBMC as well (TABLE 4). Currently we are performing more comprehensive studies on the numbers of CFU-C, BFU-E, CFU-E, and CFU-GEMM at various intervals of human and macaque nylon screen LTBMC.

Human and monkey LTBMC can be established on a stratum of fetal human fibroblasts but this matrix will not support the growth of rat LTBMC. The fetal cells reach a

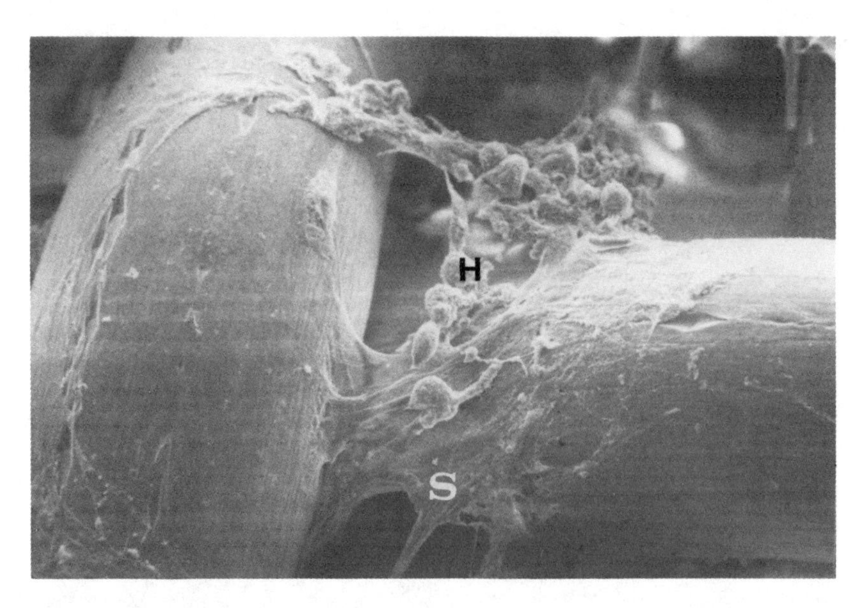

FIGURE 2. SEM of nylon screen culture (human) 2 d after the second (hematopoietic cell) inoculation. Stromal cells (S) and their processes envelop each nylon strand and hematopoietic cells (H) adhere to these. 380×.

stage of subconfluence which will allow the inoculation of marrow cells much faster than marrow stroma. When cynomologous bone marrow is grown on a bed of fetal fibroblasts the phenotypic profile of the adherent zone shows that more cells react with the Plt-1 antibody than in the other cultures we studied but the other hematologic lineages are represented also (TABLE 3). It is not known to what extent this finding reflects cross-reactivity of the antibody or a shift in the cell population of the adherent zone mediated by the fetal cells. We expect that variability will decrease as more cultures are studied and the number of monoclonal antibodies used to determine the profiles are expanded.

When confluent stromal cell monolayers are co-cultured with nylon screen LTBMC, both hematopoiesis and stromal cell growth on the suspended culture are inhibited (TABLE 5) as compared to LTBMC suspended in flasks without adherent stroma ($p < 0.01$) or with stromal cells at approximately 50% confluence ($p < 0.02$). The most pronounced effects are seen in the hematopoietic cells of the 28-d cultures, whose numbers diminish by almost 50% in the presence of the confluent stromal monolayer. Stromal cell growth on the nylon screen LTBMC is affected as well, with cell counts decreasing by approximately one quarter over the same time period. Stromal monolayers at about 50% confluence had neither an inhibitory nor a stimulatory influence on suspended nylon screen LTBMC (TABLE 5). Observations with the inverted phase microscope revealed that co-culture with confluent stromal monolayers results in the detachment and release of mesh-associated stromal cells into the nonadherent zone. In addition, hematopoietic colonies coalesce and cease growing. If the LTBMC is transferred to a new flask, recovery of hematopoiesis is seen by 3–5 d in most cultures.

DISCUSSION

Nylon screen can be employed as a template for the establishment not only of rat LTBMC[16] but for human and nonhuman primate LTBMC as well. The Dexter LTBMC system sustains the self-renewal of murine pluripotent stem cells although the committed stem cells become predominantly myeloid in character[20] after several weeks. This system can be modulated to favor lymphopoiesis[21] or erythropoiesis[22] but substantial numbers of each of these cell types are not produced by single cultures.[2] LTBMC of human[23] and nonhuman primate[24] cells has been achieved using a modification of the Dexter technique. Thus, continued proliferation of myeloid progenitors (CFU-C) was reported in a human LTBMC of 20 wk duration[23] and in a prosimian culture system for at least 25 wk.[24] Studies of the adherent zone in human monolayer-type LTBMC revealed the persistence of myeloid (CFU-C) and erythroid (BFU-E) progenitors, although their numbers diminished by approximately 90% between wks 2 and 7 of culture.[25] Hematopoiesis in LTBMC systems appears to be dependent upon regulatory and microenvironmental factors provided by the stromal support cells. A number of regulatory factors have been reported to originate from stromal cells. These include CSF which can be produced by fibroblasts,[26] adipocyte-like cells[27] and endothelia,[7] interleukin-1 which originates from macrophages and induces CSF production by fibroblasts[28] and endothelia,[29] erythropoietin which is elaborated and/or stored by certain macrophages,[30] and other as yet uncharacterized stimulatory factors.[12]

Stromal factors which inhibit the growth and/or differentiation of hematopoietic cells have been reported also. In this regard, macrophages, which stimulate erythroid progenitor growth while in a quiescent state,[31] produce inhibitors to erythroid maturation when activated.[32] The monokine, tumor necrosis factor (TNF), has been found to inhibit the growth of BFU-E, CFU-E, and CFU-GEMM in a dose-dependent fashion.[33] Restriction of expression due to inhibitors like TNF may reflect a normal regulatory activity; receptors for TNF have been discovered on hematopoietic cells.[34] A recent study provides

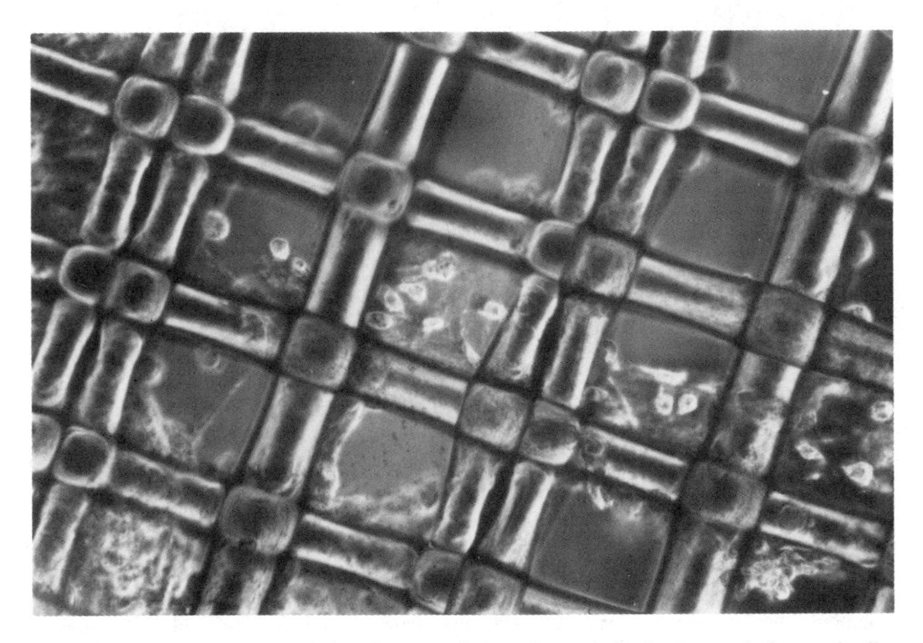

FIGURE 3. Human hematopoietic cells appear light against a darker background of stromal cells to which they are attached. 5 d following second inoculation. Inverted phase contrast. Wratten filter. 200×.

evidence that another broadly acting regulatory molecule, transforming growth factor β1, can inhibit colony formation by factor-dependent (interleukin-3 or GM-CSF) hematopoietic progenitor cells.[35] Medullary stromal elements also may regulate hematopoiesis by restricting differentiation.[36,37] In this regard, murine marrow stromal cells have been reported to sustain the proliferation of lymphoblastoid lines while inhibiting the growth of more differentiated cell lines.[36] In addition to regulatory factors, stromal cells contribute a number of extracellular matrix components to the hematopoietic microenvironment including glycosaminoglycans,[38] proteoglycans,[39] and several types of collagens.[6,9] Glycosaminoglycan-depleted hematopoietic cells fail to adhere to the stromal matrix in LTBMC.[38] Analysis of adherent zones of LTBMC reveals that chondroitin sulfate,[39] heparan sulfate,[40] and dermatan sulfate[40] are the predominant proteoglycans present. Perturbations in these components may contribute to hematologic disease. In this regard,

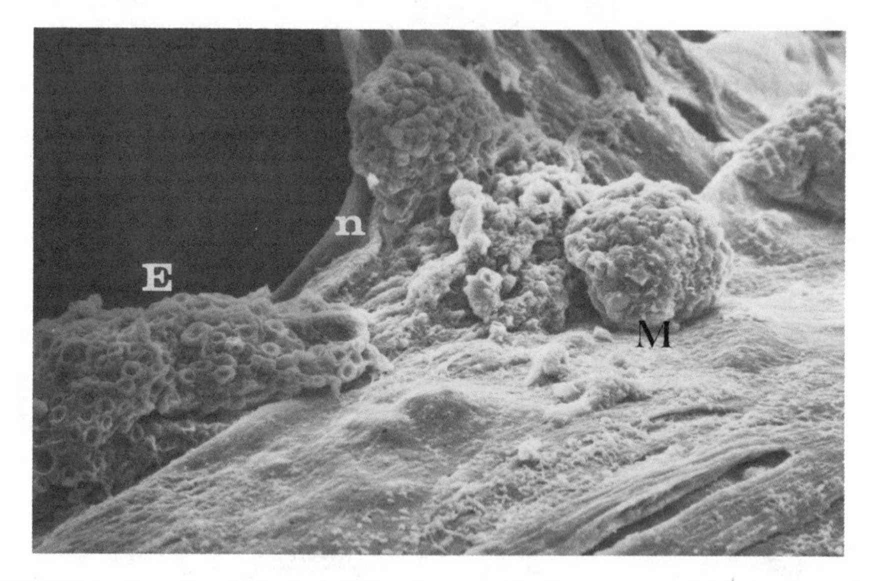

FIGURE 4. Several myeloid colonies (M) and an erythroid colony (E) are seen in this SEM of a human culture at 30 d after second inoculation. A portion of a nylon strand (n) is visible also. 1560×.

alterations in the levels of chondroitin sulfate and hyaluronic acid in the spleens of Sl/S^d mice may contribute to their anemia.[41] Changes in the levels of acid mucopolysaccharides in spleens of polycythemic mice have been reported also.[42] Collagen is an important component of the hematopoietic microenvironment as well. Marrow-derived adherent cells produce type I and type III collagen, reportedly in a ratio of 5:1.[9] The presence of collagen in the extracellular matrix was indispensable to the establishment of a stratum capable of supporting LTBMC of murine cells.[43] Stromal cells which support different types of hematopoiesis may be qualitatively diverse. In this regard, medullary stromal elements which were associated with myelopoiesis stained positively for alkaline phosphatase[44] whereas erythropoietic areas of the marrow were acid phosphatase positive. Erythropoiesis in LTBMC is prolonged if hematopoietic cells are inoculated onto a matrix consisting of fetal liver-derived stroma,[45] which also are acid phosphatase positive. Both

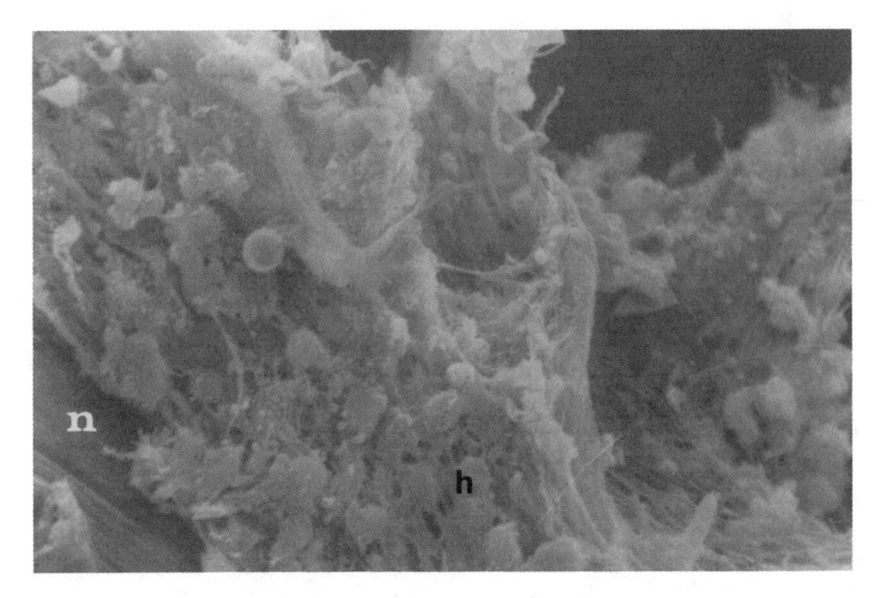

FIGURE 5. Nylon screen LTBMC of cynomologous macaque cells 6 wk following second inoculation. Hematopoietic cells (h) are associated with a complex stratum of stromal cells and their processes, some of which are seen enveloping a nylon fiber (n). 400×.

acid and alkaline phosphatase positive stromal cell populations are expressed in LTBMC.[46] The relative contributions of these stromal cell subpopulations to the microenvironment which enable them to support different types of hematopoiesis has not as yet been determined.

TABLE 1. Cell Content[a] of the Nonadherent Zone

Time in Culture (Wk)	Differential (%)[b]									
	Human					Cynomologous Macaque				
	MY	E	L	MAC/STR	Other	MY	E	L	MAC/STR	Other
0	63.9	19.0	10.8	3.6	2.7	64.8	14.2	10.1	8.2	2.7
1	59.0	14.0	8.6	14.9	3.5	60.2	16.0	6.8	12.9	4.1
2	48.5	14.4	9.9	23.7	3.9	57.4	14.9	7.5	16.4	3.8
3	51.9	9.2	9.6	24.7	4.6	49.7	12.4	10.0	23.5	4.4
4	41.2	10.4	6.1	33.9	8.4	49.7	9.9	7.9	26.2	6.3
5	41.9	12.7	10.3	29.0	6.1	43.0	10.7	6.1	32.0	8.2
6	45.2	11.2	8.0	27.2	8.4	39.2	8.0	6.0	36.7	10.1
7	39.8	10.1	6.3	34.8	9.0	—	—	—	—	—
8	38.6	9.8	6.5	37.1	8.0	38.8	4.3	8.4	39.2	9.8
9	40.3	5.6	6.6	38.4	9.1	27.6	7.7	8.6	46.1	10.0
10	35.9	5.5	6.8	40.6	11.2	33.5	6.2	7.7	42.0	10.6
11	31.3	6.7	5.4	43.2	13.4	—	—	—	—	—
12	30.1	5.0	4.1	44.6	16.2	35.4	6.0	6.9	39.2	12.5

[a]Results reflect an average of 3–5 cultures. Each culture contained one $3.6 = cm^2$ nylon screen.
[b]MY = myeloid; E = erythroid; L = lymphoid; MAC/STR = macrophages, monocytes, and fibroblastic cells; other = megakaryocytes and unidentified blasts.

Nylon screen provides a greater relative surface area for cell growth as compared to the bottom of a culture flask. However, if the nylon is not treated prior to inoculation, stromal cell attachment and growth follows a somewhat random pattern. Perhaps this is due to the heterogeneous distribution of covalent binding sites on this material.[47] Whereas mild acidification of the mesh prior to coating with type IV mouse collagen provided a homogeneous matrix for the attachment of rat stromal cells, primate marrow cells would not adhere to this substrate unless it is treated with polylysine and/or FBS as well. Even so, stromal seeding efficiency was only 20–30%. Although hematopoietic colonization occurs rapidly, stromal seeding appears to be the rate limiting step for hematopoiesis in this type of system, since hematopoietic cells from the second inoculum seed only those areas where a support matrix is present. Colonization occurs in the natural interstices formed by the partially developed stromal layers and is seen also on the outermost surfaces of the nylon screen culture (FIG. 5). The surface colonies are somewhat smaller than those observed within the meshwork and appear at times to be part of the nonadherent zone. Actually, they are loosely attached and remain after feeding. These cells, which are found consistently in monolayer-type LTBMC, have been termed the "pseudoadherent layer."[25]

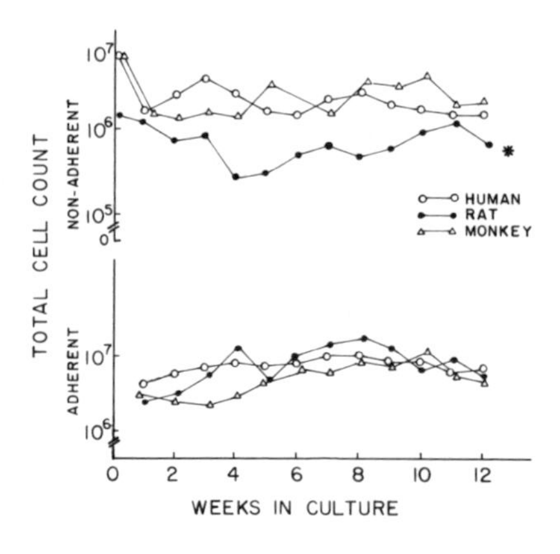

FIGURE 6. Mean numbers of cells in the adherent and nonadherent zones at various intervals of LTBMC of human, cynomologous macaque, and rat cells. *Although data for rat LTBMC is reported for 12.5 wk for comparison, the cultures were hematopoietically active for 39 wk.[16]

In our experience, nylon screen LTBMC supports the expression of several hematologic lineages as evidenced by differential counts of the nonadherent and enzyme-dissociated adherent zones (TABLES 1 & 2) of human, macaque, and rat cells. Cytofluorographic analysis of the adherent zones of human and nonhuman primate LTBMC using monoclonal antibodies reactive against various cell lines revealed the presence of early and late myeloid precursors, mature granulocytes, B and T lymphocytes, megakaryocytes/platelets, and monocytes/macrophages (TABLE 3). Although supportive of the data obtained from differential counts, the results were highly variable, presumably because the flow cytometry and cytospin data were derived from different cultures. It is likely that variation in hematopoietic expression in nylon screen LTBMC is a result of the initial inoculation, when random populations of stromal cells were used to form the support matrix. Likewise, the number of progenitors located in the adherent zone was highly variable. In the rat, the numbers of CFU-C dropped by approximately 30% between wks 2 and 26 of LTBMC and fell 60% by 39 wk, when the cultures were lost to a contamination[16] (FIG. 7). Although mean numbers of BFU-E fell by approximately 60%, these progenitors were still present in human cultures of 10 wk duration (TABLE 4).

TABLE 2. Cell Content of the Adherent Zone[a]

Time in Culture (Wk)	Differential (%)[b]							
		Human Hematopoietic				Cynomologous Macaque Hematopoietic		
	Stromal	E	MY	Other	Stromal	E	MY	Other
1	66.4	6.2	20.4	7.0	53.1	8.0	35.7	3.2
2	60.0	5.4	26.4	8.2	66.0	8.3	19.2	6.5
3	54.2	6.6	29.2	10.0	68.6	7.4	18.1	5.9
4	62.6	6.8	24.5	6.1	57.0	5.1	29.2	8.7
5	65.1	2.7	25.2	7.0	56.6	5.8	27.5	10.1
6	65.4	6.1	21.6	6.9	63.1	3.9	24.0	9.0
7	59.7	7.7	25.4	7.2	—	—	—	—
8	64.3	5.1	24.0	6.6	68.1	4.8	20.2	6.9
9	72.9	2.7	18.4	6.0	59.3	4.0	27.3	9.4
10	73.2	3.7	17.7	5.4	70.0	4.4	17.3	8.3
11	71.3	3.0	19.6	6.1	—	—	—	—
12	74.7	2.9	17.4	5.0	65.3	4.2	21.9	8.6

[a]Cells of the adherent zone were disaggregated by enzyme treatment.

[b]Stromal includes fibroblasts, macrophages, adipocytelike cells, and endothelia; E = erythroid; MY = myeloid; other = lymphoid, thromboid, and unidentified blasts.

Likewise, the adherent zone of human LTBMC continued to support substantial numbers of CFU-C (FIG. 7).

In the human LTBMC, hematopoiesis diminished as the relative numbers of stromal cells increased (TABLE 2), but multilineage expression was still evident after 12.5 wk of culture. Hematopoiesis may be dependent upon growth-related activities of the support cells. Thus, when actively hematopoietic nylon mesh LTBMCs are floated in flasks which contain a confluent monolayer of stromal cells, an inhibition of both hematopoiesis and stromal cell growth is observed on the nylon mesh, as compared to those cultures suspended in flasks without confluent stromal monolayers (TABLE 5). This suggests that stromal cell products influence not only hematopoietic cells, but other stromal elements as well. Separate populations of marrow stromal cells which either stimulate or inhibit hematopoiesis have been postulated.[48] Overgrowth of inhibitory stroma may account for the decline in hematopoiesis observed in LTBMC with time. Some investigators have

TABLE 3. Mean Percent Reactivity[a] (± 1 SEM) of Uncultured Bone Marrow and Cells from LTBMC with Various FITC-Labelled Monoclonal Antibodies

Sample[b]	Monoclonal Antibodies				
	B-1	T-3	Plt-1	Mo-1	MY-9
Human					
2 wk LTBMC[c]	10.20 ± 1.43	18.64 ± 1.88	4.40 ± 1.33	20.10 ± 1.04	3.98 ± 0.26
7 wk LTBMC	6.76 ± 0.98	11.18 ± 1.86	8.08 ± 0.92	17.26 ± 2.29	3.70 ± 0.68
10.5 wk LTBMC	22.73 ± 1.37	13.01 ± 1.84	17.05 ± 4.10	20.98 ± 1.41	3.46 ± 0.25
uncultured marrow	11.96 ± 1.13	9.90 ± 0.64	8.72 ± 1.83	15.60 ± 0.84	1.46 ± 0.54
Macaque					
7 wk LTBMC[d]	31.01	18.13	46.50	40.87	21.64
uncultured marrow	8.37 ± 0.99	11.56 ± 2.10	8.53 ± 1.09	24.64 ± 2.25	5.49 ± 0.83

[a]Calculated by subtracting nonspecific labelling with Ms-IgM-FITC control.

[b]Results reflect data from 4–5 cultures.

[c]Times listed are following the second inoculation.

[d]Mean of 2 cultures inoculated onto fetal human fibroblasts.

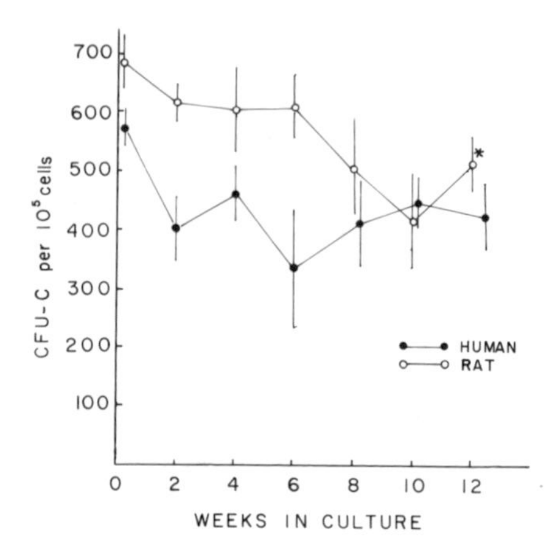

FIGURE 7. Mean number of CFU-C per 10^5 cells in the human and Long-Evans rat LTBMC systems. *Vertical bars* through the means indicate ± one standard error of the mean. *Rat cultures were followed for 39 wk in a separate study.[16]

reported that irradiation of stroma increases CSF elaboration,[7,12,48] possibly by destroying these inhibitory cells. Alternately, single populations of stromal cells may produce factors which stimulate growth in newly inoculated cultures but inhibit it in cultures nearing confluence. Two factors, TNF and TGF-B1, which can inhibit the growth of both tissue cells[49] and hematopoietic progenitors,[33,35] have been identified. The most likely explanation of our finding that cellular proliferation on nylon screen LTBMC declined when co-cultured with confluent stromal monolayers is that an inhibitor(s) was made by the confluent cells. Supportive of this hypothesis is a study demonstrating the failure to intensify hematopoiesis in a declining human LTBMC by recharging with fresh bone marrow.[50] In contrast, no stimulatory effect was noted in nylon screen LTBMC which was co-cultured with semiconfluent monolayers. One possible explanation of these phenomena is that whereas stimulatory factors in LTBMC may act locally,[51,52] inhibitor(s) may have a broader scope of influence.

Human fetal lung fibroblasts support the growth of human and nonhuman primate bone marrow cells. One advantage of the fetal cells is that they achieve the level of subconfluence necessary for the second inoculation much faster than stromal cells derived from the bone marrow (4–7 d vs 7–13 d, respectively). Although we have not as yet attempted to apply viral or other markers to these cells, it seems that the fetal fibroblasts are displaced continually into the nonadherent zone by the stromal cells present in the marrow (second) inoculum. In this regard, it is not unusual to see macrophages attacking the support cells and to find excessive numbers of fibroblastoid cells in the medium. Rather than inhibiting the growth of cells in the marrow inoculum, this activity stimulates

TABLE 4. BFU-E in the Adherent Zone at Various Intervals of Human LTBMC

Time of Culture (Wk)	Numbers of BFU-E[a]
Uncultured marrow	19 ± 6
2	14 ± 4
4	12 ± 5
7	8 ± 3
9	11 ± 6
10	8 ± 3

[a]Colonies per 10^5 cells; mean of 3–4 plates ± SEM.

it. The number of fibroblasts in the nonadherent zone diminishes as the mesh is ''taken over'' by marrow-derived stromal cells. Hematologic engraftment after bone marrow transplantation (BMT) may be dependent upon the simultaneous (or prior) engraftment of cells which form the microenvironment.[8] Thus, an evaluation of LTBMC established using cells derived from BMT patients revealed that stromal cells were primarily of donor origin.[8] In a recent study, Da *et al*[15] demonstrated that the numbers of CFU-F were diminished for 21 d following BMT but recovered by 42 d. Although radiation damage to the medullary stroma[51] of the BMT recipients in this study probably accounts for the initial decline in CFU-F, it is unclear whether the enhanced CFU-F activity seen in the last wk of the trial is due to recovery of recipient stromal cells or engraftment of donor stroma.

TABLE 5. The Effect[a] of Stromal Monolayers at Approximately 50% and 100% Confluence on Cellular Proliferation in a Suspended Nylon Screen LTBMC[b] in the rat.

	Time of Culture[c]	50% Confluent	100% Confluent
Stromal cells[d]	7 d	0 ± 2.5	−13.5 ± 4.7
	15 d	+3.3 ± 2.0	18.0 ± 5.1
	28 d	+1.7 ± 0.9	−24.3 ± 4.0
Hematopoietic cells[e]	7 d	−1.0 ± 4.1	−17.3 ± 3.2
	15 d	+6.3 ± 3.4	−30.8 ± 7.7
	28 d	+2.7 ± 1.9	−49.9 ± 10.2

[a]Results are expressed as mean percent difference (+/−) ± 1 SEM as compared to LTBMC grown in the absence of adherent cells on the bottom of the flask.

[b]Nylon screen bone marrow culture at 2 wk following the second inoculation (with hematopoietic cells).

[c]Time after introduction of the nylon screen LTBMC into a flask containing adherent cells at either approximately 50% or 100% confluence.

[d]Includes fibroblasts, macrophages, adipocytelike cells, and endothelia.

[e]Includes blasts and late stage precursors of all lineages.

CONCLUSIONS

1. The nylon screen LTBMC system supports multilineage expression of hematologic cells.
2. Cell growth on the nylon mesh follows a three-dimensional pattern allowing for the establishment of many localized microenvironments. Indicative of this is the observation of closely apposed myeloid and erythroid colonies in the 210-μm mesh openings.
3. Because of the large surface area of the nylon screen, stromal cells must actively divide for an extended period of time prior to reaching confluence. The growth of hematopoietic cells in this system is related to the mitotic activity of the underlying stroma.
4. Confluent stromal cell monolayers inhibit the proliferation of both hematopoietic and stromal cells in nylon mesh LTBMC suspended in the same flask. This may be attributed to the release of an inhibitor from the confluent cells.
5. Subconfluent stromal cell monolayers failed to enhance proliferative activity in nylon screen LTBMC suspended in the same flask.
6. Human fetal lung fibroblasts support LTBMC of human and nonhuman primate but not rat cells. Marrow stromal cells of the second inoculum (donor) appear to displace the fetal fibroblasts over time.

7. Use of human fetal fibroblasts to establish the support matrix hastens hematopoietic colonization.

ACKNOWLEDGMENTS

The authors wish to thank Glen Feye and Ernest Campany for their technical assistance.

REFERENCES

1. Puck, T. T. & P. I. Marcus. 1955. A rapid method for viable cell titration and clone production with HeLa cells in tissue culture: The use of X-irradiated cells to supply conditioning factors. Proc. Natl. Acad. Sci. USA **41:** 432–437.
2. Dexter, T. M., T. D. Allen & L. G. Lajtha. 1977. Conditions controlling the proliferation of haematopoietic stem cells *in vitro*. J. Cell. Physiol. **91:** 335–344.
3. Bradley, T. R. & D. Metcalf. 1966. The growth of mouse bone marrow cells *in vitro*. Aust. J. Exp. Biol. Med. Sci. **44:** 287–300.
4. Bessis, M. & J. Breton-Gorius. 1959. Nouvelle observations sur l'ilot erythroblastique et la rhopheocytose de la ferritin. Rev. Hematol. **14:** 165–174.
5. Broudy, V. C., K. S. Zuckerman, S. Jetmelani, J. H. Fitchen & G. C. Bagby. 1986. Macrophages stimulate fibroblastoid bone marrow stromal cells to produce multilineage hematopoietic growth factors. Blood **68:** 530–534.
6. Bentley, S. A., O. Alabaster & J.-M. Foidart. 1981. Collagen heterogeneity in normal human bone marrow. Br. J. Haematol. **48:** 287–291.
7. Quesenberry, P. J. & M. A. Gimbrane. 1980. Vascular endothelium as a regulator of granulopoiesis: Production of CSA by human endothelial cells. Blood **56:** 1060–1067.
8. Keating, A., J. W. Singer, P. D. Killen, G. E. Striker, A. C. Salo, E. D. Thomas, D. Thorning & P. J. Fialkow. 1982. Donor origin of the *in vitro* haematopoietic microenvironment after marrow transplantation in man. Nature **298:** 280–283.
9. Bentley, S. A. & J.-M. Foidart. 1980. Some properties of marrow derived adherent cells in tissue culture. Blood **56:** 1006–1012.
10. Allen, T. D. & T. M. Dexter. 1976. Cellular interrelationships during in vitro granulopoiesis. Differentiation **6:** 191–194.
11. Lichtman, M. A. 1984. The relationship of stromal cells to hemopoietic cells in marrow. *In* Long-Term Bone Marrow Culture. D. G. Wright & J. S. Greenberger, Eds. Kroc Foundation. Series Vol. **18:** 3–29. A.R. Liss. New York, NY.
12. Gualtieri, R. J., R. K. Shadduck, D. G. Baker & P. J. Quesenberry. 1984. Hematopoietic regulatory factors produced in long-term murine bone marrow cultures and the effect of in vitro irradiation. Blood **64:** 516–525.
13. Westen, H. & D. F. Bainton. 1979. Association of alkaline phosphatase positive reticulum cells in bone marrow with granulocytic precursors. J. Exp. Med. **150:** 919–937.
14. Patt, H. M. & M. S. Maloney. 1975. Bone marrow regeneration after local injury. Exp. Hematol. **3:** 135–148.
15. Da, W.-M., D. D. F. Ma & J. C. Biggs. 1986. Studies of hemopoietic stromal fibroblastic colonies in patients undergoing bone marrow transplantation. Exp. Hematol. **14:** 266–270.
16. Naughton, B. A., R. A. Preti & G. K. Naughton. 1987. Hematopoiesis on nylon mesh templates. I. Long-term culture of rat bone marrow cells. J. Med. **18:** 219–250.
17. Griffin, J. D., J. Ritz, L. M. Nadler & S. F. Schlossman. 1981. Expression of myeloid differentiation antigens on normal and malignant cells. J. Clin. Invest. **68:**932–941.
18. Cashman, J., D. Henkelman, C. Eaves & A. Eaves. 1983. Individual BFU-E in polycythemia vera produce both erythropoietin dependent and independent progeny. Blood **61:** 876–884.
19. Boswell, H. S., P. R. Albrecht, R. E. Shupe, D. E. Williams & J. Burgess. 1987. Role of stromal populations in hemopoietic stem cell proliferation. I. Physically distinct subpop-

ulations of hemopoietic stem cells and stromal progenitors determine long-term culture hemopoiesis. Exp. Hematol. **15:** 46–53.

20. DEXTER, T. M., E. G. WRIGHT, F. KRIZSA & L. G. LAJTHA. 1977. Regulation of haemopoietic stem cell proliferation in long-term bone marrow cultures. Biomedicine **27:** 344–349.
21. SCHRADER, J. W. & S. SCHRADER. 1978. *In vitro* studies on lymphocyte differentiation. I. Long-term *in vitro* culture of cells giving rise to functional lymphocytes in irradiated mice. J. Exp. Med. **148:** 823–828.
22. ELIASON, J. F., N. G. TESTA & T. M. DEXTER. 1979. Erythropoietin-stimulated erythropoiesis in long-term bone marrow culture. Nature **281:** 382–384.
23. GARTNER, S. & H. S. KAPLAN. 1980. Long-term culture of human bone marrow cells. Proc. Natl. Acad. Sci. USA **77:** 4756–4759.
24. MOORE, M. A. S. & A. P. SHERIDAN. 1979. Pluripotential stem cell replication in continuous human, prosimian, and murine bone marrow culture. Blood Cells **5:** 297–311.
25. COULOMBEL, L., A. C. EAVES & C. J. EAVES. 1983. Enzymatic treatment of long-term human marrow cultures reveals the preferential location of primitive hemopoietic progenitors in the adherent layer. Blood **62:** 291–297.
26. BROCKBANK, K. G. M., J. P., DEJONG, A. H. PIERSMA & J. S. A. VOERMAN. 1986. Hematopoiesis on purified bone marrow-derived reticular fibroblasts *in vitro*. Exp. Hematol. **14:** 386–394.
27. HINES, D. L. 1983. Lipid accumulation and production of colony-stimulating activity by the 266AD cell line derived from mouse bone marrow. Blood **61:**397–402.
28. ZUCALI, J. R., H. E. BROXMEYER, C. A. DINARELLO, M. A. GROSS & R. S. WEINER. 1987. Regulation of early human hematopoietic (BFU-E and CFU-GEMM) progenitor cells in vitro by interleukin-1 induced fibroblast-conditioned medium. Blood **69:** 33–37.
29. BAGBY, G. C., C. A. DINARELLO, P. WALLACE, C. WAGNER, S. HEFENEIDER & E. MCCALL. 1986. Interleukin-1 stimulates granulocyte macrophage colony-stimulating activity release by vascular endothelial cells. J. Clin. Invest. **78:** 1316–1323.
30. NAUGHTON, G. K., B. A. NAUGHTON & A. S. GORDON. 1985. Erythropoietin production by macrophages in the regenerating liver. J. Surg. Oncol. **30:** 184–197.
31. KURLAND, J. J., P. A. MEYERS & M. A. S. MOORE. 1980. Synthesis and release of erythroid colony and burst promoting activities by purified populations of murine peritoneal macrophages. J. Exp. Med. **151:** 839–846.
32. ROODMAN, G. D., V. W. HORADAM & T. L. WRIGHT. 1983. Inhibition of erythroid colony formation by autologous bone marrow adherent cells from patients with the anemia of chronic disease. Blood **62:** 406–412.
33. BROXMEYER, H. E., D. E. WILLIAMS, L. LU, S. COOPER, S. L. ANDERSON, G. S. BEYER, R. HOFFMAN & B. Y. RUBIN. 1986. The suppressive influences of human tumor necrosis factors on bone marrow hematopoietic progenitor cells from normal donors and patients with leukemia: Synergism of tumor necrosis factor and interferon γ. J. Immunol. **12:** 4487–4495.
34. ROODMAN, G. D., A. BIRD, D. HUTZLER & W. MONTGOMERY. 1987. Tumor necrosis factor-alpha and hematopoietic progenitors: Effects of tumor necrosis factor on the growth of erythroid progenitors CFU-E and BFU-E and the hematopoietic cell lines K562, HL60, and HEL. Exp. Hematol. **15:** 928–935.
35. OHTA, M., J. S. GREENBERGER, P. ANKLESARIA, A. BASSOLS & J. MASSAGUE. 1987. Two forms of transforming growth factor-β distinguished by multipotential haematopoietic progenitor cells. Nature **329:** 539–541.
36. ZIPORI, D. 1980. In vitro proliferation of mouse lymphoblastoid cell lines: Growth modulation by various populations of adherent cells. Cell Tissue Kinet. **13:** 287–298.
37. ZIPORI, D. & T. SASSON. 1980. Adherent cells from mouse marrow inhibit the formation of colony stimulating factor (CSF) induced myeloid colonies. Exp. Hematol. **8:** 816–817.
38. DEL ROSSO, M., R. CAPPALLETTI, G. DINI, G. FIBBI, S. VANNUCCHI, V. CHIARUGI & C. GUAZELLI. 1981. Involvement of glycosaminoglycans in detachment of early myeloid precursors from bone marrow stromal cells. Biochim. Biophys. Acta **676:** 129–136.
39. WIGHT, T. N., M. G. KINSELLA, A. KEATING & J. W. SINGER. 1986. Proteoglycans in human long term bone marrow cultures: Biochemical and ultrastructural analyses. Blood **67:** 1333–1343.
40. GALLAGHER, J. T., E. SPOONCER & T. M. DEXTER. 1983. Role of extracellular matrix in

haemopoiesis. I. Synthesis of glycosaminoglycans by mouse bone marrow cultures. J. Cell Sci. **63:** 155–171.

41. McCuskey, R. S. & H. A. Meinke. 1973. Studies of the hemopoietic microenvironment. III. Differences in the splenic microvascular system and stroma between $S1/S^d$ and W/W^v anemic mice. Am. J. Anat. **137:** 187–197.
42. Schrock, L. M., J. T. Judd, H. A. Meinke & R. S. McCuskey. 1973. Differences in concentrations of acid mucopolysaccharides between normal and polycythemic CF_1 mice. Proc. Soc. Exp. Biol. Med. **144:** 593–595.
43. Zuckerman, K. S., R. K. Rhodes, D. D. Goodrum, V. R. Patel, B. Sparks, J. Wells, M. S. Wicha & L. A. Mayo. 1985. Inhibition of collagen deposition in the extracellular matrix prevents the establishment of stroma supportive of hematopoiesis in long term murine bone marrow cultures. J. Clin. Invest. **75:** 970–975.
44. Westen, H. & D. F. Bainton. 1979. Association of alkaline phosphatase-positive reticulum cells in bone marrow with granulocytic precursors. J. Exp. Med. **150:** 919–937.
45. Slapen-Cortenbach, I., R. Ploemacher & B. Lowenberg. 1987. Different effects of human bone marrow and fetal liver stromal cells on erythropoiesis in long-term bone marrow culture. Blood **69:** 135–139.
46. Gordon, M. Y., M. Aguado & D. Grennan. 1982. Human marrow stromal cells in culture: Changes induced by T lymphocytes. Blut **44:** 131–139.
47. Weiss, A. J. & L. A. Blankstein. 1987. Membranes as a solid phase for clinical diagnostic assays. Am. Clin. Prod. Rev. **6:** 8–19.
48. Naparstek, E., J. Pierce, D. Metcalf, R. Shadduck, J. Ihle, A. Leder, M. Sakakeeny, K. Wagner, J. Falco, T. J. Fitzgerald & J. S. Greenberger. 1986. Induction of growth alterations in factor-dependent hematopoietic progenitor cell lines by co-cultivation with irradiated bone marrow stromal cell lines. Blood **67:** 1395–1403.
49. Sporn, M. B., A. B. Roberts, L. M. Wakefield & R. K. Assoian. 1986. Transforming growth factor-B: Biological function and chemical structure. Science **233:** 532–534.
50. Moore, M. A. S., H. E. Broxmeyer, A. P. C. Sheridan, P. A. Meyers, N. Jacobsen & R. J. Winchester. 1980. Continuous human bone marrow culture: Ia antigen characterization of probable pluripotent stem cells. Blood **55:** 682–690.
51. Greenberger, J. S., V. Klassen, K. Kase, R. K. Shadduck & M. Sakakeeny. 1984. Effects of low dose rate irradiation on plateau phase bone marrow stromal cells *in vitro:* Demonstration of a new form of non-lethal physiologic damage to support of hematopoietic stem cells. Int. J. Rad. Oncol. Biol. Phys. **10:** 1027–1037.
52. Zipori, D. 1981. Cell interactions in the bone marrow microenvironment: Role of endogenous colony-stimulating activity. J. Supramol. Struct. Cell. Biochem. **17:** 347–357.

Evidences That Fibroblasts and Epithelial Cells Produce a Specific Type of Macrophage and Granulocyte Inducer, Also Known as Colony-Stimulating Factor, and That Monocyte-Macrophages Can Produce Another Factor with Proliferative Inducing Activity on Myeloid Cells and Differentiative Activity on Macrophages[a]

ISAAC R. ZAMBRANO, JULIO R. CACERES,
JORGE F. MENDOZA, EDELMIRO SANTIAGO, LOURDES M.
MORA, MARIA G. MORALES, MARIA T. CORONA, AND
BENNY WEISS-STEIDER[b]

Escuela Nacional de Estudios Profesionales Zaragoza
Laboratorio de Diferenciacion Celular y Cancer
UNAM
Mexico D.F.

INTRODUCTION

Since the discovery of an assay to measure the factor responsible for the proliferation of myeloid precursors 23 years ago by the group of Leo Sachs,[1] a powerful tool for *in vitro* research was made available to study the mechanisms of myeloid cell differentiation. Even though this group was the first to report the formation of colonies from hematopoietic precursors using a double semisolid culture media, it was the group of Donald Metcalf[2] that one year later by using a similar culture system but with bone marrow cells instead of spleen cells, established the basis for the assays that have been in use all these years to study the differentiation of granulocytes and macrophages. It was this last group that several years later coined the name colony-stimulating factor (CSF)[3] for the molecule responsible for this property. Even though 2 years later the Sachs group named the same factor macrophage and granulocyte inducer (MGI),[4] and another group colony-stimulating activity (CSA),[5] the term CSF is the most widely accepted, and the majority of the work done in this area uses this name.[6–20]

[a]This work was supported in part by the Dirección Adjunta de Formación de Recursos Humanos (DAFRHU) of Consejo Nacional de Ciencia y Tecnología (CONACyT), Departmento de Becas.

[b]Correspondence address: Dr. Benny Weiss, ENEP-Zaragoza, UNAM, Apartado Postal 9-020, Mexico 15000 D.F., Mexico.

The MGI school postulates the existence of a family of proliferation-inducing molecules called MGI-1, and another with differentiation properties called MGI-2 produced by the cells in response to MGI-1.[21–23] On the other hand, the CSF school maintains that hematopoiesis is regulated by a limited series of growth factors which are progressively restricted in their biological activities and target cells.[24–27] It is important to mention that the best known CSFs (CSF-1, GM-CSF and multi-CSF) were originally produced from cells that do not belong to the myeloid compartment, and that on the other hand the differentiating molecules MGI-2 were obtained strictly from proliferating myeloid cells. Thus we think there is some ground to suppose that the proliferation of the myeloid cells is controlled by cells from outside the myeloid compartment, while the differentiation is autoregulated by the cells themselves.

In this work we have attempted by using liquid cultures to provide evidence that there are cell-specific CSFs produced by nonmyeloid cells (one from fibroblasts and another from epithelial cells) which are excellent proliferative factors, while there is at least one CSF molecule produced by the myeloid cells themselves (from macrophages), with strong differentiative capabilities. For these purposes we have produced conditioned media from fibroblasts, from epithelial cells and from bone marrow cells activated for proliferation, and determine by gel chromatography the molecular weights of the different proliferative factors. The differentiation-inducing property was evaluated by determining the induction of Fc receptors on resident and induced peritoneal macrophages.

MATERIALS AND METHODS

Mice

Mice of either sex, strain CD-1, were used at 7–10 days of age as donors of lungs and kidneys, and at 6–8 weeks as donors of fibroblasts, epithelial cells, resident and induced macrophages and bone marrow cells.

Cell Culture

All the cells were cultured under a 10% CO_2 atmosphere, at 37°C and with 95% relative humidity in minimal essential medium (MEM) (Gibco Labs., Grand Island, NY) supplemented with either 10% fetal bovine serum (FBS) (Microlab, Mexico), or horse serum (HS) (Microlab) previously inactivated at 56°C for 30 min. Streptomycin 100 mcg/ml, penicillin G 100 U/ml, and sodium bicarbonate 3.7 g/liter were added to the MEM before culture. All the cultures were done using 60 × 15-mm tissue culture dishes with a final volume of 5 ml.

Cells

Lung fibroblasts were obtained after an enzymatic dispersion with 0.05% of collagenase type IV in a phosphate-buffered saline solution (PBS) at 37°C. On the other hand, to obtain epithelial cells from kidneys and lungs, the organs were subjected to enzymatic dispersion with 0.25% of crude trypsin from pig pancreas, in PBS at room temperature. Finally the cells were always resuspended in culture media containing FBS. To obtain resident macrophages from the peritoneal cavity, the mice were sacrificed and the cells removed by flushing the cavity with 10 ml of PBS. This operation was repeated two more times with 5 ml of PBS to recover as many cells as possible. The cells were then washed three times with PBS at 500 g for 5 min, and finally 8×10^6 cells were seeded in culture media containing HS. To obtain induced macrophages from the peritoneal cavity, the mice were injected ip with 3 ml of sodium caseinate (Difco Labs., Detroit, MI) at 10% in PBS. After 4 days the animals were sacrificed and the peritoneal exudates recovered by flushing the cavity with 20 ml of PBS. The cells were washed three times and 8×10^6 cells were seeded with culture media. In order to enrich the macrophage populations, the cells were incubated for 1 h at 37°C, followed by the removal of those cells that did not adhere to the culture substrate.

Conditioned Media

The sources for the CSF's were the media conditioned (CM) by the different cell types employed. CSF from fibroblasts and epithelial cells were produced by cultures in their second passages, either by seeding 2.7×10^5 lung fibroblasts and collecting the CM after 12 days, or 3×10^5 lung epithelial cells also after 12 days, or 1.5×10^5 kidney epithelial cells after 11 days. The media conditioned by 8×10^6 bone marrow cells stimulated with 2 ml of either the CM from fibroblasts or epithelial cells, as a source of CSF, were collected after 4 days in culture. All the CM were stored at −20°C until use.

Gel Chromatography

A total of 2 ml of CM from either fibroblasts, epithelial cells or bone marrow cells, were chromatographed in a Sephadex G-100 (Pharmacia Fine Chemicals, Uppsala, Sweden) in a 2.5 × 90-cm column, and eluted with 75 mM Tris–HCl buffer, pH 7.7, at 4°C and with a flow rate of 4 cm/h. Fractions of 5.5 ml were collected and stored at −20°C until use. Transferrin, ovalbumin, myoglobin, half hemoglobin and cytochrome C were used as external standards. For the biological assays, aliquots of 0.5 ml from each fraction were employed.

CSF Assay

For the colony-stimulation assay the double agar layer technique was employed.[1] Briefly in a 60 × 15-mm Petri dish a first layer with 0.6% of agar was added to the medium to be tested, and a second layer with 0.3% of agar was overlayered with approximately 5×10^5 bone marrow cells. After 7 days of incubation, all the colonies with more than 20 cells were counted using an inverted microscope.

Cell Proliferation Assay in Liquid Cultures

For the cell proliferation assay the liquid culture technique was employed. Briefly, in a 35 × 10-mm Petri dish 0.25 ml of the medium to be tested was added to approximately 5×10^6 bone marrow cells and 10% of HS. After 4 days of incubation, the number of cells was evaluated in each culture by means of an hemocytometer.

Assay for Fc Receptors

The Fc receptors were measured by the rosette technique.[28] Briefly, IgG against sheep red blood cells (SRBC) (Cordis, Labs., Miami, FL) was diluted 1:1600 in PBS and mixed with a nonagglutinating concentration of SRBC previously washed in PBS and incubated at 37°C for 30 min. The erythrocytes coated with antibody (EA) were washed three more times in PBS to get rid of free IgG and stored in PBS for a maximum of 4 days at 4°C until used in the Fc receptor assay. The cells to be tested for Fc receptors were resident and elicited peritoneal macrophages that were cultured during 4 days with and without the test media. For the assay 1 ml of EA was added to the culture dishes diluted 1:1 with PBS, and incubated for 30 min at 37°C. Finally, the nonattached erythrocytes were removed and the dishes fixed and dyed with May Greenwald Giemsa. The percentage of rosettes was evaluated by determining the macrophages with more than 4 erythrocytes attached to them. A minimum of 300 cells were counted for each determination.

Miscellaneous

The results are the averages of duplicates of two independent assays that did not depart more than 15% from each other. A culture without CM was always added as control. All the chemicals were from Sigma Chemical Co., St. Louis, MO, unless otherwise specified.

RESULTS

Production of Proliferation Factors for Myeloid Cells by Epithelial Cells and Fibroblasts

Conditioned media (CM) from cultures of kidney epithelial cells and from lung fibroblasts were chromatographed and the different fractions evaluated for myeloid cell proliferation and colony-stimulating activity. Half a milliliter of each fraction was employed either to activate proliferation in liquid cultures of 5×10^6 bone marrow cells during 4 days, or to induce colony formation in 5×10^5 during 7 days. When using epithelial cells CM we obtained for both activities a single peak that corresponded to a mol wt of 22,000 d (FIG. 1), while for the fibroblasts CM another single peak of 70,000 d (FIG. 2). Taking into consideration that until today there have been no reports in the scientific literature of indirect CSF inducing agents with these mol wts, that on the other hand there are other well-known CSFs with mol wts of 70,000 d and 22,000 d (CSF-1 and GM-CSF),[6,27] and that the 70,000-d molecule is known to be produced by a fibroblast cell line, we think that the colony-stimulating activities measured in this work represent true CSFs. In order to determine if the 22,000-d CSF molecule found in the CM of epithelial cells from the kidneys is also produced by similar cells from other organs, we evaluated the presence of a proliferative activity in a CM obtained from epithelial cells from lungs. Using the same methods we found a single peak of activity with 22,000 d (FIG. 3).

Production of Proliferation Factors by Bone Marrow Cells Induced by CM Containing CSF from Fibroblasts and Epithelial Cells

Once we had determined that fibroblasts and epithelial cells can secrete specific CSFs, we used CM containing these molecules to evaluate whether bone marrow cells could be induced to secrete new proliferating molecules. For this purpose we induced for proliferation 8 × 10^6 bone marrow cells with 2.0 ml of each CM, and after 4 days in culture we collected the new CM. We found 4 peaks with proliferative activity in the chromatographic fractions from the CM of the bone marrow cells that were induced by fibroblast CM. Once again the proliferation-inducing activity also corresponded in mol wt to the peaks with colony-stimulating activity (FIG. 4). Apart from the 70,000-d peak similar in mol wt to the fibroblasts CSF, we obtained three new ones, with 45,000, 30,000 and 17,000 d, respectively. Taking into consideration that we have previously found that a macrophage cell line produces a CSF with 45,000 d,[29] that alveolar and peritoneal macrophages are known to produce a CSF with a similar mol wt,[30] and that fibroblast CSF

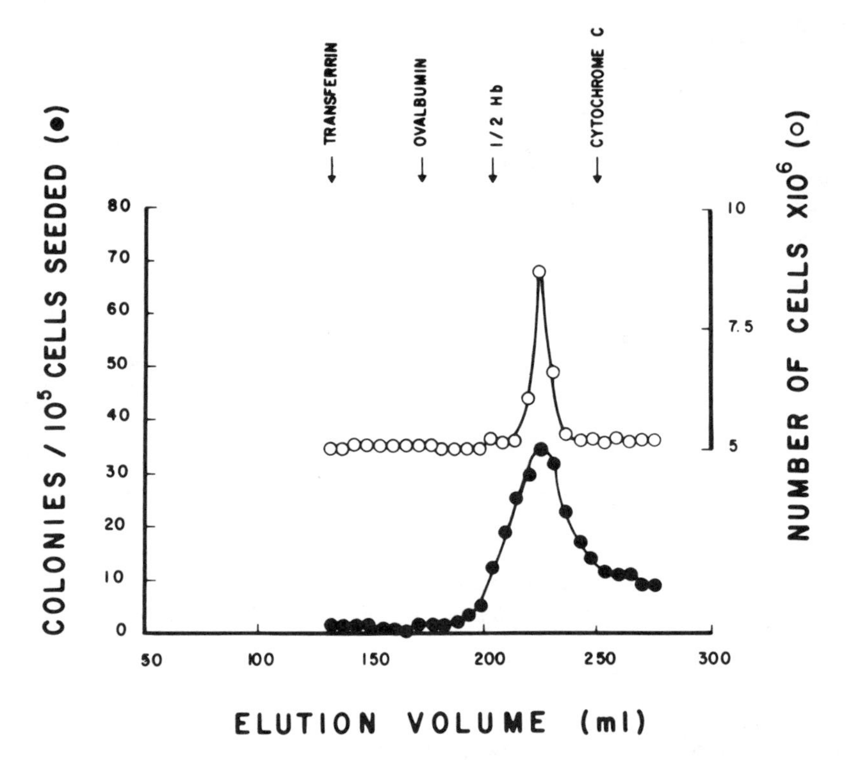

FIGURE 1. Elution profile of the media conditioned by kidney epithelial cells. Colony formation on bone marrow cells (●). Proliferation of bone marrow cells in liquid cultures (○).

is known to induce monocyte-macrophage proliferation, we think that this 45,000-d molecule was secreted by the proliferating monocytes in the bone marrow cultures. We do not know at this stage whether the 30,000- and the 17,000-d molecules represent true CSF molecules, or if they are factors that could have been induced indirectly. In fact there is a 17,000-d molecule known as IL-1,[31] which is secreted by macrophages[32] and can induce the secretion of CSF.[33–35]

When under similar experimental conditions epithelial cell CM was used as an activator for bone marrow cell proliferation, we obtained three peaks of activity (FIG. 5). Once again the 45,000-, the 30,000- and the 17,000-d molecules were present with both a proliferative and a colony-stimulating activity.

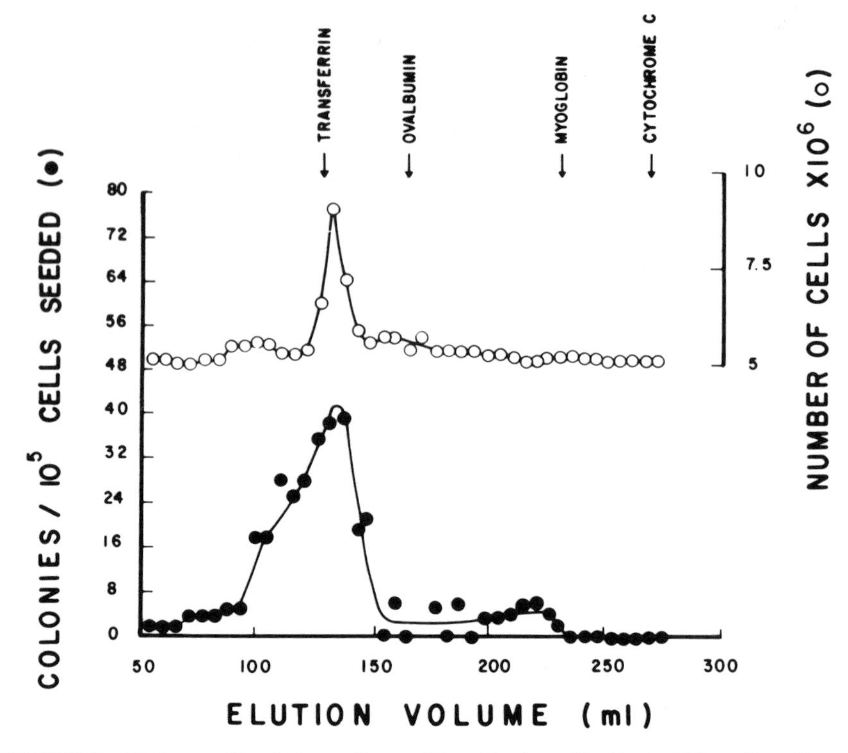

FIGURE 2. Elution profile of the media conditioned by lung fibroblasts. Colony formation on bone marrow cells (●). Proliferation of bone marrow cells in liquid cultures (○).

Production of Proliferating Factors by Bone Marrow Cultures Induced by the 45,000-d Proliferating and Colony-Stimulating Factor

In order to determine whether all the three new peaks of proliferative and colony-stimulating activities found in bone marrow cultures were induced by fibroblasts CSF (MGI-1), or by the 45,000-d molecule (MGI-2) induced by the 70,000-d containing fibroblast CM, we proceeded to activate bone marrow cells to proliferate by 1.0 ml of a pool of the fractions containing the 45,000-d molecule (FIG. 6) during 5 days in culture. When evaluating the presence of new proliferating and colony-stimulating activities in the chromatographed fractions of this new CM, we found apart from the 45,000-d peak the 30,000- and 17,000-d ones. Thus we interpret the 45,000-d molecule to be the inducer of these new molecules.

The 45,000 Proliferative and Colony-Stimulating Activity Induces the Expression of Fc Receptors on Resident and Elicited Peritoneal Macrophages

In order to determine if the 45,000-d molecule could represent a true MGI-2 molecule, that is, with strong differentiating properties, we evaluated the property of inducing Fc receptor expression by adding 0.5 ml of the 45,000-d fraction pool to either 1×10^6

elicited or resident peritoneal macrophages during 4 days in culture. We observed that both groups of macrophages were strongly induced for Fc receptor expression (TABLE 1). Thus our results provide evidence that the 45,000-d molecule has, aside from a proliferative activity, a strong differentiative one.

DISCUSSION

Two different phenomena have to occur in order to maintain a continuous supply of mature blood cells. There has to be first a mechanism that assures the existence of sufficient numbers of a particular cell type, and second a mechanism for the differentiation of these cells into functional ones. In fact taking into consideration that proliferation of precursor cells goes simultaneously with their maturation, in such a way that in normal cells the more the maturation level achieved the less the proliferative capabilities, then we can expect that these two mechanisms are simultaneously regulated. In consequence proliferation has to be linked to differentiation, because proliferation without differentiation would give rise to leukemia, while differentiation without proliferation would give rise to anemia.

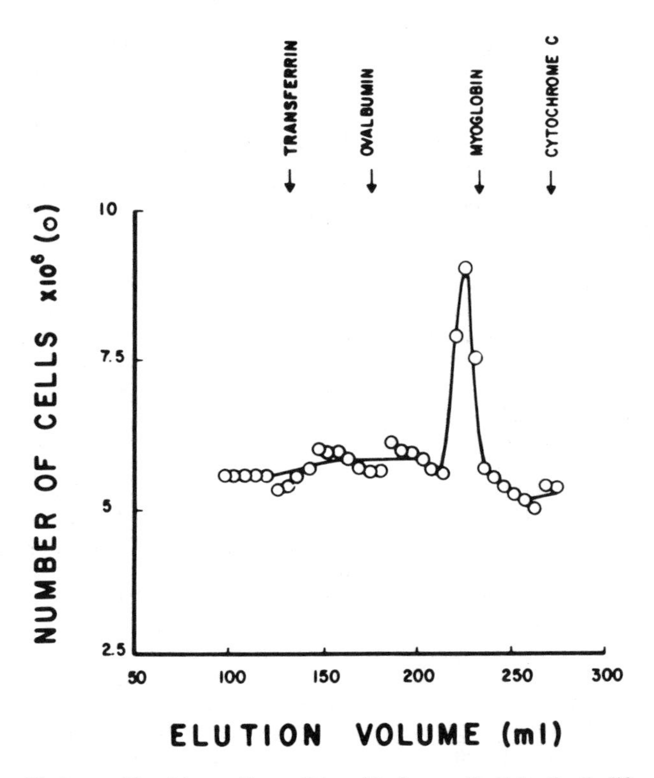

FIGURE 3. Elution profile of the media conditioned by lung epithelial cells. Proliferation of bone marrow cells in liquid cultures (0).

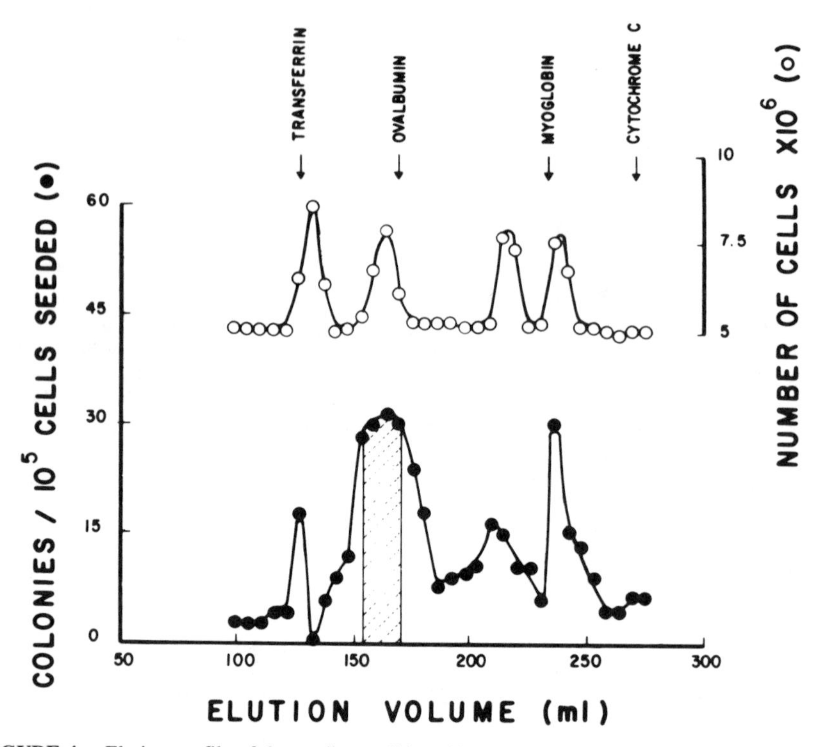

FIGURE 4. Elution profile of the media conditioned by bone marrow cells induced by the media conditioned by cultures of lung fibroblast. Colony formation on bone marrow cells (●). Proliferation of bone marrow cells in liquid cultures (○). The 45,000-d pool was taken from the fractions represented by the *shaded area.*

We can think of two mechanisms by which differentiation could be linked to proliferation. First, that the proliferating or mitogenic inducing factors induce the expression of receptors to differentiation factors, thus making the cells susceptible to respond to exogenous regulatory molecules for the expression of mature properties. And second, that proliferation induces the production of endogenous factors that in turn induce the cell to present a mature phenotype.

We consider that there are today two main schools of thought that have provided considerable data to support each one of the ideas with regard to the way in which proliferation could be linked to the differentiation of granulocyte and macrophages. The MGI school maintains that endogenous factors (MGI-2) are produced in response to exogenous mitogenic factors (MGI-1) that are responsible for cell differentiation.[21–23] On the other hand, the CSF school affirms that four different exogenous factors (multi-CSF, GM-CSF, CSF-1 and G-CSF) control the proliferation and differentiation of these blood cells, depending on a programmed expression of membrane receptors to each one of them.[24] Whether the MGI or CSF school turns out to be correct, or whether they complement each other, we think that a consensus has to be obtained to assign names to these, and to all the other hematopoietic factors. First, with regards to the CSF nomenclature, we have to take into consideration that there are many other factors that also induce the formation of hematologic colonies *in vitro,* like erythropoyetin,[36] interleukin-2,[37] interleukin-3,[25] and megakaryopoietin,[38] and that there is no clear evidence of colony formation *in vivo*. In fact, as shown in this work, the proliferative capabilities of CSF can also be measured in liquid cultures where there is obviously no colony formation. On the other hand, and as far as the MGI terminology is concerned, there is strong evidence that one type of hemopoietic factor, once thought to be specific for one cell type, can contribute to the differentiation and proliferation of another one. Thus we feel that there is a need to redefine terms by employing a more uniform nomenclature. There is a terminology using the interleukin nomenclature that does not offer a solution to this problem, because obviously it does not account for the proliferating and differentiation factors that are produced by other cell types aside from leukocytes.

In this work we have attempted to provide some evidence that the CSFs are cell specific, and that one cell type is capable of producing and secreting only one type of CSF, and different from the one produced by another cell type. We think that under certain circumstances the cells could be induced to produce other factors, but that under normal physiological conditions the majority of the CSF production is of one type. We have provided evidence that fibroblasts from different organs secrete a 70,000-d CSF molecule similar in mol wt to the already known CSF-1 also from a fibroblasts cell line.[6] We have also shown that epithelial cells from different organs produce a CSF with 22,000 d similar in mol wt to the known GM-CSF.[27] Finally evidence is provided that monocytes produce a 45,000-d CSF molecule similar in mol wt to the already known molecule from a macrophage cell line,[29] from peritoneal and alveolar macrophages,[30] and from human monocytes.[39] Throughout this paper we have stated that our molecules were similar in mol

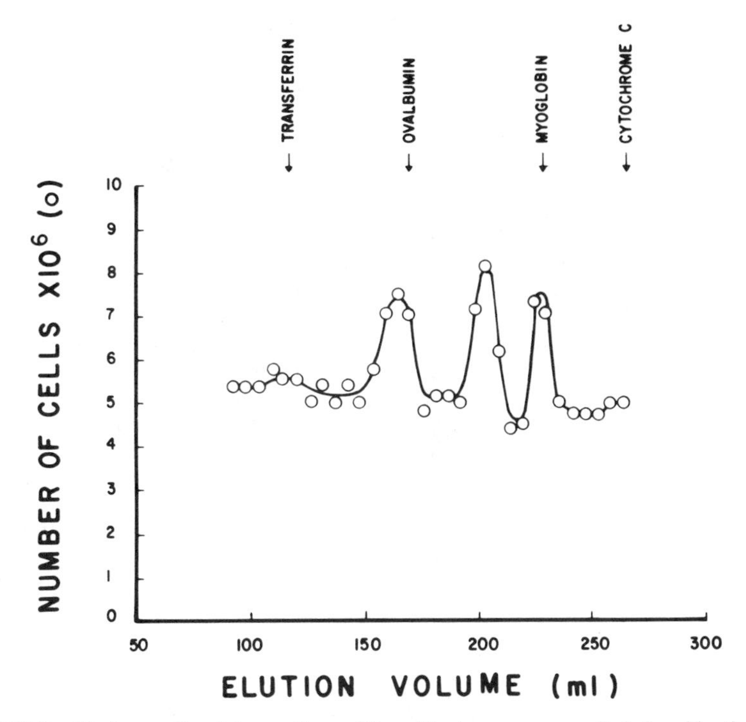

FIGURE 5. Elution profile of the media conditioned by bone marrow cells induced by the media conditioned by cultures of kidney epithelial cells. Proliferation of bone marrow cells in liquid cultures (○).

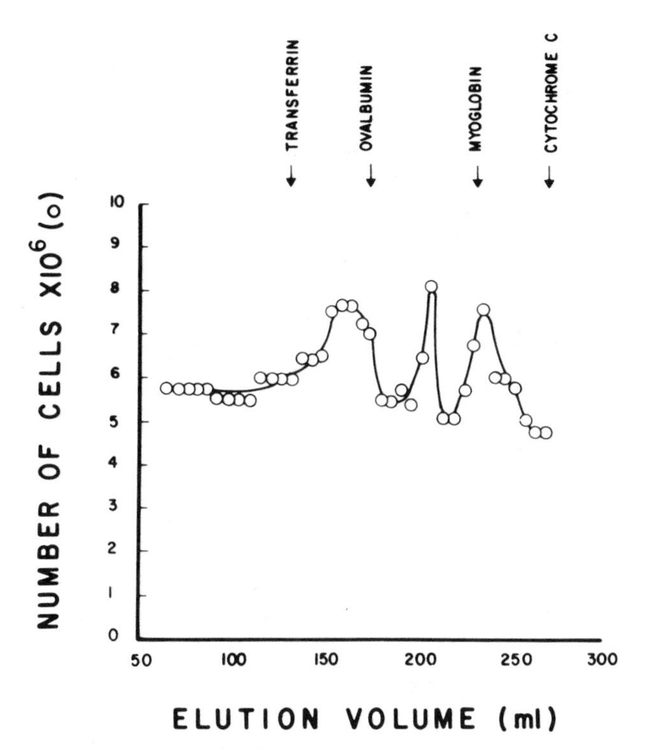

FIGURE 6. Elution profile of the media conditioned by bone marrow cells induced by the pooled fractions containing the 45,000-d peak. Proliferation of bone marrow cells in liquid cultures (○).

wt and not identical to the already existing CSFs, because we have not made cross experiments with antibodies to the already existing factors to prove identity.

We feel that liquid cultures should be more extensively used to complement the data of colony formation, to evaluate aside from an indication of colony-forming units, the proliferating and differentiating inducing properties of the different factors. We have previously shown[29] that CSF can induce colony formation on bone marrow cells but was unable to induce the expression of Fc receptors, a mature cell property, while another factor was capable of inducing Fc receptors and not colony formation. Thus we think that there is a family of molecules some, with proliferative properties and others with differ-

TABLE 1. Induction of Fc Rosettes on 5×10^5 Peritoneal Cells after 4 Days in Culture

	Resident Macrophages[a]				Induced Macrophages[a]			
	−		+		−		+	
	Exp. 1	Exp. 2	Exp. 1	Exp. 2	Exp. 1	Exp. 2	Exp. 1	Exp. 2
Percentage of Fc rosettes	30	29	54	59	42	46	61	59

[a](−) Cultures without inducer; (+) cultures with 10% of the pool containing the 45,000-d molecules.

entiative ones that contribute to the production of functional mature blood cells. In this paper we have shown that differentiative factors can be produced by cells in response to proliferative ones, and have provided evidence that this is an endogenous phenomena in accordance with the MGI-2 theory. It has to be mentioned that MGI-2 has been associated with G-CSF,[40] thus if the 45,000-d molecule described in this work is a true monocyte-macrophage-secreted CSF, we can speculate that macrophages secrete under the action of exogenous mitogens (CSF-1 or GM-CSF), endogenous factors that are aside from mitogens to other cell types, strong autodifferentiating ones (FIG. 7). Even though we have shown that the 45,000-d molecule induces cell proliferation and the secretion by bone marrow cells of other colony-stimulating activities, we do not know at this stage the target cells for all these factors, or whether the induced colony-stimulating activities represent true CSFs, or if they are the product of indirect inducing mechanisms like the one

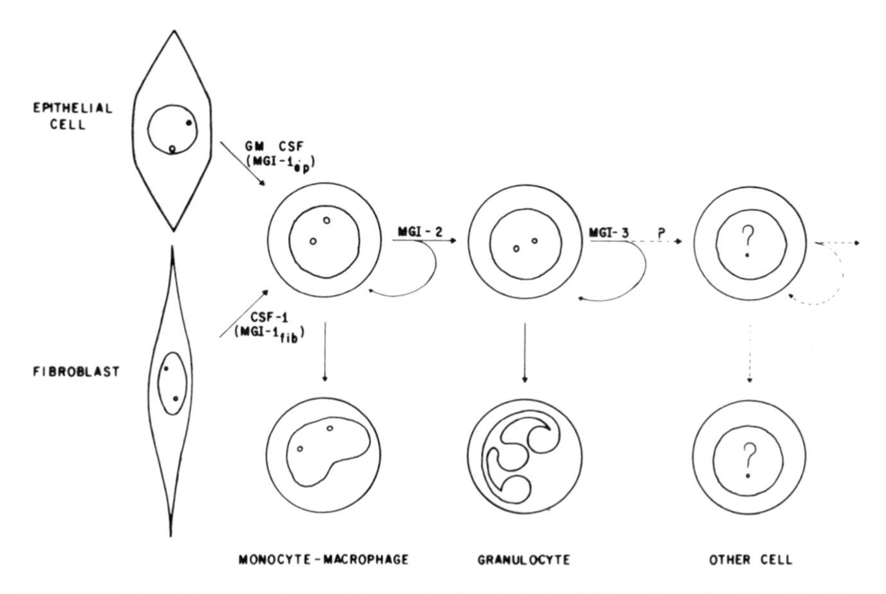

FIGURE 7. Model representing a chain reaction in which the CSF produced by either fibroblasts or epithelial cells (MGI-1) can induce a myeloid precursor to produce another CSF (MGI-2) with proliferative and autodifferentiative inducing properties.

associated with IL-1. Thus we do not know if there is a chain reaction in which one exogenous mitogen induces the appearance of an endogenous mitogen on another cell type (FIG. 8), or if this secondary molecule only contributes to the proliferation and differentiation of the cell that produces it (FIG. 7).

If the hypothesis presented in this work concerning the specificity of CSF production by cell types is correct, then we can offer an alternative for the nomenclature of this family of molecules. We propose that the name be associated with the cell that produces the factor and not with the target cell. Thus proliferating and differentiating factors would be classified according to their origin and not to their different functions. Finally, we would like to mention that even though the myeloid precursors are the targets for fibroblast- and epithelial-produced CSF we do not know if they act on the same cells or in different subgroups of myeloid precursors. Obviously much work would have to be done to prove the evidences discussed here, but we think that they offer an alternative explanation for what is known about the mechanisms involved in myeloid cell differentiation.

SUMMARY

Molecules with the property to induce proliferation of bone marrow cells in liquid cultures, and with colony-stimulating activity, were found on media conditioned (MC) by

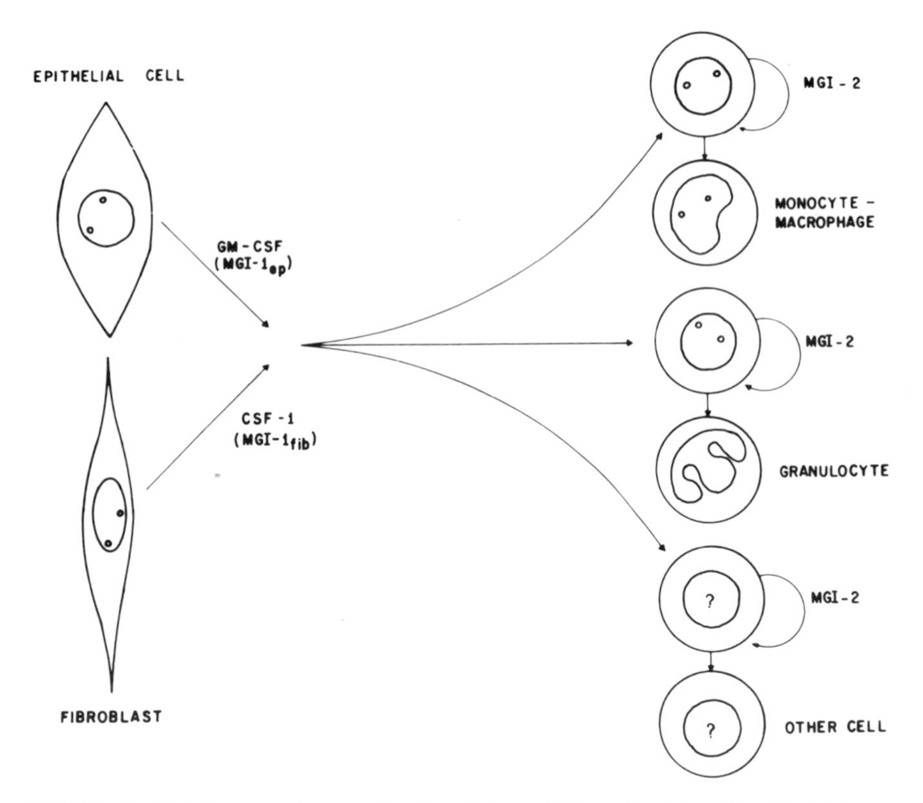

FIGURE 8. Model representing a system in which the CSF produced by either fibroblasts or epithelial cells (MGI-1) can induce several myeloid precursors to produce other CSF (MGI-2) with autodifferentiative inducing properties.

lung fibroblasts and kidney epithelial cells. These factors presented an apparent mol wt of 70,000 and 22,000 d respectively. Also when MC by epithelial cells from lungs was tested for the induction of proliferation of bone marrow cells a molecule with 22,000 d was detected. These molecules are thought to be CSF because they induce colony formation, and they are also similar in mol wt to two of the already known CSF. In fact the GM-CSF obtained from endotoxic lungs with a large epithelial cell content has a mol wt of 22,000 d, and the CSF-1 produced by a fibroblast cell line has 70,000. When the MC by fibroblast was used to induce bone marrow cells to proliferate, three new molecules with colony-stimulating activity were secreted. These molecules with apparent mol wts of 45,000, 30,000 and 17,000 d were also found in the MC by bone marrow cells when induced to proliferate with MC by epithelial cells. When the 45,000-d molecule was used to induce bone marrow cells to proliferate, once again the 30,000- and the 17,000-d molecules were secreted. Evidence is also provided that the 45,000-d molecule is produced by the monocyte-macrophage cells, and that it can induce Fc receptors on resident and elicited peritoneal macrophages. The possibility that the production of CSF is cell specific is discussed together with two models to explain the way in which these molecules can participate as proliferative (MGI-1) and differentiative (MGI-2) function in normal myeloid cell differentiation. Finally, a new terminology is proposed to classify this family of molecules.

ACKNOWLEDGMENTS

We thank Dr. Mario Calcagno and Dr. Guillermo Mendoza for help in the biochemistry techniques, and Ranulfo Pedraza and Jose G. Chavarria for excellent technical assistance.

REFERENCES

1. PLUZNIK, D. H. & L. SACHS. 1965. The cloning of normal "mast" cells in tissue culture. J. Cell. Comp. Physiol. **66:** 319–324.
2. BRADLEY, T. R. & D. METCALF. 1966. The growth of mouse bone marrow cells in vitro. Aust. J. Exp. Biol. Med. Sci. **44:** 287–299.
3. METCALF, D., M. A. S. MOORE & N. L. WARNER. 1969. Colony formation in vitro by myelomonocytic leukemic cells. J. Natl. Cancer Inst. **43:** 983–1001.
4. LANDAU, T. & L. SACHS. 1971. Characterization of the inducer required for the development of macrophage and granulocyte colonies. Proc. Natl. Acad. Sci. USA **68:** 2540–2544.
5. AUSTIN, P. E., E. A. MCCULLOCH & J. E. TILL. 1971. Characterization of the factor in L-cell conditioned medium capable of stimulating colony formation by mouse bone marrow cells in culture. J. Cell. Physiol. **77:** 121–134.
6. BURGESS, A. W., J. CAMAKARIS & D. METCALF. 1977. Purification and properties of colony-stimulating factor from mouse lung conditioned medium. J. Biol. Chem. **252**(6): 1988–2003.
7. GOUGH, N. M., J. GOUGH, D. METCALF, A. KELSO, D. GRAIL, N. A. NICOLA, A. W. BURGESS & A. R. DUNN. 1984. Molecular cloning of cDNA encoding a murine haemapoietic growth regulator, granulocyte-macrophage colony-stimulating factor. Nature **309**(5970): 763–767.
8. STANLEY, E. R. & L. J. GUILBERT. 1981. Methods for the purification assay, characterization and target cell binding of a colony stimulating factor (CSF-1). J. Immunol. Methods **42**(3): 253–284.
9. GASSON, J. L., R. H. WEISBART, S. E. KAUFMAN, S. C. CLARK, R. M. HEWICH, G. G. WONG & W. GOLDE. 1984. Purified human granulocyte-macrophage colony-stimulating factor: Direct action on neutrophils. Science **226**(4680): 1339–1342.
10. KAWASAKI, E. S., M. B. LADNER, A. M. WANG, J. VAN ARSELL, M. K. WARREN, M. Y. COYNE, D. L. SCHWICKART, M. LEE, K. J. WILSON, A. BOOSMAN, E. R. STANLEY, P. RALPH & D. F. MARK. 1985. Molecular cloning of a complementary DNA encoding human macrophage-specific colony-stimulating factor (CSF-1). Science **230**(4723): 291–296.
11. DELAMARTER, J. F., J. J. MERMOD, C. M. LIANG, S. F. ELIASON & D. THATCHER. 1985. Recombinant murine GM-CSF from E. coli has biological activity and is neutralized by a specific antiserum. EMBO J. **4**(10): 2575–2578.
12. METCALF, D. & N. A. NICOLA. 1983. Proliferative effects of purified granulocyte colony-stimulating factor (G-CSF) on normal mouse hemopoietic cells. J. Cell. Physiol. **116:** 198–206.

13. WAHEED, A. & R. K. SHADDUCK. 1979. Purification and properties of L-cell-derived colony-stimulating factor. J. Lab. Clin. Med. **94**(1): 180–189.
14. SHADDUCK, R. K., G. PIGOLI, C. CARAMATTI, G. DEGLIONTONI, V. RIZZOLI, A. PORCELLINI, A. WAHEED & L. SCHIFFER. 1983. Identification of hemopoietic cells responsive to colony-stimulating factor by autoradiography. Blood **62**(6): 1197–1202.
15. DONAHUE, R. E., E. A. WANG, D. K. STONE, R. KAMEN, G. G. WONG, P. K. SEHGAL, D. G. NATHAN & S. C. CLARK. 1986. Stimulation of haemopoiesis in primates by continuous infusion of recombinant human GM-CSF. Nature **321**(6073): 872–875.
16. TSUDA, H., L. M. NECKERS & D. H. PLUZNIK. 1986. Colony stimulating factor-induced differentiation of murine M1 myeloid leukemia cells is permissive in early G1 phase. Proc. Natl. Acad. Sci. USA **83**(12): 4317–4321.
17. LE BEAU, M. M., C. A. WESTHROOK, M. O. DIAZ, R. A. LARSON, J. D. ROWLEY, J. C. GASSON, D. W. GOLDE & C. J. SHERR. 1986. Evidence for the involvement of GM-CSF and FMS in the deletion (5q) in myeloid disorders. Science **231**(4741): 984–987.
18. DONAHUE, R. E., S. G. EMERSON, E. A. WANG, G. G. WONG, S. C. CLARK & D. G. NATHAN. 1985. Demonstration of burst-promoting activity at recombinant human GM-CSF on circulating erythroid progenitors using an assay involving the delayed addition of erythropoietin. Blood **66**(6): 1476–1481.
19. WANG, S. Y., H. CASTRO-MALASPINA, L. LU & M. A. MOORE. 1985. Biological characterization of a granulopoietic enhancing activity derived from cultured human lipid-containing macrophages. Blood **65**(5): 1181–1190.
20. QUESENBERRY, P. J., J. N. IHLE & E. MCGRATH. 1985. The effect of interleukin 3 and GM-CSA-2 on megakaryocyte and myeloid clonal colony formation. Blood **65**(1): 214–217.
21. SACHS, L. 1980. Constitutive uncoupling of pathways of gene expression that control growth differentiation in myeloid leukemia: A model for the origin and progression of malignancy. Proc. Natl. Acad. Sci. USA **77**(10): 6152–6156.
22. SYMONDS, G. & L. SACHS. 1983. Synchrony of gene expression and the differentiation of myeloid leukemic cells: Reversion from constitutive to inducible protein synthesis. EMBO J. **2**(5): 663–667.
23. LOTEM, J. & L. SACHS. 1982. Mechanisms that uncouple growth and differentiation in myeloid leukemia: Restoration of requirement for normal growth-inducing protein without restoring induction of differentiation-inducing protein. Proc. Natl. Acad. Sci. USA **79**(14): 4347–4351.
24. METCALF, D. 1986. Review. The molecular biology and functions of the granulocyte-macrophage colony-stimulating factor. Blood **62**(2): 257–267.
25. IHLE, J. N., J. KELLER, L. HENDERSON, F. KLEIN & E. W. PALASZYNSKI. 1982. Procedures for the purification of interleukin 3 to homogeneity. J. Immunol. **129**(6): 2431–2436.
26. STANLEY, E. R. & P. M. HEARD. 1977. Factors regulating macrophage production and growth: Purification and some properties of colony stimulating factor from medium conditioned by mouse L. cells. J. Biol. Chem. **252**(12): 4305–4312.
27. SPARROW, L. G., D. METCALF, M. W. HUNKAPILLER, L. E. HOOD & A. W. BURGESS. 1985. Purification and partial amino acid sequence of asialo murine granulocyte-macrophage colony stimulating factor. Proc. Natl. Acad. Sci. USA **82**(2): 292–296.
28. BIANCO, C., R. PATRICK & V. NOSSENZWEIG. 1970. A population of lymphocytes bearing a membrane receptor for antigen-antibody complement complexes. I. Separation and characterization. J. Exp. Med. **132**(4): 702–720.
29. CALCAGNO, M., J. R. PEREZ, M. G. WALDO, G. CABRERA & B. WEISS STEIDER. 1982. Evidence of the existence of a factor that induces Fc receptors on bone marrow cells. Blood **59**(4): 756–760.
30. LOTEM, J., J. H. LIPTON & L. SACHS. 1980. Separation of different molecular forms of macrophage and granulocyte inducing proteins for normal and leukemic myeloid cells. Int. J. Cancer **25**(6): 763–771.
31. TOCCI, M. J., N. I. HUTCHINSON, P. M. CAMERON, K. E. KURK, D. J. NORMAN, J. CHIN, E. A. RUPP, G. A. LIMJUCO, J. M. B. ARGUDO & J. A. SCHMIDT. 1987. Expression in Escherichia coli of fully active recombinant human IL 1 beta. Comparison with native human IL 1 beta. J. Immunol. **138**(4): 1109–1114.
32. NATHAN, C. F. 1987. Secretory products of macrophages. J. Clin. Invest. **79**(2): 319–326.

33. ZSEBO, K. M., V. N. YUSCHENKOFF, S. SCHIFER, D. CHANG, E. MCCALL, C. A. DINARELLO, M. A. BROWN, B. ALFROCK & G. C. BAGBY. 1988. Vascular endothelial cell and granulopoiesis: Interleukin-1 stimulates release of G-CSF and GM-CSF. Blood **17**(1): 99–103.
34. SEGAL, G. M., E. MCCALL, T. STUEVE & G. C. BAGBY. 1987. Interleukin-1 stimulates endothelial cell to release multilineage human colony-stimulating activity. J. Immunol. **138**(6):1772–1778.
35. SEIFF, C. A., S. TSAI & D. V. FALLER. 1987. Interleukin-1 induced cultured human endothelial cell production of granulocyte macrophage colony-stimulating factor. J. Clin. Invest. **79**(1): 48–51.
36. DONAHUE, R. E., S. G. EMERSON, E. A. WANG, G. G. WONG, S. C. CLARK & D. G. NATHAN. 1985. Demonstration of burst-promoting activity of recombinant human GM-CSF on circulating erythroid progenitors using an assay involving the delayed addition of erythropoietin. Blood **66**(6): 1479–1481.
37. JOURDAN, M., T. COMMER & B. KLEIN. 1985. Control of human T-colony by interleukin-2. Immunology **54**(2): 249–253.
38. HOFFMAN, R., H. H. YANG, E. BRUNO & J. E. STRANEVA. 1985. Purification and partial characterization of a megakaryocyte colony-stimulating factor from human plasma. J. Clin. Invest. **74**(4): 1174–1182.
39. SULLIVAN, R., E. W. LIPKIN, R. BELL, N. E. LARSEN & L. A. MCCARROLL. 1985. The kinetics of the production of granulocyte-monocyte colony-stimulating activity (GM-CSA) by isolated human monocytes: Response to bacterial endotoxin. Prog. Clin. Biol. Res. **184:** 173–187.
40. DEXTER, T. M. 1984. Blood cell development: The message in the medium. Nature **309**(28): 746–747.

PART V. COLONY-STIMULATING FACTORS FOR MEGAKARYOCYTOPOIESIS AND GRANULOPOIESIS

Development of a Radioimmunoassay for Human Macrophage Colony-Stimulating Factor (CSF-1)[a]

RICHARD K. SHADDUCK AND ABDUL WAHEED

Department of Medicine
Montefiore Hospital
University of Pittsburgh
School of Medicine
3459 Fifth Avenue
Pittsburgh, Pennsylvania 15213

INTRODUCTION

The *in vitro* growth of granulocyte-macrophage colonies is stimulated by certain glycoprotein hemopoietic growth factors termed colony-stimulating factors.[1,2] At this time four distinct CSFs are recognized.[3] These include macrophage CSF (M-CSF or CSF-1), granulocyte-macrophage CSF (GM-CSF), granulocyte CSF (G-CSF) and multi-CSF or interleukin-3. These factors induce either single or multilineage colonies after 7 to 14 days of incubation with bone marrow cells *in vitro*.

Originally, such factors were described as activities that could be found in the serum, urine or media conditioned by the growth of a variety of cell lines. Assay systems for these activities utilized *in vitro* growth of colonies in a semisolid medium;[1,2] however, such assays are hampered by low sensitivity and susceptibility to a variety of inhibitors. Such inhibitors include high molecular weight lipoprotein substances in the serum[4] as well as low molecular weight dialyzable materials that are found in conditioned media.[5,6]

Recently each of these CSFs has been purified and the amino acid sequences determined.[7–13] cDNA clones have been identified such that many of the factors are now produced by recombinant techniques.[14–17] Use of such factors and the corresponding antibodies should provide the basis for developing sensitive radioimmunoassay, radioreceptor assay and ELISA techniques for detection of minute quantities of these factors. Such assays should prove useful in the study of potential physiologic roles of the CSFs and should also serve as tools for study of the pharmacokinetics following injection of these materials.

The present report describes the development of a radioimmunoassay for human urinary CSF or human CSF-1. The human urinary CSF was purified using a cross-reactive antibody produced against L cell CSF,[18,19] a form of murine CSF-1. This report describes the preparation of antibody to human urinary CSF, iodination of the CSF and the development and validation of a radioimmunoassay.

[a]Supported by National Institutes of Health Grant R01CA15237.

METHODS

Human urinary CSF was concentrated and purified as described previously.[19] In essence, 10 liter pools of this material were concentrated 200-fold by ultrafiltration and dialyzed against 0.1 M Tris, 0.3% polyethylene glycol (mw 4000) (PEG) buffer. The concentrated material was applied to a column of an IgG fraction containing anti-CSF with selective retention of the CSF activity. After extensive rinsing, the purified material was released under low pH, high molarity conditions. This material was purified further on Con A-Sepharose with elution using alpha methylglucoside. The final specific activity was approximately 2×10^7 U/mg. After iodination, this material showed a single band of radioactivity in SDS-acrylamide gel.

In order to prepare antiserum to human CSF-1, two rabbits were injected with 20 μg of the purified HU-CSF emulsified in complete Freund's adjuvant with 2.5 mg/ml of added H37 RA desiccated mycobacteria (DIFCO). The animals were injected on one occasion in ~30–40 intradermal sites. Serum samples were obtained starting 6 weeks after immunization. The antisera were assayed using varying dilutions; 0.1 ml of each dilution was incorporated in a standard agar gel bioassay.[20]

The purified HU-CSF was iodinated using a chloramine-T technique with addition of DMSO to protect the methionine residues[21] (TABLE 1). After incubation at 0°C for 15 min

TABLE 1. Iodination of HU-CSF

1. Pure HU-CSF (0.4–0.6 μg)	10 μl
2. $Na^{125}I$ (1 mCi)	2 μl
3. DMSO 1/10 in 0.2 M phosphate, pH 7.0	1.4 μl
4. Chloramine T (0.18 μg)	5 μl
5. Mix and hold at 0°C, 15 min	
6. Reaction terminated by:	
Potassium iodide (20 μg)	1.2 μl
Sodium metabisulfite (0.13 μg)	1.3 μl
7. Separate on Sephadex G 25	

the reaction was terminated by the addition of potassium iodide and sodium metabisulfite. The reaction mixture was taken up in 100 μl of Tris-PEG (polyethylene glycol 4000) buffer pH 7.5. The vial was rinsed with an additional 50 μl of buffer and the combined iodination mixture was layered on a 1 × 15-cm column of Sephadex G25. The early or high molecular weight fractions represented the iodinated CSF. The iodinated fractions were assayed for colony-stimulating activity for intervals up to 22 days after iodination; unlabeled material served as a control.

The radioimmunoassay was modeled after our previous assay for murine CSF-1[22,23] (TABLE 2). Unknown samples or the standards were mixed with phosphate buffer in the presence of EDTA and dilute first antibody or rabbit anti-CSF serum. The iodinated CSF was added at that time or addition was delayed for 24 hr in order to increase the sensitivity of the assay. All samples were held at 4°C for 48 hr at which time a second antibody, *i.e.*, sheep antirabbit IgG serum was added in order to precipitate the ^{125}I CSF-antibody complex. After 3 hr at 4°C, the precipitate was removed by centrifugation.

Initially it was necessary to determine the capability of the rabbit anti-CSF serum to precipitate the iodinated CSF. Varying dilutions of the rabbit antibody were mixed in the RIA for 48 hr. That dilution of antiserum that caused 33–40% precipitation of maximum precipitable counts was then used in subsequent assays.

Standard curves were constructed using known quantities of pure human urinary CSF.

TABLE 2. Radioimmunoassay Procedure

Step	Amount
1. Unknown or standard	0.025–0.1 ml
2. Phosphate buffer with 2.5% rabbit serum	0.6 ml
3. Na EDTA (0.1 M), pH 7.6	0.1 ml
4. ^{125}I CSF (~12,000 cpm) delayed 24 hr	0.2 ml
5. Rabbit anti-CSF serum diluted 1:30,000	0.1 ml
6. Incubate 48 hr at 4°C	
7. Sheep anti-rabbit IgG serum	0.25 ml
8. Incubate 3 hr at 4°C	
9. Centrifuge and count precipitate	

In most assays 0.3 to 150 units were used. Several different antibodies were tested as was the delayed tracer technique, *i.e.*, addition of the iodinated CSF 24 hr after preparing the assay tubes. Specificity of the radioimmunoassay was evaluated by adding a number of human and murine hormones and growth factors in place of the purified human urinary CSF. These included recombinant human GM-CSF (Immunex Corporation, Seattle, WA), recombinant human IL-1, IL-2 and IL-3 (Genzyme Corporation, Boston, MA), human follicle-stimulating hormone (FSH) (National Institutes of Health, Bethesda, MD.), luteotropic hormone (LH) (National Institutes of Health), recombinant human erythropoietin (Amgen Biologicals, Thousand Oaks, CA), murine GM-CSF (Immunex Corporation), and IL-3 (Biogen Corporation, Cambridge, MA).

Serum and urine samples were obtained from normal volunteers. These were held at room temperature, 4°C or −20°C for at least 1 day prior to assay. Most serum samples were then held at −20°C and urine samples were held at 4°C prior to assay. Some serum and urine samples were assessed repetitively over a 6-month interval.

In further studies of specificity a variety of other sera were substituted as unknowns in the RIA. These included mouse, rat, calf, lamb and rabbit serum. As a measure of potential untoward effects of human serum or urine on CSF-1 activity, a number of serum and urine samples were assayed simultaneously before and after the addition of 32 units of pure human urinary CSF.

Weekly serum and urine samples were obtained from a patient who was undergoing autologous bone marrow transplant for treatment of acute myelogenous leukemia.[24] This patient was in second remission of her disease at the time of bone marrow harvest. The conditioning therapy included high dose busulfan (4 mg/kg per day) for 4 days followed by cyclophosphamide (50 mg/kg intravenously) daily for 4 days. On the following day, cryopreserved marrow was infused. Serial observations were made of serum and urinary

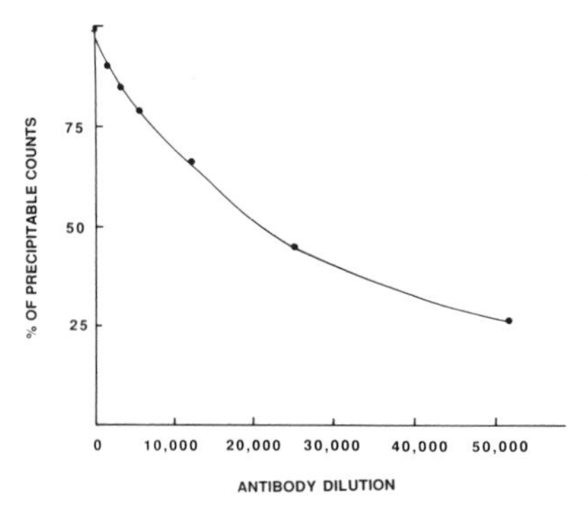

FIGURE 1. Precipitation of ^{125}I-CSF by rabbit anti-CSF serum.

CSF values prior to conditioning, during the aplastic phase and during early recovery from the transplant.

RESULTS

In a series of 14 iodinations a mean of 13.6% of the radioiodine was bound to CSF. Essentially 100% of the biologic activity was recovered after correction for a 20% preparative loss on the Sephadex column. The ^{125}I CSF retained activity for at least 3 weeks after iodination.

Antisera were obtained from 2 rabbits and tested for neutralizing activity in the agar

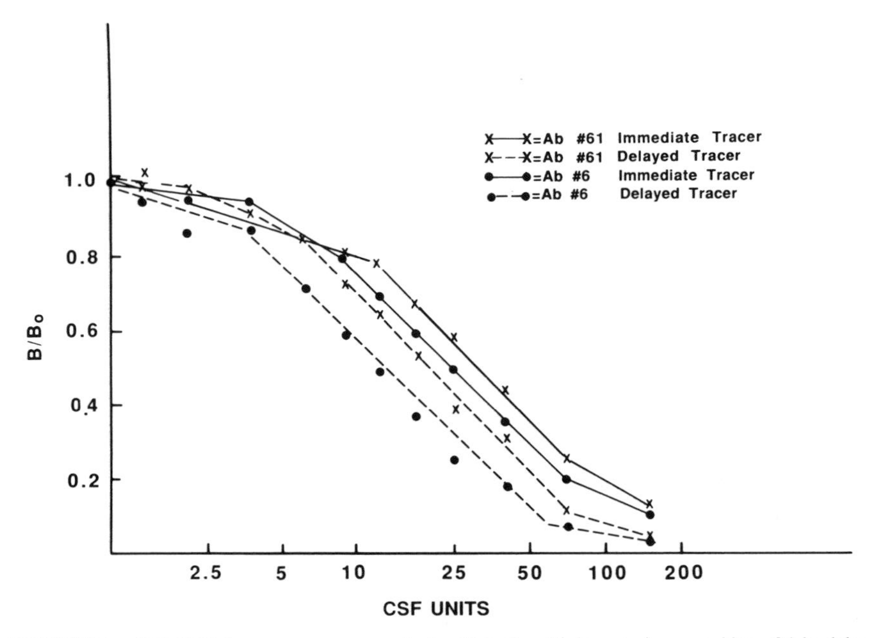

FIGURE 2. HU-CSF dose response curves in the RIA. Sensitivity was improved by a 24-hr delay in the addition of ^{125}I-CSF.

gel bioassay. Titers of anti-CSF activity ranged from 1:1500 to 1:6000. Antibody levels declined gradually over a 1-year interval. The anti-CSF titer from rabbit #6, which was used in the RIA, rose to 1:8000 after a booster immunization.

Two antisera were evaluated in the RIA. Dilutions of 1:10 to 1:100,000 were tested with the ^{125}I CSF. Both antisera yielded ~60–90% precipitation of the tracer in dilutions of 1:10 to 1:1000. The results with the most active antiserum #6 are shown in FIGURE 1. In this assay, a 1:1000 dilution yielded 90% precipitation or a bound over bound zero (B/B_0) of 0.9. This value represented the maximum precipitable counts or the 100% value. With further dilution to 1:52,800, precipitation of the tracer decreased to 28%. Based on this study a working dilution of 1:30,000 was used for this antibody.

Displacement of the ^{125}I CSF-antibody complex was assessed using known quantities of pure HU-CSF. Two antisera were evaluated both by immediate and delayed addition of the tracer (FIG. 2). The displacement curve was linear between 15 and 60 units of CSF for antiserum #61. A modest improvement in sensitivity was seen with delayed addition of the tracer. Antiserum from rabbit #6 proved more useful in the assay. Using the delayed tracer technique, the curve was linear between 5 and 50 units of CSF. In more recent studies linearity has been observed to ~2 units of CSF activity.

The specificity of the RIA was tested in several ways. There was no cross-reactivity with human IL-3 (0.008 to 2.0 ng) or with human GM-CSF (FIG. 3) including doses as high as 300 ng per sample. A variety of other hormones including human recombinant IL-1 (0.06–2 ng), IL-2 (0.04–10 ng), erythropoietin (0.01–0.2 ng), FSH (0.3–300 ng), LH (0.2–50 ng), murine GM-CSF (0.3–300 ng) and IL-3 (0.008–2 ng) were also inactive in this assay. Specificity was tested further using sera from a number of species. Virtually

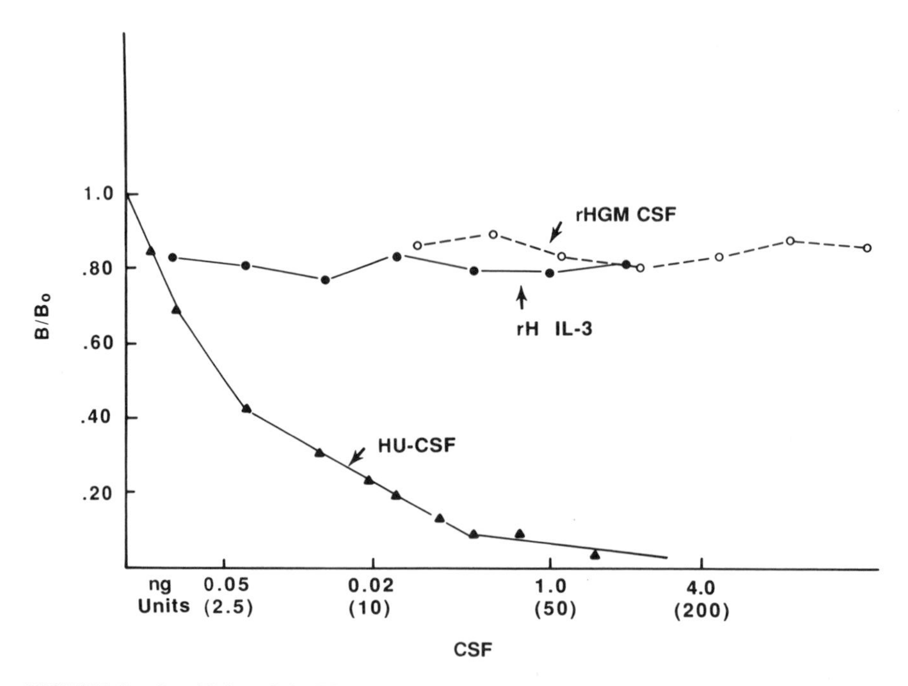

FIGURE 3. Specificity of the RIA. Neither recombinant human GM-CSF or IL-3 were cross-reactive in the assay.

no CSF activity was detected with normal mouse, rat, calf, lamb or rabbit serum (NRS) (TABLE 3). Results with 2 dose levels of NRS were inconsistent and suggested an alteration in RIA conditions. Further addition of NRS to the assay probably exceeded the capacity of the second antibody to precipitate the rabbit IgG carrier, thereby providing a false estimate of displacement by CSF.

The activity of 2 human sera was evaluated using 5–100 μl per assay. The 100-μl samples yielded a B/B_0 of less than 0.1, thereby providing inaccurate estimates of CSF activity. Dilution curves were, however, linear with sample volumes of 5–50 μl (FIG. 4).

To determine whether human serum or urine would interfere with the RIA, a known quantity of pure HU-CSF was added to each of 15 normal sera and 11 normal urine samples. This assay was conducted early in the development phase when serum samples

TABLE 3. Serum Cross-Reactivity in the Radioimmunoassay[a]

Serum	Units/Sample	
	25 μl	100 μl
Mouse	0.1	0.9
Rat	0.2	1.0
Calf	0.5	1.2
Lamb	0.6	1.6
Rabbit	-0-	13.0

[a]Values represent units/sample using the indicated volumes of serum. Activity was barely discernible except with 100 μl of rabbit serum.

yielded greater activity than at the present time. This may have resulted from assigning improper unitage to the CSF standards. Despite this reservation, addition of 32 units of pure HU-CSF was associated with reasonably accurate recovery of this additional CSF. The observed increment averaged 35–36 units with the sera and 26–29 units with the urine samples (TABLE 4).

The effect of temperature on 4 serum and urinary CSF values is shown in TABLE 5. All samples were held at room temperature, 4°C or −20°C for 24 hr after they were obtained and they were then converted to −20°C for serum and 4°C for urine. There was no degradation of CSF at room temperature and seemingly no advantage to storage at −20°C.

Further serum and urine samples were obtained from 10 normal laboratory personnel. They were stored at −20°C (serum) and 4°C (urine) as above. The results of 5 sequential assays performed over a 6-month interval is shown in TABLE 6. There was a considerable drift in serum CSF activity but no appreciable change in urinary CSF values in these repetitive assays. Studies are underway to determine the reason for this drift by systematically reexamining the various assay conditions. Although 24-hr urine samples were not collected from the normal subjects, an estimated daily output of ~1000 ml would yield total urinary CSF values of ~85,000 to 104,000 units.

To determine whether serum and urinary CSF values might increase with bone marrow aplasia, serial samples were obtained from a patient undergoing an autologous bone

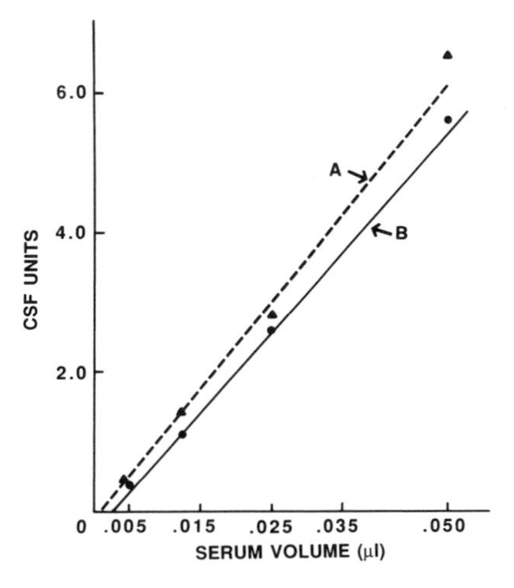

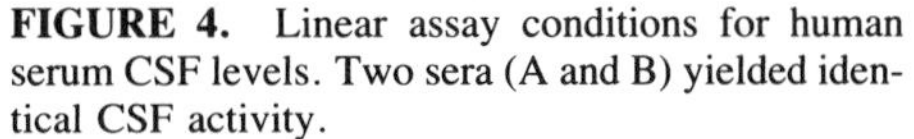
FIGURE 4. Linear assay conditions for human serum CSF levels. Two sera (A and B) yielded identical CSF activity.

TABLE 4. Recovery of Pure HU-CSF after Addition to Serum and Urine Samples[a]

	Units/ml		
Source	Baseline Activity	Sample + HU-CSF	CSF Increment
Serum 12.5 μl	7.4 ± 0.4	42.2 ± 0.7	34.8 ± 0.7
25 μl	12.3 ± 0.9	48.3 ± 1.9	36.0 ± 2.0
Urine 25 μl	9.3 ± 1.8	37.8 ± 2.1	28.5 ± 1.3
50 μl	16.8 ± 3.6	42.7 ± 2.4	25.8 ± 2.8

[a]Fifteen normal sera and 11 urine samples were assayed alone or admixed with 100 μl of pure HU-CSF. RIA of HU-CSF alone yielded 32 units/100μl. Values are mean ± 1 SE.

marrow transplant for treatment of acute myelogenous leukemia (TABLE 7). All samples were evaluated in the same assay. The total leukocyte count fell to 0 after the 8-day conditioning regimen with busulfan and cyclophosphamide. The baseline serum CSF values ranged from 204 to 282 u/ml. CSF activity rose by day 10 after the transplant reaching a peak value of 988 u/ml during the aplasia. Serum CSF values returned toward baseline as the leukocyte count recovered. The 24-hr urinary CSF ranged from 24,500 to 82,700 units/day in the baseline period. This rose greater than 10-fold to 727,000–1,251,000 units/day during the aplastic phase. Urinary CSF values tended to parallel the serum activity by decreasing in concert with the rising leukocyte count.

DISCUSSION

These studies show that purified human urinary CSF (HU-CSF) or CSF-1 can be measured effectively by a sensitive and specific radioimmunoassay. This depended on several key factors, namely, that HU-CSF elicited high titer antibodies and that the CSF could be radioiodinated without loss of biologic or immunologic reactivity.

The standard technique for measurement of CSF is by the *in vitro* bioassay.[20] Generally concentration of urine samples is sufficient to detect this form of CSF in the murine assay.[25] However, crude urinary concentrates contain inhibitors to colony formation which may lead to underestimates of activity.[26] Using a technique with adsorption to and elution from calcium phosphate gels, Stanley *et al.*[27] described a cumbersome technique for quantification of urinary CSF. Total normal 24-hr urinary excretion of CSF was estimated as 93,700 U/day, which compares favorably with the values estimated in this study.

In contrast, serum CSF has been difficult to measure with reliability. Most sera contain high molecular weight lipoprotein inhibitors that decrease colony formation.[4,28] For this reason there are few indepth studies of serum CSF. Additional problems with *in vitro* bioassays are the time required (7–14 days), the relative insensitivity to low levels of CSF and the relative contribution of other types of CSF to colony growth. Bioassays

TABLE 5. Effect of Temperature on CSF Activity

	Units/ml[a]	
Temperature	Serum	Urine
23°C	101 ± 11	88 ± 24
4°C	100 ± 13	113 ± 38
−20°C	88 ± 10	85 ± 24

[a]Mean ± 1 SE, 4 samples.

TABLE 6. Normal Serum and Urine CSF Values

RIA	Units/ml[a] Serum	Urine
A	91 ± 9	89 ± 17
B	112 ± 7	97 ± 17
C	138 ± 9	105 ± 25
D	93 ± 10	72 ± 16
E	132 ± 13	98 ± 24

[a]Mean + 1 SE; 10 samples.

require 7 days with murine target cells and 10–14 days with human colony assays. These assays are less sensitive with loss of linearity below ~20 colonies or units per 0.1-ml sample. The RIA described herein can be completed in 3 working days with sensitivity to 2–5 u/sample. Although the effect of inhibitors has not been rigorously excluded, the recovery studies wherein serum and urine samples were spiked with a known quantity of CSF ruled against significant inhibitory effects in the RIA. Moreover, a detailed study with the murine radioimmunoassay for CSF-1 indicated no interference by either serum- or tissue-derived inhibitors.[23]

The radioimmunoassay described here appears to be specific for HU-CSF or human CSF-1. A variety of other hormones and hemopoietic growth factors were not crossreactive in the assay. Moreover, no appreciable CSF was detected in sera from other species. Murine serum is known to contain ~500–1000 U of CSF/ml in the murine RIA but was essentially inactive in the present assay. This would appear to confirm a high degree of specificity in these assay systems.

Previously, Stanley[29] and we[22,23] have described several radioimmunoassays for murine CSF-1. These assays showed the same degree of specificity and virtually the same dose response curves as the present RIA for human CSF-1. Das *et al.*[30] described a similar approach to the assay of human CSF-1. In this technique, partially purified CSF was radioiodinated and bound to a macrophage cell line with receptors for CSF. The material that was eluted at low pH was presumed to represent pure CSF and was used in conjunction with a low titer rabbit antiserum to develop the radioimmunoassay. This assay was

TABLE 7. Changes in Leukocyte Values, Serum and Urine CSF Values after Bone Marrow Transplantation[a]

Days after BMT	Leukocytes/μl	Neutrophils/μl	CSF Serum	Urine
−8	3,800	2200	204	24,583
−6	5,300	4450	282	82,728
0	0.062	ND	230	76,800
7	-0-	-0-	268	727,533
14	0.002	-0-	465	736,512
21	0.070	ND	505	805,500
28	0.040	ND	988	1,250,930
35	0.300	ND	338	483,700
42	0.180	ND	368	769,500
49	1,100	264	257	383,800
56	2,400	456	260	313,484

[a]Values represent serial leukocyte and neutrophil counts, serum CSF (units/ml) and 24 hr urine CSF excretion. Observations ranged from 8 days prior to BMT to 56 days after BMT.

useful in determining that there are different types of CSF; however, no details were provided regarding the sensitivity or specificity for assay of human serum and urine samples. The present assay extends those previous observations using a source of pure CSF and high titer antibody. As shown, it is specific for human CSF-1 and is useful for assessment of serum and urinary CSF values. Further use of this assay will be expected to aid in our understanding of the role of CSF-1 in human hemopoietic cell production. Use of such a specific radioimmunoassay will allow further definition of the kinetics of recombinant CSF molecules as they are used in clinical medicine.

SUMMARY

Purified human urinary CSF-1 was used for production of polyclonal CSF antibodies in rabbits. The purified CSF was iodinated by a modified chloramine-T technique with retention of biologic activity. Dilutions of anti-CSF were reacted with 15,000 cpm of ^{125}I-CSF in EDTA-phosphate buffer for 48 hr. Sheep antirabbit serum was added for 3 hr to precipitate the tracer-anti-CSF complex. A 1:1000 dilution of anti-CSF caused 60–90% precipitation of tracer; optimal conditions were observed with a 1:30,000 dilution. Linear displacement curves were obtained with 2–50 U of pure CSF-1. Related hormones did not cross-react in the assay; no displacement was seen with human GM-CSF, IL-1, IL-2, IL-3, EP, LH or FSH. Reactivity was also not observed with murine GM-CSF or IL-3. Ten normal human sera yielded CSF values of 91–138 U/ml in 5 assays. Urine values were 72–105 U/ml. When 32 U of pure CSF-1 was added to normal serum and urine samples, quantitative recovery was observed. Serial assays revealed a rise in serum and urinary CSF during marrow aplasia in a patient undergoing autologous BMT; CSF values returned to normal during the recovery phase. This sensitive and specific radioimmunoassay should prove useful in the further study of CSF-1 responses *in vivo*.

ACKNOWLEDGMENTS

The authors gratefully acknowledge the excellent technical assistance of Mrs. Darlene DePasquale, Mrs. Florence Boegel and Mrs. Donna Hodge. The recombinant human and murine GM-CSF were kindly provided by Immunex Corporation, Seattle, WA. Murine IL-3 was also kindly provided by Biogen Corporation, Cambridge, MA.

REFERENCES

1. Wing, E. J. & R. K. Shadduck. 1985. Colony stimulating factor. *In* Biological Response Modifiers. P. F. Torrence, Ed.: 219–243. Academic Press. New York, NY.
2. Metcalf, D. 1985. The granulocyte-macrophage colony stimulating factors. Science, **229:** 16–22.
3. Shadduck, R. K. & A. Waheed. 1984. The role of colony stimulating factor in the regulation of granulopoiesis. Blood Cells **10:** 163–176.
4. Zidar, B. L. & R. K. Shadduck. 1978. Serum colony stimulating and colony inhibitory activity in response to neutropenia. J. Lab. Clin. Med. **91:** 584–591.
5. Metcalf, D. 1971. Inhibition of bone marrow colony formation in vitro by dialysable products of normal and neoplastic haemopoietic cells. Aust. J. Exp. Biol. Med. Sci. **49:** 351–363.
6. Shadduck, R. K. 1976. Leukocyte colony stimulating factor and inhibitor release. J. Lab. Clin. Med. **87:** 1041–1049.

7. Waheed, A. & R. K. Shadduck. 1979. Purification and properties of L-cell derived colony-stimulating factor. J. Lab. Clin. Med. **94:** 180–194.
8. Stanley, E. R. & P. M. Heard. 1977. Factors regulating macrophage production and growth: Purification and properties of the colony-stimulating factor from medium conditioned by mouse L cells. J. Biol. Chem. **252:** 4305–4312.
9. Ben-Avram, C. M., J. E. Shively, R. K. Shadduck, A. Waheed, T. Rajavashisth & A. J. Lusis. 1985. Amino-terminal amino acid sequence of murine colony-stimulating factor 1. Proc. Natl. Acad. Sci. USA **82:** 4486–4489.
10. Burgess, A. W., J. Camakaris & D. Metcalf. 1977. Purification and properties of colony-stimulating factor from mouse lung-conditioned medium. J. Biol. Chem. **252:** 1998–2003.
11. Sparrow, L. G., D. Metcalf, M. W. Hunkapillar, L. E. Hood & A. W. Burgess. 1985. Purification and partial amino acid sequence of asialo murine granulocyte-macrophage colony-stimulating factor. Proc. Natl. Acad. Sci. USA **82:** 292–297.
12. Nicola, N. A., D. Metcalf, M. Matsumato & G. R. Johnson. 1983. Purification of a factor inducing differentiation in murine myelomonocytic leukemia cells: Identification as a granulocyte colony-stimulating factor (G-CSF). J. Biol. Chem. **258:** 9019–9023.
13. Ihle, J. N., J. Keller, L. Henderson, F. Klein & E. Palaszynski. 1982. Procedure for the purification of interleukin-3 to homogeneity. J. Immunol. **129:** 2431–2436.
14. Rajavashisth, T. B., R. Eng, R. K. Shadduck, A. Waheed, C. M. Ben-Avram, J. E. Shively & A. J. Lusis. 1987. Cloning and tissue-specific expression of mouse macrophage colony-stimulating factor. Proc. Natl. Acad. Sci. USA **84:** 1157–1161.
15. Tsuchiya, M., S. Asano, Y. Kaziro & A. Nagata. 1986. Isolation and characterization of the cDNA for murine granulocyte colony-stimulating factor. Proc. Natl. Acad. Sci. USA **83:** 7633–7637.
16. Gough, N. M., J. Gough, D. Metcalf, A. Kelso, D. Grail, N. A. Nicola, A. W. Burgess & A. R. Dunn. 1984. Molecular cloning of cDNA encoding a murine hematopoietic regulator, granulocyte-macrophage colony-stimulating factor. Nature **309:** 763–767.
17. Fung, M. C., A. J. Hapel, S. Ymer, D. R. Cohen, R. M. Johnson, H. D. Campbell & I. G. Young. 1984. Molecular cloning of cDNA for murine Interleukin-3. Nature **307:** 233–237.
18. Shadduck, R. K. & D. Metcalf. 1975. Preparation and neutralization characteristics of an anti-CSF antibody. J. Cell. Physiol. **86:** 247–252.
19. Waheed, A. & R. K. Shadduck. Purification of human urine colony stimulating factor by affinity chromatography. Exp. Hematol. **17.** In press.
20. Shadduck, R. K. & N. G. Nagabhushanam. 1971. Granulocyte colony stimulating factor. I. Response to acute granulocytopenia. Blood **38:** 559–568.
21. Shadduck, R. K., A. Waheed, A. Porcellini, V. Rizzoli & G. Pigoli. 1979. Physiologic distribution of colony stimulating factor in vivo. Blood **54:** 894–905.
22. Shadduck, R. K. & A. Waheed. 1979. Development of a radioimmunoassay for colony stimulating factor. Blood Cells **5:** 421–434.
23. Boegel, F., A. Waheed & R. K. Shadduck. 1981. Evaluation of radioimmunoassay and in vitro colony assay techniques for determination of colony stimulating factor and inhibitory activity in murine serum and tissue. Blood **58:** 1141–1147.
24. Rosenfeld, C. S., K. F. Mangan, R. K. Shadduck & O. M. Colvin. 1987. Hemopoietic recovery after bone marrow transplantation with 4-hydroperoxycyclophosphamide purged marrows. *In* Autologous Bone Marrow Transplantation. K. A. Dicke, G. Spitzer & S. Jagannath, Eds.: 119–123. The University of Texas M.D. Anderson Hospital and Tumor Institute. Houston, TX.
25. Robinson, W. A., E. R. Stanley & D. Metcalf. 1969. Stimulation of bone marrow colony growth in vitro by human urine. Blood **33:** 396–399.
26. Stanley, E. R. & D. Metcalf. 1969. Partial purification and some properties of the factor in normal and leukemic human urine stimulating mouse bone marrow colony growth in vitro. Aust. J. Exp. Biol. Med. Sci. **47:** 467–483.
27. Stanley, E. R., D. Metcalf, J. S. Maritz & G. F. Yeo. 1972. Standardized bioassay for bone marrow colony stimulating factor in human urine: Levels in normal man. J. Lab. Clin. Med. **79:** 657–668.
28. Chan, S. H., D. Metcalf & E. R. Stanley. 1971. Stimulation and inhibition by normal

human serum of colony formation in vitro by bone marrow cells. Br. J. Haematol. **20:** 329–341.

29. STANLEY, E. R. 1979. Colony-stimulating factor (CSF) radioimmunoassay: Detection of a CSF subclass stimulating macrophage production. Proc. Natl. Acad. Sci. USA **76:** 2969–2973.
30. DAS, S. K., E. R. STANLEY, L. J. GUILBERT & L. W. FORMAN. 1981. Human colony-stimulating factor (CSF) radioimmunoassay: Resolution of three subclasses of human colony-stimulating factors. Blood **58:** 630–641.

Molecular Characterization of Colony-Stimulating Factors and Their Receptors: Human Interleukin-3

DAVID L. URDAL, VIRGINIA PRICE,
HELMUT M. SASSENFELD, DAVID COSMAN, STEVEN GILLIS,
AND LINDA S. PARK

Immunex Corporation
51 University Street
Seattle, Washington 98101

INTRODUCTION

The proliferation and differentiation of hematopoietic cells is controlled by a number of distinct protein hormones. The proteins which dictate the development of cells of the granulocyte and macrophage cell lineages are called the colony-stimulating factors (CSF) and are comprised of a family of proteins that have been identified by their capacity to promote the growth *in vitro* of colonies of cells having distinct phenotypes.[1,2] Macrophage-CSF (M-CSF, CSF-1) for example, stimulates the growth of macrophage colonies. In contrast, granulocyte-CSF (G-CSF) influences the differentiation of precursor cells into granulocytes, while the presence of granulocyte-macrophage-CSF (GM-CSF) in culture gives rise to the appearance of colonies composed of both granulocytes and macrophages. Interleukin-3 (multi-CSF) appears to act on hematopoietic precursor cells common to several lineages since it can influence the appearance of colonies containing cells of several distinct phenotypes.

The number of proteins involved in the control of hematopoiesis has grown with the recent demonstration that other proteins, among them interleukin-1 (IL-1),[3–5] interleukin-4 (IL-4, B cell stimulating factor),[6] interleukin-5 (IL-5, eosinophil-differentiating factor)[7] and interleukin-6 (IL-6, B cell-stimulating factor-2, interferon β_2)[8] are also active in regulating hematopoiesis either directly, by potentiating the activity of the colony-stimulating factors, or indirectly, by inducing the production of colony-stimulating activity by other cells.

Like other polypeptide hormones, the colony stimulating factors interact with cells through specific glycoprotein receptors found on the cell surface. The availability of purified natural or recombinant colony-stimulating factors has made the study of these molecules possible.[9] Consequently, our understanding of the interactions between the ever increasing number of hormones, receptors, and cells involved in the regulation of hematopoiesis is advancing rapidly.

This understanding is currently at a stage where the molecules involved are being cataloged. All of the protein factors mentioned above have been cloned and recombinant proteins are available for study and application to the clinic. Receptor studies are one short step behind. The human CSF most recently cloned was interleukin-3.[10] Murine IL-3 was cloned earlier,[11,12] but identification of the human homologue remained elusive until just recently. The gene encoding human interleukin-3 appears to be polymorphic, with at least two alleles being present in the population. Expression of recombinant human interleukin-3 (rhu IL-3) in yeast has made possible the purification of rhu IL-3 in amounts sufficient

```
                                 C                                   GAT
     GCT CCC ATG ACC CAG ACG ACG TCC TTG AAG ACC AGC TGG GTT AAC TGC TCT AAC ATG ATC
1    ---------+---------+---------+---------+---------+---------+   60
     CGA GGG TAC TGG GTC TGC TGC AGG AAC TTC TGG TCG ACC CAA TTG ACG AGA TTG TAC TAG

     Ala Pro Met Thr Gln Thr Thr Ser Leu Lys Thr Ser Trp Val Asn Cys Ser Asn Met Ile
                                 Pro                             Asp

     GAT GAA ATT ATA ACA CAC TTA AAG CAG CCA CCT TTG CCT TTG CTG GAC TTC AAC AAC CTC
61   ---------+---------+---------+---------+---------+---------+   120
     CTA CTT TAA TAT TGT GTG AAT TTC GTC GGT GGA AAC GGA AAC GAC CTG AAG TTG TTG GAG

     Asp Glu Ile Ile Thr His Leu Lys Gln Pro Pro Leu Pro Leu Leu Asp Phe Asn Asn Leu

     AAT GGG GAA GAC CAA GAC ATT CTG ATG GAA AAT AAC CTT CGA AGG CCA AAC CTG GAG GCA
121  ---------+---------+---------+---------+---------+---------+   180
     TTA CCC CTT CTG GTT CTG TAA GAC TAC CTT TTA TTG GAA GCT TCC GGT TTG GAC CTC CGT

     Asn Gly Glu Asp Gln Asp Ile Leu Met Glu Asn Asn Leu Arg Arg Pro Asn Leu Glu Ala

                                              G
     TTC AAC AGG GCT GTC AAG AGT TTA CAG AAC GCA TCA GCA ATT GAG AGC ATT CTT AAA AAT
181  ---------+---------+---------+---------+---------+---------+   240
     AAG TTG TCC CGA CAG TTC TCA AAT GTC TTG CGT AGT CGT TAA CTC TCG TAA GAA TTT TTA

     Phe Asn Arg Ala Val Lys Ser Leu Gln Asn Ala Ser Ala Ile Glu Ser Ile Leu Lys Asn
                                         Asp

     CTC CTG CCA TGT CTG CCC CTG GCC ACG GCC GCA CCC ACG CGA CAT CCA ATC CAT ATC AAG
241  ---------+---------+---------+---------+---------+---------+   300
     GAG GAC GGT ACA GAC GGG GAC CGG TGC CGG CGT GGG TGC GCT GTA GGT TAG GTA TAG TTC

     Leu Leu Pro Cys Leu Pro Leu Ala Thr Ala Ala Pro Thr Arg His Pro Ile His Ile Lys

     GAC GGT GAC TGG AAT GAA TTC CGG AGG AAA CTG ACG TTC TAT CTG AAA ACC CTT GAG AAT
301  ---------+---------+---------+---------+---------+---------+   360
     CTG CCA CTG ACC TTA CTT AAG GCC TCC TTT GAC TGC AAG ATA GAC TTT TGG GAA CTC TTA

     Asp Gly Asp Trp Asn Glu Phe Arg Arg Lys Leu Thr Phe Tyr Leu Lys Thr Leu Glu Asn

     GCG CAG GCT CAA CAG ACG ACT TTG AGC CTC GCG ATC TTT TAG
361  ---------+---------+---------+---------+--   402
     CGC GTC CGA GTT GTC TGC TGA AAC TCG GAG CGC TAG AAA ATC

     Ala Gln Ala Gln Gln Thr Thr Leu Ser Leu Ala Ile Phe End
```

FIGURE 1. Sequence of the complementary DNA encoding human interleukin-3. Nucleotides are numbered from the mature NH_2-terminus of the molecule. The C at position 22 was found in four cDNA clones isolated from a library prepared from human peripheral blood T cells and results in the coding of proline at residue 8 of the mature protein. The codons at residues 15 and 70 were changed such that Asp could be encoded in place of Asn, precluding potential glycosylation at these positions following expression in yeast.

to contemplate clinical trials and to initiate studies on the characterization of the receptor specific for this growth factor.

The Gene Encoding hu IL-3 is Polymorphic

Human peripheral blood T cells were purified and stimulated *in vitro* with PHA and PMA. Messenger RNA was extracted from the cells and cDNA was prepared from the polyA plus messenger RNA. The cDNA was packaged into λgT10, and oligonucleotides complementary to both 5' and 3' regions of the published sequence for human IL-3[10] were then used to identify clones to which they hybridized. Eleven clones were isolated and sequencing of four of these cDNAs revealed a C at position 22 instead of the T reported in the published sequence. This base change results in an amino acid change from serine to proline at residue eight of the molecule (FIG. 1).

In order to test the relative abundance of the pro^8 versus the ser^8 gene or, alternatively, to test if it represented a true gene and not an artifact of cDNA construction, we examined the DNA isolated from the peripheral blood cells of 13 individuals for the presence of the two DNAs (FIG. 2). The polymerase chain reaction[13] was used to amplify the DNA containing the segment of the IL-3 gene around position 8. Each of the thirteen individuals examined (eight are depicted in FIG. 2) contained DNA encoding proline at position 8 of the interleukin-3 gene. Three were also positive for serine at this position, suggesting that the gene encoding human IL-3 is polymorphic and that of the two alleles identified, the proline-8 gene is the most prevalent (in this limited survey).

Yeast Expression of rhu IL-3

The cDNA encoding IL-3 was engineered into a yeast expression vector which places the DNA under transcriptional control of the alcohol dehydrogenase 2 promoter. This promoter is repressed when the yeast are fermented in the presence of glucose as a carbon source and derepressed when glucose is depleted from the medium. The secretion of the protein is realized by the juxtaposition of the cDNA encoding rhu IL-3 to DNA encoding the leader sequence of the yeast pheromone, α-factor (FIG. 3).

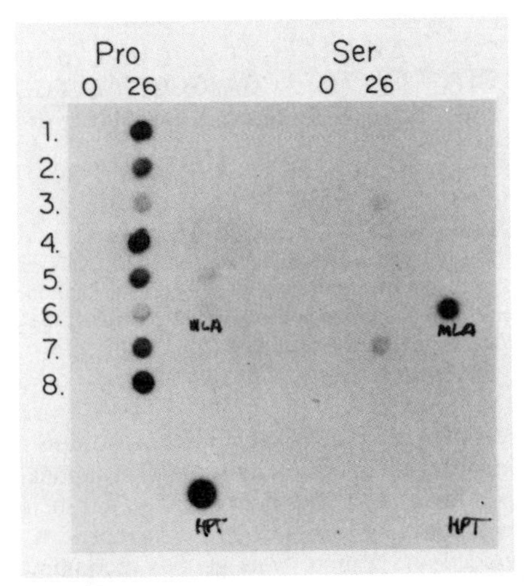

FIGURE 2. Prevalence of IL-3 ser^8 and IL-3 pro^8 genes in DNA from different individuals. DNA was extracted from the peripheral blood leukocytes of eight individuals and subjected to 0 or 26 cycles of the polymerase chain reaction as described previously,[13] using synthetic oligonucleotides complementary to regions 5' and 3' of residue 8. Radiolabeled synthetic oligonucleotides complementary to the ser^8 and pro^8 forms of IL-3 were then used to probe the amplified DNA that had been dotted to nitrocellulose. Positive controls included DNA extracted from MLA 144 cells which is known to contain the gene encoding ser^8 gibbon ape IL-3,[10] and human peripheral T cell DNA which was the source of the pro^8 human IL-3 gene described in FIGURE 1. All eight individuals were positive for the pro^8 IL-3 gene, two were also positive for the ser^8 gene indicating that both genes are represented in the population.

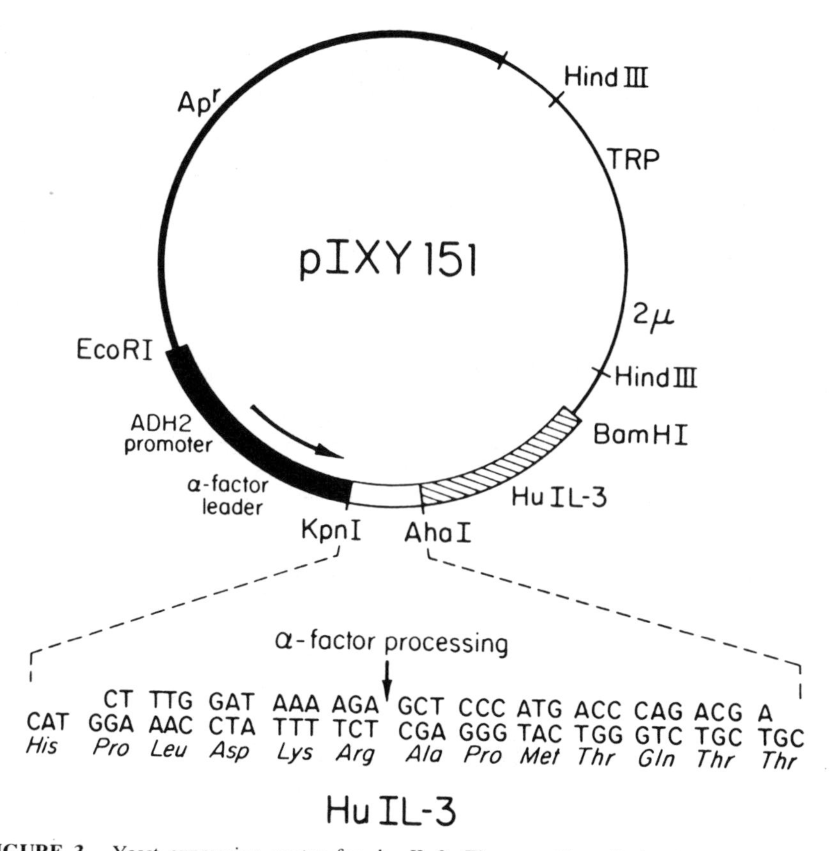

FIGURE 3. Yeast expression vector for rhu IL-3. The yeast/*E. coli* shuttle vector pIXY151 contains sequences from pBR322 (*thick lines*) that allow selection (Ap^r) and replication in *E. coli* and the yeast TRP1 gene and 2 μ origin of replication for selection and autonomous replication in a trpl yeast strain (*thin line*). Expression of foreign genes is under control of the glucose-repressible ADH2 promoter and α-factor leader as previously described.[14]

Expression of proteins in yeast can result in the addition of carbohydrate to the recombinant protein. In order to minimize this posttranslational modification, the asparagine residues at positions 15 and 70 which could signal the addition of sugar, were changed to aspartic acid residues by site-directed mutagenesis. As a result, glycosylation of the molecule is precluded and a more homogenous recombinant protein is obtained. These changes did not affect the specific activity of IL-3 (TABLE 1).

Purification of rhu IL-3

Recombinant hu IL-3 was purified from the yeast beer by reversed-phase HPLC as described previously for other recombinant proteins expressed in yeast.[14] Briefly, the yeast beer was filtered and pumped directly onto a C-4 reversed-phase column (Vydac, Separations Group, Hespernia, CA). Bound rhu IL-3 was eluted with a gradient of acetonitrile in 0.1% trifluoracetic-acid (TFA) and fractions containing rhu IL-3 were pooled, diluted in 0.1% TFA and pumped onto a C-18 reversed-phase column (Vydac). Bound rhu IL-3 was again eluted with a gradient of acetonitrile in 0.1% TFA. Fractions containing rhu IL-3 were pooled and the pH was adjusted to 3.6 by the addition of β-alanine (0.05 M final concentration). This pool was then applied to an S-Sepharose column (Pharmacia) equilibrated in 0.05 M β-alanine pH 4.0. The column was washed and then the rhu IL-3 was eluted with trishydroxymethylaminomethane pH 7.4. FIGURES 4 and 5 display the results of the analysis of the purified protein by SDS polyacrylamide gel electrophoresis (FIG. 4) and by analytical reversed-phase HPLC (FIG. 5).

Characterization of Human IL-3 Receptors

Interleukin-3 receptors on murine cells have been well characterized previously,[15,16] but the inability of murine IL-3 to bind to human cells has precluded the analysis of human

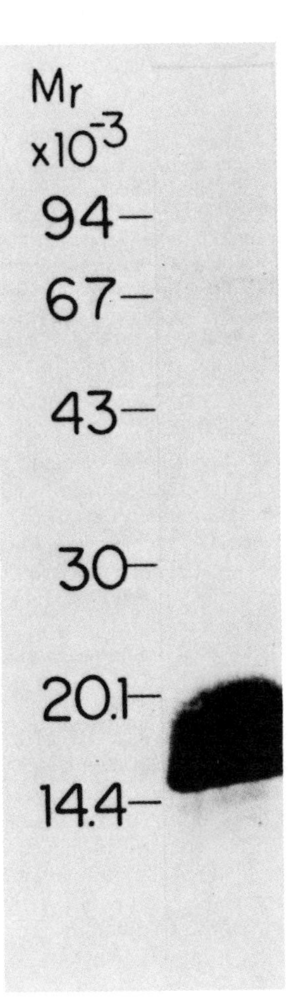

FIGURE 4. Analysis of purified recombinant human IL-3 by SDS polyacrylamide gel electrophoresis and silver staining. Recombinant human IL-3 was purified from yeast beer as described in the text and analyzed by SDS-PAGE according to the procedure described by Laemmli.[19] Silver staining was performed as previously described.[20]

IL-3 receptors. The availability of purified recombinant human IL-3 has made the study of the human IL-3 receptor possible.

The cDNA sequence depicted in FIGURE 1 reveals that only a single tyrosine is present in the human IL-3 molecule. In order to facilitate the radiolabeling of rhu IL-3, advantage was taken of a procedure in use in the laboratory to 'tag' proteins with a small immunogenic octapeptide. Briefly, interleukin-3 was engineered to contain this octapeptide at the NH_2-terminus. When expressed in the yeast expression system described above, rhu IL-3 is secreted into the yeast beer and can be affinity purified by virtue of the presence of the antigenic peptide at the NH_2-terminus. The octapeptide used for this purpose happens to contain a tyrosine residue, the presence of which allows the radiolabeling of this form of rhu IL-3 to a higher specific activity using the enzymobead procedure as previously described.[17]

Binding studies were performed using ^{125}I-labeled rhu IL-3 and monocytes purified from human peripheral blood. Free (unbound) and cell bound counts are converted to molecules of IL-3 bound and the parameters summarized in TABLE 2 readily determined. For example, in order to establish the conditions required to achieve equilibrium binding of ^{125}I labeled rhu IL-3, association experiments were performed in which ^{125}I-rhu IL-3 at different concentrations was incubated with monocytes at 37°C or at 4°C. The level of

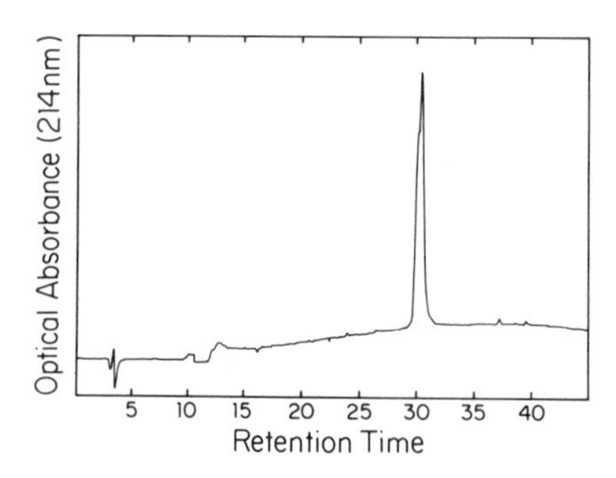

FIGURE 5. Analytical reversed-phase high performance liquid chromatography of purified rhu IL-3. Recombinant human IL-3 (20 μg) was injected onto a C4 reversed-phase HPLC column (Vydac, Separation Group, Hespernia, CA) and eluted with a gradient of acetonitrile in trifluoroacetic acid (0.1%) at a flow rate of 1 ml/min and rate of change of 2% per minute using equipment previously described.[20]

^{125}I-rhu IL-3 binding achieved at equilibrium and the rate at which maximum binding was achieved were dependent on the initial concentration of ^{125}I-rhu IL-3 in the medium. Analysis of these data allow the prediction of the association rate constant k_1, and dissociation rate constant k_{-1} depicted in TABLE 2. Binding at 37°C achieved equilibrium within 30 minutes while that at 4°C took over eight hours, making rhu IL-3 similar in binding to rhu GM-CSF, G-CSF and M-CSF, each of which binds much more rapidly at 37°C than at 4°C.

The results of equilibrium binding experiments for ^{125}I-rhu IL-3 binding to monocytes at 37°C is depicted in FIGURE 6. A display of the data in the Scatchard coordinate system yielded a straight line, suggesting that a single class of binding sites for ^{125}I-rhu IL-3 exists on these cells. The apparent K_a was calculated to be $8.1 \pm 0.9 \times 10^9$ M^{-1} with an average of 168 ± 4 binding sites detected on the surface of human monocytes.

Unlabeled rhu IL-3 was used to establish a quantitative inhibition curve from which the binding affinity (K_I) of the unlabeled rhu IL-3 could be determined. This value was $1.68 \pm 0.22 \times 10^{10}$ M^{-1} and was very close to the affinity constant estimated from equilibrium binding studies, suggesting that little loss of binding activity was realized upon radiolabeling the rhu IL-3. Furthermore, the octapeptide 'tag' at the NH_2-terminus of radiolabeled IL-3 does not appear to affect the interaction of rhu IL-3 with its receptor.

TABLE 1. rhu IL-3 Biological Activity[a]

IL-3 Gene	Specific Activity
Wild type rhu IL-3	7,500 U/μg
D^{15} D^{70} rhu IL-3[b]	8,300 U/μg

[a]Recombinant human IL-3 was purified from yeast beer and tested for the capacity to stimulate the proliferation of human bone marrow cells. One unit is defined as the amount of rhu IL-3 required to stimulate half maximal incorporation of ^{3}H-thymidine into bone marrow cultures. Protein concentrations were determined by amino acid analysis of the purified protein.

[b]$D^{15}D^{70}$ rhu IL-3 represents the rhu IL-3 in which the asparagine at positions 15 and 70 was changed to aspartic acid.

Inhibition of ^{125}I-rhu IL-3 binding was also used to establish the specificity with which rhu IL-3 bound to its receptor. The results depicted in TABLE 3 suggest that of a panel of purified cytokines and polypeptide hormones, only rhu IL-3 had the capacity to compete with ^{125}I-rhu IL-3 for binding to monocytes.

Interestingly, rhu GM-CSF inhibited the binding of ^{125}I-rhu IL-3 to monocytes to a small, but reproducible extent suggesting either a partial cross-reactivity of IL-3 and GM-CSF for the same receptor on monocytes or the down modulation of IL-3 receptors by GM-CSF even in the presence of sodium azide which is present in the binding media to minimize internalization of receptors. The partial inhibition of ^{125}I-rhu IL-3 binding by GM-CSF was not seen on the pre B cell line, JM-1, which unlike monocytes, does not bind GM-CSF.

Cell Surface Display of IL-3 Receptors

Radiolabeled rhu IL-3 has been used to determine the tissue distribution of IL-3 receptors. TABLE 4 summarizes the results of experiments in which a number of human cell lines and primary human cells, including leukemia cells, were examined for the capacity to bind ^{125}I-rhu IL-3. Radiolabeled rhu IL-3 binding, like that of the other CSFs (G-CSF and GM-CSF), is very restricted in the number of cell types to which it binds. Unexpectedly, none of four leukemia cell lines having a monocytic or myelogenous phenotype bound ^{125}I-rhu IL-3. In contrast, three cell lines of pre B cell phenotype bound IL-3. In addition, monocytes from normal individuals and blast cells collected from patients with acute myeloblastic leukemia bound ^{125}I-rhu IL-3, while cells isolated from patients with acute lymphoblastic leukemia failed to bind the radiolabeled protein.

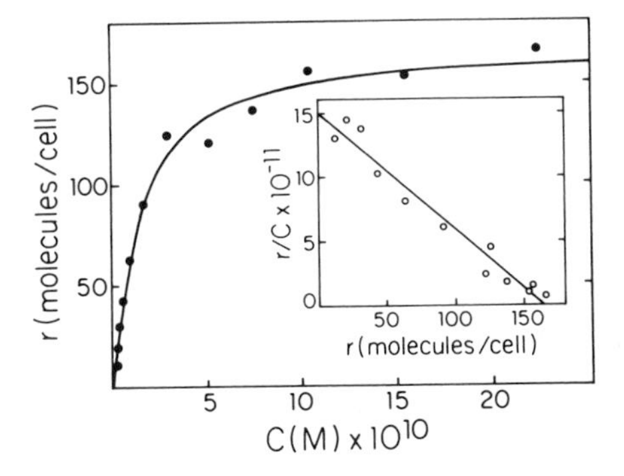

FIGURE 6. Equilibrium binding of ^{125} I-labeled rhu IL-3. Human monocytes (3.3×10^7 cell/ml) were incubated with ^{125}I-rhu IL-3 for one hour at 37°C. Data are corrected for nonspecific binding (2.2×10^{11} molecules/cell/M) measured in the presence of a 200-fold molar excess of unlabeled IL-3. The *inset* shows a Scatchard representation of specific binding.

TABLE 2. Summary of rhu IL-3 Receptor Binding Parameters at 37°C[a]

Parameter	Value
k_1 (M^{-1} min^{-1})	1.8×10^8
k_{-1} (min^{-1})	6×10^{-2}
K_a (M^{-1})	1×10^{10}
K_I (M^{-1})	1.7×10^{10}
Specific Activity (cpm/mmole)	5×10^{15}

[a]Recombinant human IL-3 was radiolabeled with ^{125}I by the enzymobead procedure as previously described.[17] Association (k_1), dissociation (k_{-1}), equilibrium binding (K_a) and inhibition (K_I) constants were determined as previously described.[17]

Molecular Characterization of Cytokine Receptors

The molecular characterization of cytokine receptors is hampered by the low abundance of these molecules on cells. Thus, further characterization of the human IL-3 receptor may have to await the discovery of a cell line which expresses substantially more receptor molecules than those described in TABLE 4.

The receptor for murine IL-3 has, in contrast, been more completely characterized[15,16] due to the availability of cell lines dependent on murine IL-3 that express a greater abundance of receptors than found on human cells. Chemical cross-linking studies have been performed in the murine system in which ^{125}I murine IL-3 is bound to cells and then

TABLE 3. Specificity of ^{125}I-IL-3 Binding to Human Monocytes[a]

Inhibitor	^{125}I-IL-3 Bound Molecules/Cell
None	120.2 ± .4
hu IL-3	25.9 ± .3
hu GM-CSF	77.1 ± 1.7
hu G-CSF	112.7 ± 3.4
hu CSF-1	119.9 ± 1.5
hu IL-4	98.7 ± 8.1
hu IL-1α	108.0 ± 3.5
hu IL-1β	115.7 ± 2.6
hu IL-2	107.6 ± .1
EGF	98.8 ± 12.3
FGF	102.9 ± 14.1
NGF	117.6 ± 4.1
HLH	109.0 ± .2
HGH	126.5 ± 13.2
TSH	112.1 ± .1
FSH	111.1 ± .3

[a]Specificity of ^{125}I-IL-3 binding to human monocytes and JM-1 cells. Monocytes and JM-1 cells (3.3×10^7 cells/ml) were incubated with ^{125}I-IL-3 (2.5×10^{-10} M) and the following unlabeled proteins at the concentrations indicated: rhu IL-3, 1×10^{-7} M; rhu GM-CSF, 7.5×10^{-7} M; rhu G-CSF, 1×10^{-7} M; rhu CSF-1, 1×10^{-7} M, rhu IL-4, 5×10^{-7} M; rhu IL-1α, 5×10^{-7} M; rhu IL-1β, 5×10^{-7} M; rhu IL-2, 5×10^{-7} M; epidermal growth factor (EGF), 3 μg/ml; fibroblast growth factor (FGF), 1 μg/ml; nerve growth factor (NGF), 2 μg/ml; human luteinizing hormone (HLH), 1 μg/ml; human growth hormone (HGH), 1.7×10^{-7} M; thyroid-stimulating hormone (TSH), 1 μg/ml; follicle-stimulating hormone (FSH), 1 μg/ml. Concentrations of partially pure hormone preparations are given in micrograms per milliter total protein. Incubation was for 1 hr at 37°C.

chemically cross-linked to the receptor molecule. Detergent extracts of cells are then analyzed by SDS polyacrylamide gel electrophoresis and autoradiography. The murine IL-3 receptor is estimated to have a molecular mass of 72,500.[16]

The ultimate molecular characterization of the IL-3 receptor, of murine or human origin, will come about with the cloning of the genes encoding these molecules. Recent advances in the purification of cytokine receptor molecules[18] and in the cloning of these molecules (J. Sims, unpublished results) will make this goal attainable. Our understanding of the protein hormones and receptors involved in the control of hematopoiesis is thus increasing rapidly as each is identified and characterized at the molecular level.

TABLE 4. Distribution of Cytokine Receptors on Human Cells[a]

Cells	Characteristics	Molecules Bound/Cell ^{125}I-rhu IL-3
Human cell lines		
Jurkat	T-lymphoma	<10
HSB2	T-lymphoma	<10
PEER	T-leukemia	<10
RAJI	B-lymphoma	<10
ARH77	B-leukemia	<10
CESS	B-lymphoblast	<10
BMB	B stem cell tumor	120 ± 30
Nalm-6	pre-B lymphoma	60 ± 30
JM-1	pre-B lymphoma	90 ± 60
HL-60	promyelocytic	<10
U937	monocytic tumor	<10
KG-1	myelogenous	<10
K562	myelogenous	<10
HPT	carcinoma	<10
HBT5637	bladder carcinoma	<10
A375	melanoma	<10
Human primary cells		
	neutrophils	<10
	monocytes	125 ± 85
	AML patient 1	120 ± 30
	AML patient 2	35 ± 15
	AML patient 3	20
	ALL patient 4	<10
	ALL patient 5	<10
	ALL patient 6	<10

[a]Molecules bound/cell for IL-3 were determined by Scatchard analysis of complete sets of binding data and represent total sites/cell.

ACKNOWLEDGMENTS

The authors gratefully acknowledge the excellent assistance of Dirk Anderson, Noel Balantac, Billy Clevenger, Della Friend, Ralph Klinke, and Betsy McGrath and thank Linda Troup for the preparation of the manuscript. The authors would also like to thank Tom Hopp, Kathryn Prickett, and Irwin Bernstein for their help in the project.

REFERENCES

1. METCALF, D. 1986. Blood **67:** 257–267.
2. CLARK, S. & R. KAMEN. 1987. Science **236:** 1229–1237.
3. MOCHIZUKI, D., J. EISENMAN, P. CONLON, A. LARSEN & R. TUSHINSKI. 1987. Proc. Natl. Acad. Sci. USA **84:** 5267–5271.
4. MOORE, M. & D. J. WARREN. 1987. Proc. Natl. Acad. Sci. USA **84:** 7134–7138.
5. STANLEY, E. R., A. BARTOCCI, D. PATINKIN, M. ROSENDAHL & T. R. BRADLEY. 1986. Cell **45:** 667–674.
6. PESCHEL, C., W. E. PAUL, J. OHARA & I. GREEN. 1987. Blood **70:** 254–263.
7. SANDERSON, C., A. O'GARRA, D. WARREN & G. KAUS. 1986. Proc. Natl. Acad. Sci. USA **83:** 437–440.
8. GARMAN, R., K. JACOBS, S. CLARK & D. RAULET. 1987. Proc. Natl. Acad. Sci. USA **84:** 7629–7633.
9. URDAL, D. & L. PARK. 1988. Behring Inst. Mitt. **83:** 27–39.
10. YANG, Y.-C., A. B. CIARLETTA, P. A. TEMPLE, M. P. CHUNG, S. KOVACIC, J. S. WITEK-GIANNOTI, A. C. LEARY, R. KRIZ, R. E. DONAHUE, G. C. WONG & S. C. CLARK. 1986. Cell **47:** 3–10.
11. YOKOTA, T., F. LEE, D. RENNICK, C. HALL, N. ARAI, T. MOSMANN, G. NABEL, H. CANTOR & K. ARAI. 1984. Proc. Natl. Acad. Sci. USA **81:** 1070–1074.
12. FUNG, M. C., A. J. HAPEL, S. YMER, D. R. COHEN, R. N. JOHNSON, H. D. CAMPBELL & I. G. YOUNG. 1984. Nature **307:** 233–237.
13. MULLIS, K., F. FALOONA, S. SCHARF, R. SAIKI, G. HORN & H. ERLICH. 1986. *In* Cold Spring Harbor Symp. Quant. Biol. Vol. **LI:** 263–273. Cold Spring Harbor, NY.
14. PRICE, V., D. MOCHIZUKI, C. MARCH, D. COSMAN, M. DEELEY, R. KLINKE, W. CLEVENGER, S. GILLIS, P. BAKER & D. URDAL. 1987. Gene **55:** 287–293.
15. PALASZYNSKI, E. & J. IHLE. 1984. J. Immunol. **132:** 1872–1878.
16. PARK, L. S., D. FRIEND, S. GILLIS & D. L. URDAL. 1986. J. Biol. Chem. **261:** 205–210.
17. PARK, L. S., D. FRIEND, H. SASSENFELD & D. URDAL. 1987. J. Exp. Med. **166:** 476–488.
18. URDAL, D. L., S. M. CALL, J. L. JACKSON & S. K. DOWER. 1988. J. Biol. Chem. **263:** 2870–2877.
19. LAEMMLI, U. K. 1970. Nature **227:** 680–685.
20. URDAL, D. L., D. MOCHIZUKI, P. J. CONLON, C. J. MARCH, M. L. REMEROWSKI, J. EISENMAN, C. RAMTHUN & S. GILLIS. 1984. J. Chromatogr. **296:** 171–179.

The Effects of Oxidizing Species Derived from Molecular Oxygen on the Proliferation *In Vitro* of Human Granulocyte-Macrophage Progenitor Cells[a]

HAL E. BROXMEYER,[b,c,d] SCOTT COOPER,[b,d]
AND THEODORE GABIG[b,d]

Departments of [b]Medicine (Hematology/Oncology),
[c]Microbiology and Immunology
[d]The Walther Oncology Center
Indiana University School of Medicine
Indianapolis, Indiana 46223

INTRODUCTION

The regulation of myeloid blood cell production is controlled *in vitro* by a complex network of biomolecule-cell interactions.[1,2] There is increasing evidence that the biomolecules involved in this regulation *in vitro* also have activities *in vivo* that would be expected from the information generated through studies *in vitro*.[3] Most studies *in vitro* have been carried out at ambient oxygen (O_2) tensions of approximately 20%, but it is likely that biomolecule-cell interactions *in vivo* occur at lowered O_2 tensions that may be in the 5% range. In fact, comparisons of studies performed *in vitro* at 20% verses 5% O_2 tension have demonstrated that at 5% O_2 there are more detectable colonies formed from hematopoietic progenitors,[4–11] and that these cells are more sensitive to both stimulating and inhibiting biomolecules; also increased factor production by accessory cells is evident.[12–14] Thus, studies performed *in vitro* at lowered oxygen tension may more accurately reflect the actual situation *in vivo*. In order to better understand the differences noted in the growth of granulocyte-macrophage progenitor cells (CFU-GM) at 20% verses 5% O_2 tension, we assessed the influence of superoxide dismutase, catalase, glucose oxidase and horseradish peroxidase on colony and cluster formation by normal human bone marrow CFU-GM present in a low density, or a nonadherent low density, fraction of cells cultured at 20% or 5% O_2 tension. These studies allowed us to evaluate the effects of oxidizing species derived from molecular oxygen on the proliferation of CFU-GM.

[a]These studies were supported by Public Health Service Grants CA 34646 and CA 36740 from the National Cancer Institute (to H.E.B.), Grant AI 21961 from the National Institutes of Health (to T.G.), and by a grant from the Biocyte Corporation (to H.E.B.).

MATERIALS AND METHODS

Cells and Cell Separation Procedures

Bone marrow cells were obtained by aspiration from the posterior iliac crest of healthy volunteers who had given informed consent according to guidelines established by the Human Investigation Committee of the Indiana University School of Medicine. Bone marrow cells were separated into a low density (<1.077 gm/cm^3) fraction using Ficoll-Hypaque, and a nonadherent low density fraction obtained after adherence of cells to plastic petri dishes in medium containing 10% fetal bovine serum (FBS) at 37°C for 1 1/2 hr and collection of the cells not adhering to the plastic. The nonadherent cells contained $\leq$3% nonspecific esterase staining cells.

Enzymes

Superoxide dismutase (bovine erythrocyte), catalase (bovine liver), glucose oxidase (Grade II, Aspergillas niger) and horseradish peroxidase were purchased from Boehringer-Mannheim Biochemicals, Indianapolis, IN.

Colony and Cluster Formation

Low density and nonadherent low density bone marrow cells from normal donors were plated at 10^5/ml in 35 mm standard tissue culture dishes in 1 ml of 0.3% agar (Difco Laboratories, Detroit, MI) culture medium containing McCoy's 5A medium supplemented with additional essential and nonessential amino acids, glutamine, serine, asparagine, sodium pyruvate (Gibco Laboratories, Grand Island, NY) with 10% heat-inactivated (56°C for 1 1/2 hr) FBS (Hyclone, Logan, UT) in the presence of 10% v/v medium conditioned by the human urinary bladder carcinoma cell line, 5637.[15] Culture dishes were incubated at 37°C in a humidified atmosphere flushed with 5% CO_2 at ambient (20%) or lowered (5%) O_2 tension and scored for colonies ($>$50 cells/aggregate) and clusters (5–50 cells/aggregate) after 7 days. Low O_2 tension was maintained using an oxyreducer (Reming Bioinstruments, Redfield, NY).

Statistics

The results are expressed as the mean $\pm$ 1 SEM of three to four plates per point for each experiment. Levels of significance for comparisons between samples were determined using Student *t* distribution.

RESULTS AND DISCUSSION

Increased numbers and size of colonies and clusters are apparent when bone marrow cells are cultured at lowered (5%) O_2 tension, when compared to cells grown at ambient (20%) O_2 tension.[4–11] It has been suggested by others that this increase in colony formation by erythroid progenitors (BFU-E) is mediated by monocytes and T-lymphocytes.[8] We compared the growth of CFU-GM colonies and clusters stimulated by 5637 cell line conditioned medium in the presence of either 20% or 5% O_2, and assessed the influence of removal of adherent cells on the growth of CFU-GM in the low density fraction of bone marrow (TABLE 1). In 7 experiments, there was a significant increase in both colonies, and colonies plus clusters, when low density marrow cells were grown at 5% O_2. The mean % increase was 255 ± 99 (range +31 to +750) for colonies and 55 ± 9 (range +17 to +90) for colonies plus clusters demonstrating a greater increase in colony than in cluster numbers. Removal of adherent cells eliminated the increase in colony and cluster numbers seen at 5% O_2 with % change being −10 ± 2 (range −8 to +13) for colonies, and −4 ± 2 (range −1 to −8) for colonies plus clusters in three experiments. This suggested that the enhancing effect of lowered oxygen tension might be mediated through monocytes, alone or in combination with other accessory cell types as reported by others.[8]

Oxygen toxicity is a noted, but clearly complex phenomenon.[16–18] Evidence suggests that superoxide ($O_2^{\cdot}$) is an agent of oxygen toxicity and that superoxide dismutase (SOD), which catalyzes the following reaction, provides an essential defense of the toxicity: [$O_2^{\cdot} + O_2^{\cdot} + 2H^+ \rightarrow$ hydrogen peroxide (H_2O_2) + O_2]. Others have shown that SOD

TABLE 1. Influence of Lowered Oxygen Tension and Adherent Cells on Colony Formation by Day 7 Granulocyte-Macrophage Progenitor Cells in Normal Human Bone Marrow[a]

Experiment	Low Density Bone Marrow				Nonadherent Low Density Bone Marrow			
	Colonies		Colonies & Cluster		Colonies		Colonies & Cluster	
	20% O_2	5% O_2	20% O_2	5% O_2	20% O_2	5% O_2	20% O_2	5% O_2
1	26 ± 1	51 ± 2[b]	100 ± 6	117 ± 2[b]	48 ± 5	42 ± 2	90 ± 4	91 ± 2
2	8 ± 1	44 ± 1[b]	141 ± 13	254 ± 17[b]	43 ± 1	39 ± 4	204 ± 10	187 ± 10
3	76 ± 4	108 ± 6[b]	111 ± 4	162 ± 5[b]	62 ± 6	57 ± 2	94 ± 11	90 ± 3
4	58 ± 2	76 ± 2[b]	90 ± 4	140 ± 5[b]				
5	38 ± 4	95 ± 2[b]	82 ± 7	156 ± 6[b]				
6	2 ± 0.6	17 ± 2[b]	144 ± 11	199 ± 2[b]				
7	18 ± 1	66 ± 4[b]	131 ± 8	204 ± 6[b]				

[a]Cells were plated at 10^5/ml in 0.3% agar-culture medium in the presence of 10% v/v 5637 CM and scored after 7 days of incubation for CFU-GM colonies (>50 cells/aggregate) and clusters (3–50 cells/aggregate).

[b]Significant % change when cells are grown in presence of 5% O_2 as compared to 20% O_2, $p < 0.005$; other numbers are not significantly different, $p > 0.05$.

enhances the growth of mouse bone marrow, peritoneal exudate, alveolar, and blood mononuclear cells.[9] We assessed the effects of SOD on colony formation by CFU-GM colony and cluster formation in low density and nonadherent low density human bone marrow cells grown at 20% and 5% O_2 tension (TABLE 2). Concentrations of 30–90 μg SOD/ml significantly enhanced colony and cluster formation of low density bone marrow cells grown at 20% O_2. This enhancement was not apparent when low density bone marrow cells were cultured at 5% O_2 or when nonadherent low density cells were cultured at 20% or 5% O_2. These results suggested that either $O_2^{\cdot}$ was inhibitory to colony and cluster formation at 20% O_2 or alternatively that the levels of H_2O_2 generated by SOD were stimulating. The apparent inactivity of SOD at 5% O_2 suggested that little or no $O_2^{\cdot}$ might be produced at this lowered oxygen tension. The ineffectiveness of SOD to enhance colony and cluster numbers at 20% O_2 when adherent cells were removed suggests that this effect, similar to the low O_2 tension effect, is mediated also by adherent cells. In order to test the possibility that levels of H_2O_2 were stimulating at 20% O_2 tension, catalase was

TABLE 2. Influence of Superoxide Dismutase, Catalase, and Adherent Cells on Colony Formation by Day 7 Granulocyte-Macrophage Progenitor Cells in Normal Human Bone Marrow at Normal (20%) versus Lowered (5%) Oxygen Tension[a]

	Percent Change (N)[b]							
	Low Density Bone Marrow				Nonadherent Low Density Bone Marrow			
	20% O_2		5% O_2		20% O_2		5% O_2	
	Colonies	Colonies & Clusters	Colonies	Colonies & Clusters	Colonies	Colonies & Clusters	Colonies	Colonies & Clusters
Superoxide Dismutase (30–90 μg/ml)	+82 ± 17[c] (6)	+62 ± 9[c] (6)	+7 ± 3 (6)	−6 ± 2 (6)	−5 ± 4 (2)	−6 ± 3 (2)	−10 (1)	−1 (1)
Catalase (10 μg/ml)	−6 ± 7 (6)	−12 ± 5 (7)	−36 ± 4[c] (6)	−44 ± 3[c] (6)	−10 ± 6 (2)	−1 ± 3 (2)	+14 ± 4 (2)	+5 ± 3 (2)
Superoxide Dismutase (45–90 μg/ml) + Catalase (10 μg/ml)	−16 ± 7 (5)	−13 ± 2 (5)	−47 ± 4[c] (5)	−45 ± 1[c] (5)	−6 ± 6 (2)	−3 ± 1 (2)	−5 (1)	−3 (1)

[a] 10^5 cells/ml were scored for CFU-GM colonies and clusters after 7 days of incubation in the presence of 10% v/v 5637 CM.

[b] Percent change is expressed as the mean ± 1 SEM of the number (N) of experiments shown in parentheses and refers to change in presence of superoxide dismutase and/or catalase when compared to controls plated in the absence of these enzymes. A plus sign refers to an increase and a negative sign refers to a decrease.

[c] Significant percent change from control, $p < 0.005$; other numbers are not significantly different, $p > 0.05$.

added to the cultures alone and in combination with SOD. Catalase converts H_2O_2 to H_2O. As shown in TABLE 2, when low density bone marrow cells were cultured in the presence of 20% O_2, 10 μg catalase/ml had no effect by itself, but did counteract the SOD enhancement of colony and cluster formation suggesting that the SOD enhancement might be mediated by the generation of H_2O_2 from $O_2^{\cdot}$. Interestingly, the enhanced colony and cluster formation noted when cells were grown at 5% O_2 tension in the absence of enzymes, was suppressed by the catalase (alone or in combination with SOD). This may be due to the possible generation of H_2O_2 when cells are grown at lowered O_2 tension. It is possible that catalase decreased colony and cluster formation noted at 5% O_2 by removing H_2O_2. Both the catalase counteraction of the SOD enhancement of growth at 20% O_2, and the catalase suppression of enhanced growth at 5% O_2 would appear to be mediated via adherent cells, since the catalase effects were not noted when nonadherent low density cells was plated.

In order to explore further the possibility that enhancement of cell growth was associated with H_2O_2, low density bone marrow cells were incubated at 20% O_2 tension in the absence and presence of glucose oxidase, an enzyme that generates H_2O_2 from glucose. As shown in TABLE 3 (one experiment with actual colony and cluster numbers) and TABLE 4 (two or more experiments with results expressed as a percent change), glucose oxidase had no effect at 0.01 ng/ml, induced a dose-dependent enhancement at 0.1 to 1.0 ng/ml, but had no effect at 5 ng/ml when cells were plated at 20% O_2. Not shown is data from two experiments demonstrating inhibition of colony and cluster formation with 10–15 ng glucose oxidase/ml. Complete inhibition was seen at 50 ng glucose oxidase/ml. No enhancement was seen with glucose oxidase when cells were grown at 5% O_2 (TABLES 3 and 4). The absence of an enhancing effect and the presence of an inhibitory effect with the higher concentration of glucose oxidase opens up the possibility that while low levels of H_2O_2 may have an enhancing effect, higher concentrations of H_2O_2 may be suppressive. Catalase, which as mentioned above converts H_2O_2 to H_2O, counteracts the effects of glucose oxidase (TABLES 3 and 4). Horseradish peroxidase, which converts H_2O_2 to a more toxic oxidant, hypochlorite, has an inhibitory effect on growth at 20% (not apparent in all experiments) and 5% O_2 (see especially TABLE 4). Horseradish peroxidase not only counteracts the enhancement of growth noted with glucose oxidase at 20% O_2, but suppresses growth below control values.

Although it is not possible to be certain of the mechanisms involved from studies dealing only with enzyme modulation of colony and cluster formation, the results in total are consistent with the hypothesis that low levels of H_2O_2 in combination with the presence of adherent cells (probably monocytes/macrophages), are involved in the enhancement of CFU-GM colony and cluster formation *in vitro*. This might be due to a direct stimulation of cell proliferation, or more likely to the removal of a suppressor cell or suppressor cell mechanism by H_2O_2.

SUMMARY

In order to better understand the enhancing effects of lowered oxygen (O_2) tension on the growth *in vitro* of granulocyte-macrophage progenitor cells (CFU-GM), the effects of oxidizing species derived from molecular O_2 were assessed on CFU-GM. Low density or nonadherent low density normal human bone marrow cells were plated at ambient (20%) or lowered (5%) O_2 tension in the presence of a source of colony stimulating factors, and in the absence or presence of superoxide dismutase, catalase, glucose oxidase or horseradish peroxidase, alone or in various combinations. Enhanced colony and cluster formation of CFU-GM was noted when low density cells were grown at 5% O_2, or when cells

TABLE 3. Influence of Glucose Oxidase (GO), Catalase (Cat) and Horseradish Peroxidase (HP) Alone and in Combination on Colony and Cluster Formation, by Human Marrow Granulocyte-Macrophage Progenitor Cells Cultured at Normal (20%) versus Lowered (5%) Oxygen Tension[a]

	20% O_2		5% O_2	
	Colonies	Colonies & Clusters	Colonies	Colonies & Clusters
McCoy's medium = control	18 ± 1	131 ± 8	66 ± 4	204 ± 6
GO (5 ng/ml)	20 ± 3 (+11)[b]	147 ± 14 (+12)	65 ± 3 (−2)	202 ± 15 (−1)
GO (1 ng/ml)	41 ± 3 (+127)[c]	254 ± 18 (+94)[c]	60 ± 5 (−9)	199 ± 6 (−2)
GO (0.1 ng/ml)	28 ± 3 (+56)[c]	179 ± 1 (+37)[c]		
GO (0.01 ng/ml)	18 ± 0.3 (0)	131 ± 5 (0)		
Cat (10 μg/m!)	16 ± 1 (−11)	134 ± 5 (+2)	51 ± 3 (−23)[c]	121 ± 0.3 (−41)[c]
HP (10 μg/ml)	16 ± 1 (−11)	132 ± 5 (0)	44 ± 5 (−33)[c]	133 ± 7 (−35)[c]
GO (5 ng/ml) + Cat (10 μg/ml)	13 ± 2 (−28)[c]	113 ± 8 (−14)	46 ± 4 (−30)[c]	140 ± 8 (−31)[c]
GO (1 ng/ml) + Cat (10 μg/ml)	15 ± 3 (−17)	135 ± 6 (+3)	48 ± 2 (−27)[c]	122 ± 3 (−40)[c]
GO (0.1 ng/ml) + Cat (10 μg/ml)	13 ± 1 (−28)[c]	120 ± 4 (−8)		
GO (0.01 ng/ml) + Cat (10 μg/ml)	15 ± 2 (−17)	128 ± 9 (−2)	42 ± 2 (−36)[c]	119 ± 5 (−42)[c]
GO (5 ng/ml) + HP (10 μg/ml)	10 ± 2 (−44)[c]	112 ± 6 (−15)[c]	37 ± 4 (−44)[c]	117 ± 3 (−43)[c]
GO (1 ng/ml) + HP (10 μg/ml)	13 ± 1 (−28)[c]	112 ± 7 (−15)[c]	34 ± 3 (−48)[c]	114 ± 4 (−44)[c]
GO (0.01 ng/ml) + HP (10 μg/ml)	16 ± 1 (−11)	120 ± 7 (−8)	40 ± 1 (−39)[c]	122 ± 6 (−40)[c]

[a] 10^5 low density normal human bone marrow cells were scored for CFU-GM colonies and clusters after 7 days of incubation in the presence of 10% v/v 5637 CM. A single experiment is shown.

[b] Percent change from control.

[c] Significant percent change, p at least <0.05.

were grown at 20% O_2 in the presence of superoxide dismutase or glucose oxidase. Both of these enzymes are capable of generating hydrogen peroxide (H_2O_2), although by different mechanisms. Low concentrations of glucose oxidase resulted in increased formation of colonies and clusters, but higher concentrations of glucose oxidase were inhibitory. Catalase, which converts H_2O_2 to H_2O, had no effect by itself on cells growing at 20% O_2, but it eliminated the superoxide dismutase and glucose oxidase enhancing effects. Catalase decreased colony formation of cells grown at 5% O_2. Removal of adherent cells ablated the growth-enhancing effects noted at lowered (5%) O_2 tension and also the superoxide dismutase and catalase effects at 20% or 5% O_2. Horseradish peroxidase, which converts H_2O_2 to a more toxic oxidant, hypochlorite, had a suppressive effect on colony and cluster numbers and at 20% O_2 converted the glucose oxidase effects

TABLE 4. Influence of Glucose Oxidase (GO), Catalase (Cat), and Horseradish Peroxidase (HP), Alone and in Combination, on Colony and Cluster Formation by Day 7 Granulocyte-Macrophage Progenitor Cells in Low Density Normal Human Bone Marrow Cells Cultured at Normal (20%) versus Lowered (5%) Oxygen Tension[a]

	Percent Change (N)[b]			
	Colonies		Colonies & Clusters	
	20% O_2	5% O_2	20% O_2	5% O_2
GO (5 ng/ml)	+ 11 (1)	+ 5 ± 7 (2)	− 1 ± 13 (2)	− 5 ± 4 (2)
GO (1 ng/ml)	+ 127[c] (1)	− 2 ± 8 (2)	+ 84 ± 10[c](2)	− 4 ± 2 (2)
GO (0.1 ng/ml)	+ 56[c] (1)	+ 6(1)	+ 22 ± 16[c](2)	− 8(1)
GO (0.01 ng/ml)	0 (1)	0(1)	− 2 ± 2 (2)	− 5(1)
Cat (10 μg/ml)	− 6 ± 7 (6)	− 36 ± 4[c] (6)	− 12 ± 5 (7)	− 44 ± 3[c] (6)
HP (10 μg/ml)	− 30 ± 10[c] (4)	− 33 ± 3[c] (4)	− 24 ± 9[c] (5)	− 37 ± 10[c](4)
GO (5 ng/ml) + Cat (10 μg/ml)	− 28[c] (1)	− 42 ± 12[c](2)	− 25 ± 11[c](2)	− 40 ± 9[c] (2)
GO (1 ng/ml) + Cat (10 μg/ml)	− 17[c] (1)	− 40 ± 13[c](2)	− 18 ± 2[c] (2)	− 45 ± 5[c] (2)
GO (0.1 ng/ml) + Cat (10 μg/ml)	− 28[c] (1)		− 12 ± 20 (2)	
GO (0.01 ng/ml) + Cat (10 μg/ml)	− 17[c] (1)	− 45 ± 9[c] (2)	− 15 ± 17 (2)	− 47 ± 5[c] (2)
GO (5 ng/ml) + HP (10 μg/ml)	− 44[c] (1)	− 42 ± 2[c] (2)	− 9 ± 7 (2)	− 46 ± 3[c] (2)
GO (1 ng/ml) + HP (10 μg/ml)	− 28[c] (1)	− 45 ± 4[c] (2)	− 26 ± 11[c](2)	− 47 ± 3[c] (2)
GO (0.1 ng/ml) + HP (10 μg/ml)		− 31[c](1)		
GO (0.01 ng/ml) + HP (10 μg/ml)	− 11 (1)	− 43 ± 4[c] (2)	− 21 ± 13[c](2)	− 46 ± 6[c] (2)

[a] 10^5 cells/ml were scored for CFU-GM colonies and clusters after 7 days of incubation in the presence of 10% v/v 5637CM.

[b] Percent change is expressed as the mean ± 1 SEM of the number (N) of experiments shown in parentheses and refers to change in presence of enzymes when compared to controls plated in the absence of enzymes.

[c] Significant percent change from control (no enzymes), p at least <0.05.

from stimulatory to inhibitory. The results suggest that adherent cells and low concentrations of H_2O_2 may mediate growth-enhancing effects of CFU-GM seen at lowered (5%) O_2 tension.

ACKNOWLEDGMENT

We thank Linda Cheung for typing the manuscript.

REFERENCES

1. BROXMEYER, H.E. 1986. Biomolecule-cell interactions and the regulation of myelopoiesis. Int. J. Cell Cloning **4:** 378–405.

2. Broxmeyer, H.E. & D.E. Williams. 1988. The production of myeloid blood cells and their regulation during health and disease. CRC Crit. Rev. Oncol./Hematol. **8:** 173–226.
3. Broxmeyer, H.E. & D.E. Williams. 1987. Actions of hematopoietic colony-stimulating factors *in vivo* and *in vitro*. Pathol. Immunopathol. Res. **6:** 207–220.
4. Bradley, T.R., G.S. Hodgson & M. Rosendaal. 1978. Effect of oxygen tension on hemopoietic and fibroblast cell proliferation *in vitro*. J. Cell. Physiol. **97:** 517–522.
5. Rich, I.N. & B. Kubanek. 1982. The effect of oxygen tension on colony formation of erythropoietic cells *in vitro*. Br. J. Haematol. **52:** 578–588.
6. Smith, S. & H.E. Broxmeyer. 1986. The influence of oxygen on the long term growth *in vitro* of haematopoietic progenitor cells from human cord blood. Br. J. Haematol. **63:** 29–34.
7. Lu, L. & H.E. Broxmeyer. 1985. Comparative influences of phytohemagglutin-stimulated leukocyte conditioned medium, hemin, prostaglandin E and low oxygen tension on colony formation by erythroid progenitor cells in normal human bone marrow. Exp. Hematol. **13:** 989–993.
8. Pennathur-Das, R. & L. Levitt. 1987. Augmentation of *in vitro* human marrow erythropoiesis under physiological oxygen tensions is mediated by monocytes and T-lymphocytes. Blood **69:** 899–907.
9. Lin, H.-S. & S. Hsu. 1986. Modulation of tissue mononuclear phagocyte clonal growth by oxygen and antioxidant enzymes. Exp. Hematol. **14:** 840–844.
10. Maeda, H., T. Hotta & H. Yamada. 1986. Enhanced colony formation of human hemopoietic stem cells in reduced oxygen tension. Exp. Hematol. **14:** 930–934.
11. Katahira, J. & H. Mizoguchi. 1987. Improvement of culture conditions for human megakaryocytic and pluripotent progenitor cells by low oxygen tension. Int. J. Cell Cloning **5:** 412–420.
12. Broxmeyer, H.E., S. Cooper, B.Y. Rubin & M.W. Taylor. 1985. The synergistic influence of human interferon-γ and interferon-α on suppression of hematopoietic progenitor cells is additive with the enhanced sensitivity of these cells to inhibition by interferons at low oxygen tension *in vitro*. J. Immunol. **135:** 2502–2506.
13. Broxmeyer, H.E., D.E. Williams, G. Hangoc, S. Cooper, P. Gentile, R.-N. Shen, P. Ralph, S. Gillis & D.C. Bicknell. 1987. The opposing actions *in vivo* on murine myelopoiesis of purified preparations of lactoferrin and the colony stimulating factors. Blood Cells **13:** 31–48.
14. Rich, I.N. 1986. A role for the macrophage in normal hemopoiesis. II. Effect of varying physiological oxygen tensions on the release of hemopoietic growth factors from bone-marrow-derived macrophages *in vitro*. Exp. Hematol. **14:** 746–751.
15. Broxmeyer, H.E., J. Bognacki, P. Ralph, M.H. Dorner, L. Lu & H. Castro-Malaspina. 1982. Monocyte-macrophage derived acidic isoferritins: Normal feedback regulators of granulocyte-macrophage progenitor cells *in vitro*. Blood **60:** 595–607.
16. Cerutti, P.A. 1985. Prooxidant states and tumor promotion. Science **227:** 375–381.
17. Halliwell, B. & J.M.C. Gutteridge. 1984. Oxygen toxicity, oxygen radicals, transition metals and disease. Biochem. J. **219:** 1–14.
18. DiGuiseppi, J. & I. Fridovich. 1983. The toxicity of molecular oxygen. CRC Crit. Rev. Toxicol. **12:** 315–342.

Humoral Regulation of Stem Cell Proliferation

MAKIO OGAWA, KENJI IKEBUCHI, AND ANNE G. LEARY

Veterans Administration Medical Center
and
Department of Medicine
Medical University of South Carolina
Charleston, South Carolina 29403

It is estimated that approximately two hundred billion erythrocytes and seventy billion neutrophilic leucocytes are produced and destroyed daily in a man weighing 70 kg. This process continues throughout the life of a man and shows very little sign of aging. In order to explain this remarkable cell renewal, our predecessors envisioned a population of cells in the marrow that are able to renew themselves and generate progenitors committed to individual hemopoietic lineages. They termed these cells hemopoietic stem cells. While the concept of hemopoietic stem cells was established early, the identity of the stem cells remained elusive. During the last two and a half decades, studies using clonal cell culture techniques significantly elucidated the functions of hemopoietic stem cells even though no cytochemical or immunochemical techniques became available for identification of the hemopoietic stem cells. Most recently progress in the molecular biology facilitated isolation and characterization of hemopoietic growth factors that regulate the proliferation of the hemopoietic progenitors in culture. In this review, we summarize the current understanding of the targets and the mechanisms of interactions between interleukin-3 (IL-3), granulocyte/macrophage colony-stimulating factor (GM-CSF), interleukin-6 (IL-6), interleukin-1 (IL-1), and granulocyte-CSF (G-CSF).

It is generally believed that in the steady state, the majority of the hemopoietic stem cells are dormant in cell cycle and that they begin cell division after varying time intervals. This concept of a long G_0 residence of the hemopoietic stem cells is supported by a number of investigations including ^{3}H-thymidine suicide studies of colony-forming units in spleen (CFU-S)[1] and formation of multilineage colonies in culture,[2] analysis of the sensitivity of primitive CFU-S to high dose 5-fluorouracil (5-FU)[3] and serial observation (mapping studies) of the formation of multipotential blast cell colonies in culture.[4] Available evidence also indicates that at a given time only a small number of stem cell clones provide cells in the entire lymphohemopoietic tissues. Mulligan and his associates[5] reconstituted the lymphohemopoietic system of a mouse by transferring cells from a retrovirally-labelled spleen colony, thereby demonstrating that the descendants of a single stem cell can generate a sufficient number of cells to occupy the entire lymphohemopoietic system. Follow-up studies of the retrovirally labelled animals indicated that hemopoiesis is supported by sequential activation of different stem cell clones.[6] Nakano *et al.*[7] using phosphoglycerate kinase (Pgk) isoenzymes as markers demonstrated long-term reconstitution of erythropoiesis by descendants of a single stem cell using a genetically anemic strain of W/W^v mice. Earlier, Mintz *et al.*[8] injected suspensions of liver cells from 13-day fetuses of C57B1/6 and Balb/c mice into placental circulation of W/W or W^f/W^f mice and obtained mice whose hemopoiesis was supported by more than one clone of stem cells. Serial observations of the genotypes of the blood cells revealed a complementary rise and fall in the proportion of cells of different genotypes. These experimental data supporting the "clonal succession" model of hemopoiesis also suggests that the majority of stem cells in a steady state are in G_0.

Humoral Factors Regulating Stem Cells

A number of humoral factors appear to be involved in the regulation of stem cell proliferation. Foremost, IL-3 appears to be necessary for sustained proliferation of multipotential progenitors. Investigators in many laboratories[9–12] have shown that IL-3 supports multilineage colony formation in culture. We have also shown that IL-3 supports formation of multipotential blast cell colonies in culture of murine[13] and human[14] hemopoietic cells. In the former experiment we observed that delayed addition of mouse IL-3 to cultures seven days after cell plating decreases the number of blast cell colonies to one half the number in cultures with IL-3 added at the initiation of culture. It did not, however, alter the proliferative and differentiative characteristics of late-emerging multipotential blast cell colonies.[13] Based on these observations we proposed that IL-3 does not trigger stem cells into active cell proliferation but is required for the continued proliferation of early multipotential progenitors. Subsequently, we observed that the development of multipotential blast cell colonies requires less IL-3 than the process of formation of multilineage colonies from blast cell colonies.[15] It appears that during stem cell development, the early multipotential progenitors are very sensitive to IL-3 and that as they gradually differentiate, the sensitivity to IL-3 declines. More recently, we examined the effects of human IL-3 on colony formation by purified hemopoietic progenitors in serum-free cultures. IL-3 alone was not effective in supporting colony formation except for a few eosinophil colonies.[16] Significant neutrophil colony or erythroid burst formation was seen only when IL-3 was combined with G-CSF or erythropoietin.

Granulocyte-macrophage CSF (GM-CSF) may also be a lineage nonspecific, intermediate acting factor. Earlier, Metcalf *et al.*[17] proposed that murine GM-CSF supports a few cell divisions of multipotential progenitors. Investigations in our laboratory also suggested that murine GM-CSF supports the proliferation of a subpopulation of multipotential progenitors that are responsive to IL-3.[18] Regarding human GM-CSF, several investigators reported that GM-CSF supports formation of human multipotential colonies in serum-containing culture.[19–21] Injections of human GM-CSF increased the number of neutrophils, monocytes and eosinophils in primates and man.[22,23] In our laboratory, human GM-CSF in the serum-free culture of purified hemopoietic progenitors was ineffective in support of colony formation except for a few small eosinophil colonies.[16] Only in the presence of G-CSF or erythropoietin did human GM-CSF support the formation of neutrophil colonies or erythroid bursts. These results indicated that the targets of human GM-CSF are also multipotential progenitors which probably represent a subpopulation of multipotential progenitors that are responsive to IL-3.

With regard to factors that might induce stem cells into active cell proliferation, there presently are three candidate factors, IL-1, IL-6 and G-CSF. Earlier, Stanley and his colleagues[24] presented evidence that hemopoietin-1, which was purified from the culture supernatant of the bladder carcinoma cell line, 5637,[25] possesses synergistic activity with IL-3 in supporting the proliferation of hemopoietic progenitors. More recently, Mochizuki *et al.*[26] and Moore and Warren[27] reported that IL-1 accounts for the hemopoietin-1 activity of the 5637 supernatant. In our laboratory we have identified IL-6[28] and G-CSF[29] as synergistic factors for IL-3 dependent proliferation of murine hemopoietic progenitors. Serial observations of the development of multipotential blast cell colonies from spleen cells obtained from mice that had been treated with 150 mg/kg 5-FU four days before revealed that, in the presence of the synergistic factors, the average length of the G_0 period was significantly shortened. In this review, we summarize our most recent results comparing the synergistic activities of IL-6 and G-CSF on murine and human blast cell colony formation.

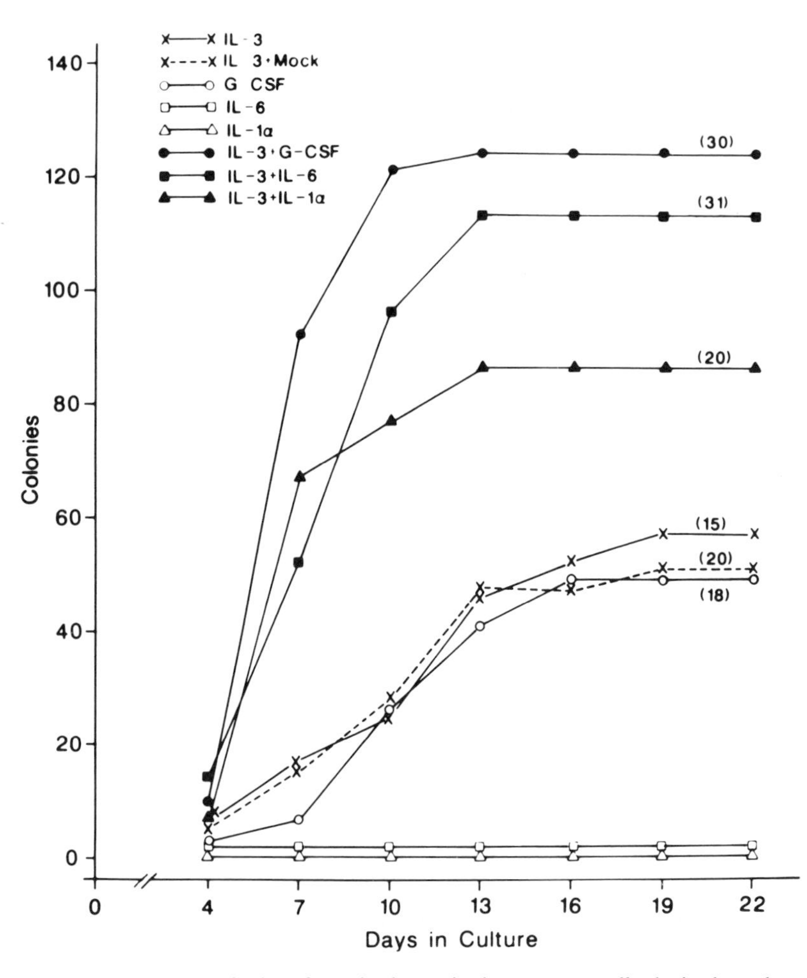

FIGURE 1. Time course of colony formation by murine bone marrow cells obtained two days after injection of 150 mg/kg 5-FU. Culture was carried out in the presence of 2 U/ml erythropoietin. The data represent total numbers of colonies in four plates each seeded with 5×10^4 marrow cells. The numbers in the *brackets* represent GEMM colonies.

The Effects of Synergistic Factors on Murine Progenitors

Murine IL-3 (approximately 5×10^6 units/mg protein) that had been purified to homogeneity from medium conditioned by WEHI-3 cells[30] was generously provided by Dr. James N. Ihle, NCI-Frederick Cancer Research Facility, Frederick, MD. The source of human IL-6 was the supernatant of Cos-1 cells transfected with cDNA coding for human IL-6 and was kindly provided by Dr. Steven C. Clark of the Genetics Institute, Cambridge, MA. *Escherichia coli*-derived human G-CSF was a generous gift from Dr. Lawrence Souza of AMGen, Thousand Oaks, CA. Recombinant human IL-1α which was produced in *Escherichia coli* was the generous gift of Dr. Yoshikatsu Hirai of Otsuka Pharmaceutical Co., Ltd., Tokushima, Japan. The concentration of the factors used in cultures were IL-3 200 units/ml, G-CSF 1×10^4 units/ml, IL-6 1:1000 dilution and IL-1α 2 ng/ml.

Earlier Stanley and his colleagues[24] demonstrated the synergistic effect of hemopoietin-1 using murine marrow cells harvested two days after injection of 5-FU. We have confirmed using BDF_1 mice that the nadir of the blast cell colony and granulocyte/erythrocyte/macrophage/megakaryocyte (GEMM) colony-forming cells in bone marrow occurs two days after injection of 5-FU. We therefore used day two post-5-FU bone marrow cells and analyzed the synergistic ability of IL-6, G-CSF and IL-1α on IL-3-dependent colony formation. 5×10^4 marrow cells harvested two days after 5-FU injection were plated in methylcellulose culture containing 2 U/ml erythropoietin and 200 U/ml IL-3, with or without a synergistic factor. Details of the culture technique have been presented elsewhere.[28,29] The results of time course analysis of colony formation are presented in FIGURE 1. IL-1α supported no colony formation. Similarly, almost no colony

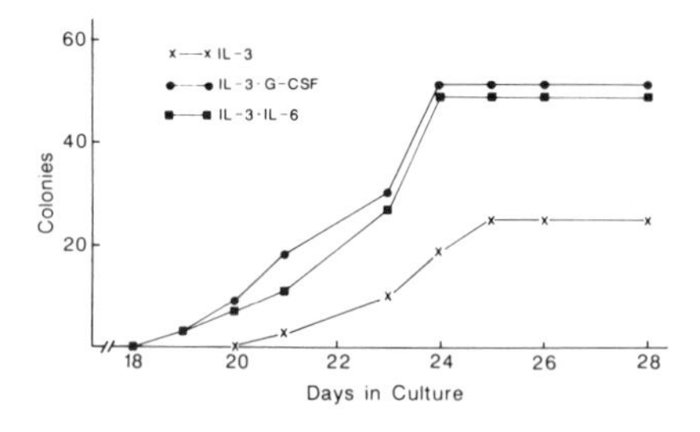

FIGURE 2. Cumulative frequency of human blast cell colonies supported by day-14 addition of IL-3 alone, IL-3 and G-CSF or IL-3 and IL-6. The data represent blast cell colony formation in a total of eight dishes.

formation was observed with IL-6 alone. G-CSF and IL-3 supported a comparable number of colonies including several GEMM colonies. The combination of IL-6, G-CSF or IL-1α with IL-3 revealed significant synergism in the time course of the development of colonies as well as the total number of colonies. The synergism between IL-1α and IL-3 was significantly less than the synergism between IL-6 or G-CSF and IL-3.

The Effects of Synergistic Factors on Human Blast Cell Colony Formation

We also compared the synergistic effects of IL-6 with that of G-CSF using a human bone marrow blast cell colony assay. Bone marrow cells obtained from healthy volunteers were enriched for progenitors by panning with monoclonal anti-My-10 antibodies and plated in methylcellulose blast cell colony assay as described previously.[31,32] Briefly, 2×10^4 My-10^+ cells were cultured in 35 mm Lux suspension culture dishes in the presence of 2% fetal calf serum and in the absence of growth factors in 5% O_2 and CO_2 in air atmosphere. On day 14 of incubation, IL-3 alone, IL-3 and IL-6, or IL-3 and G-CSF were layered over the culture dishes and the incubation continued. The dishes were then examined daily on an inverted microscope for the appearance of blast cell colonies. Previously, we documented that the majority of blast cell colonies yielded secondary colony formation upon replating.[31,32] The cumulative frequencies of the human blast cell colonies are presented in FIGURE 2. In the presence of IL-3 alone the first blast cell colony was identified on day 21 after cell plating and the colony number reached the peak on day 26 of culture. When IL-6 or G-CSF was added to the cultures containing IL-3, the identification of blast cell colonies was earlier, and approximately twice as many blast cell colonies were identified by day 24 of cell plating. These results with G-CSF were similar to the previous observations on the synergistic activity of IL-6 and IL-3 on human blast cell colony formation.[32]

DISCUSSION

Both in this paper and in our previous reports, we have shown that IL-6[28] and G-CSF[29] work as synergistic factors for IL-3-dependent proliferation of murine hemopoietic progenitors. Mapping studies that we described earlier indicated that part of the synergistic mechanisms of these factors is to shorten the G_0 period of dormant stem cells. Both IL-6 and G-CSF revealed stronger synergism than IL-1 on the multipotential progenitors. IL-1 is known to stimulate the synthesis and the release of IL-6[33] and G-CSF[34] by a variety of cell types including bone marrow fibroblasts. It is possible that the synergistic effects of IL-1 shown in the murine system were indirect and were mediated through release of factors by accessory cells. In support of this concept is our failure to detect the synergistic effect of IL-1 on human blast cell colony formation from panned My-10^+ marrow cells which are mostly devoid of accessory cells.

The genomic structures and amino acid sequences of IL-6 and G-CSF reveal significant homology.[35] It is likely that G-CSF and IL-6 are distantly related and may have evolved to assume related but distinct functions in host defense. Following infections, the host organism responds both by activation of the immune system and by an increased production of neutrophils. Antibody production is primarily regulated by IL-6 and neutrophil production by G-CSF. Production of these factors in turn may be augmented by increased production of IL-1. It may not be coincidental that these important regulators of the different aspects of host defense are also capable of shortening G_0 period of stem cells that would result in an enhanced rate of production of neutrophils (and possibly lymphocytes) and thus a more effective response to infection.

REFERENCES

1. BECKER, A. J., E. A. MCCULLOCH, L. SIMINOVITCH & J. E. TILL. 1965. The effect of differing demands for blood cell production on DNA synthesis by hemopoietic colony forming cells of mice. Blood **26:** 296–308.
2. HARA, H. & M. OGAWA. 1978. Murine hemopoietic colonies in culture containing normoblasts, macrophages, and megakaryocytes. Am. J. Hematol. **4:** 23–34.
3. HODGSON, G. S. & T. R. BRADLEY. 1979. Properties of haematopoietic stem cells surviving 5-fluorouracil treatment: Evidence for a pre-CFU-S cell? Nature **281:** 381–382.
4. SUDA, T., J. SUDA & M. OGAWA. 1983. Proliferative kinetics and differentiation of murine blast cell colonies in culture: Evidence for variable G_0 periods and constant doubling rates of early pluripotent hemopoietic progenitors. J. Cell. Physiol. **117:** 308–318.
5. WILLIAMS, D. A., I. R. LEMISCHKA, D. G. NATHAN & R. C. MULLIGAN. 1984. Introduction of new genetic material into pluripotent haematopoietic stem cells of the mouse. Nature **310:** 476–480.
6. LEMISCHKA, I. R., D. H. RAULET & R. C. MULLIGAN. 1986. Developmental potential and dynamic behavior of hematopoietic stem cells. Cell **45:** 917–927.
7. NAKANO, T., N. WAKI, H. ASAI & Y. KITAMURA. 1987. Long-term monoclonal reconstitution

of erythropoiesis in genetically anemic W/W^v mice by injection of 5-fluorouracil-treated bone marrow cells of Pgk-1^b/Pgk-1^a mice. Blood **70:** 1758–1763.

8. MINTZ, B., K. ANTHONY & S. LITWIN. 1984. Monoclonal derivation of mouse myeloid and lymphoid lineages from totipotent hematopoietic stem cells experimentally engrafted from fetal hosts. Proc. Natl. Acad. Sci. USA **81:** 7835–7840.
9. PRYSTOWSKY, M. B., G. OTTEN, M. F. NAUJOKAS, J. VARDIMAN, J. N. IHLE, E. GOLDWASSER & F. W. FITCH. 1984. Multiple hemopoietic lineages are found after stimulation of mouse bone marrow precursor cells with interleukin 3. Am. J. Pathol. **117:** 171–179.
10. IHLE, J.N., J. KELLER, S. OROSZLAN, L. E. HENDERSON, T. D. COPELAND, F. FITCH, M. B. PRYSTOWSKY, E. GOLDWASSER, J. W. SCHRADER, E. PALASZYNSKI, M. DY & B LEBEL. 1983. Biologic properties of homogeneous interleukin 3. I. Demonstration of WEHI-3 growth factor activity, mast cell growth factor activity, P cell-stimulating factor activity, colony-stimulating factor activity and histamine-producing cell-stimulating factor activity. J. Immunol. **131:** 282–287.
11. RENNICK, D. M., F. D. LEE, T. YOKOTA, K. ARAI, H. CANTOR & G. J. NABEL. 1985. A cloned MCGF cDNA encodes a multilineage hematopoietic growth factor: Multiple activities of interleukin 3. J. Immunol. **134:** 910–914.
12. MESSNER, H. A., K. YAMASAKI, N. JAMAL, M. M. MINDEN, Y.-C. YANG, G. G. WONG & S. C. CLARK. 1987. Growth of human hemopoietic colonies in response to recombinant gibbon interleukin-3: Comparison with human recombinant granulocyte-macrophage colony-stimulating factor. Proc. Natl. Acad. Sci. USA **84:** 6765–6769.
13. SUDA, T., J. SUDA, M. OGAWA & J. N. IHLE. 1985. Permissive role of interleukin 3 (IL-3) in proliferation and differentiation of multipotential hemopoietic progenitors in culture. J. Cell. Physiol. **124:** 182–190.
14. LEARY, A. G., Y.-C. YANG, S. C. CLARK, J. C. GASSON, D. W. GOLDE & M. OGAWA. 1987. Recombinant gibbon interleukin-3 (IL-3) supports formation of human multilineage colonies and blast cell colonies in culture: Comparison with recombinant human granulocyte-macrophage colony-stimulating factor (GM-CSF). Blood **70:** 1343–1348.
15. KOIKE, K., J. N. IHLE & M. OGAWA. 1986. Declining sensitivity to interleukin 3 of murine multipotential hemopoietic progenitors during their development: Application to a culture system that favors blast cell colony formation. J. Clin. Invest. **77:** 894–899.
16. SONODA, Y., Y.-C. YANG, G. G. WONG, S. C. CLARK & M. OGAWA. 1988. Analysis in serum-free culture of the targets of recombinant human hemopoietic factors: Interleukin-3 and granulocyte/macrophage colony-stimulating factors are specific for early developmental stages. Proc. Natl. Acad. Sci. USA **85:** 4360–4364.
17. METCALF, D., G. R. JOHNSON & A. W. BURGESS. 1980. Direct stimulation by purified GM-CSF of the proliferation of multipotential and erythroid precursors. Blood **55:** 138–147.
18. KOIKE, K., M. OGAWA, J. N. IHLE, T. MIYAKE, T. SHIMIZU, A. MIYAJIMA, T. YOKOTA & K. ARAI. 1987. Recombinant murine granulocyte-macrophage (GM) colony-stimulating factor supports formation of GM and multipotential blast cell colonies in culture: Comparison with the effects of interleukin-3. J. Cell. Physiol. **131:** 458–464.
19. Sieff, C. A., S. G. Emerson, R. E. Donahue, D. G. Nathan, E. A. Wang, G. G. Wong & S. C. CLARK. 1985. Human recombinant granulocyte-macrophage colony-stimulating factor: A multilineage hemopoietin. Science **230:** 1171–1173.
20. GABRILOVE, J. L., K. WELTE, P. HARRIS, E. PLATZER, L. LU, E. LEVI, R. MERTELSMANN & M. A. S. MOORE. 1986. Pluripoietin α: A second human hematopoietic colony-stimulating factor produced by the human bladder carcinoma cell line 5637. Proc. Natl. Acad. Sci. USA **83:** 2478–2482.
21. KAUSHANSKY, K., P. J. O'HARA, K. BERKNER, G. M. SEGEL, F. S. HAGEN & J. W. ADAMSON. 1986. Genomic cloning, characterization, and multilineage growth-promoting activity of human granulocyte-macrophage colony-stimulating factor. Proc. Natl. Acad. Sci. USA **83:** 3101–3105.
22. DONAHUE, R. E., E. A. WANG, D. K. STONE, R. KAMEN, G. G. WONG, P. K. SEHGAL, D. G. NATHAN & S. C. CLARK. 1986. Stimulation of haematopoiesis in primates by continuous infusion of recombinant human GM-CSF. Nature **321:** 872–875.
23. GROOPMAN, J. E., R. T. MITSUYASU, M. J. DELEO, D. H. OETTE & D. W. GOLDE. 1987.

Effect of recombinant human granulocyte-macrophage colony-stimulating factor on myelopoiesis in the acquired immunodeficiency syndrome. N. Engl. J. Med. **317:** 593–598.

24. STANLEY, E. R., A. BARTOCCI, D. PATINKIN, M. ROSENDAAL & T. R. BRADLEY. 1986. Regulation of very primitive, multipotent, hemopoietic cells by hemopoietin-1. Cell **45:** 667–674.
25. JUBINSKY, P. T. & E. R. STANLEY. 1985. Purification of hemopoietin 1: A multilineage hemopoietic growth factor. Proc. Natl. Acad. Sci. USA **82:**2764–2768.
26. MOCHIZUKI, D. Y., J. A. EISENMAN, P. J. CONLON, A. D. LARSEN & R. J. TUSHINSKI. 1987. Interleukin 1 regulates hematopoietic activity, a role previously ascribed to hemopoietin 1. Proc. Natl. Acad. Sci. USA **84:** 5267–5271.
27. MOORE, M. A. S. & D. J. WARREN. 1987. Synergy of interleukin-1 and granulocyte colony-stimulating factor: *In vivo* stimulation of stem-cell recovery and hematopoietic regeneration following 5-fluorouracil treatment of mice. Proc. Natl. Acad. Sci. USA **84:** 7134–7138.
28. IKEBUCHI, K., G. G. WONG, S. C. CLARK, J. N. IHLE, Y. HIRAI & M. OGAWA. 1987. Interleukin-6 enhancement of interleukin-3-dependent proliferation of multipotential hemopoietic progenitors. Proc. Natl. Acad. Sci. USA **84:** 9035–9039.
29. IKEBUCHI, K., S. C. CLARK, J. N. IHLE, L. M. SOUZA & M. OGAWA. 1988. Granulocyte colony-stimulating factor enhances interleukin-3-dependent proliferation of multipotential hemopoietic progenitors. Proc. Natl. Acad. Sci. USA **85:** 3445–3449.
30. IHLE, J. N., J. KELLER, L. HENDERSON, F. KLEIN & E. PALASZYNSKI. 1982. Procedures for the purification of interleukin 3 to homogeneity. J. Immunol. **129:** 2431–2436.
31. LEARY, A. G. & M. OGAWA. 1987. Blast cell colony assay for umbilical cord blood and adult bone marrow progenitors. Blood **69:** 953–956.
32. LEARY, A. G., K. IKEBUCHI, Y. HIRAI, G. G. WONG, Y.-C. YANG, S. C. CLARK & M. OGAWA. 1988. Synergism between interleukin-6 and interleukin-3 in supporting proliferation of human hemopoietic stem cells: Comparison with interleukin-1α. Blood **71:** 1759–1763.
33. KISHIMOTO, T. & T. HIRANO. 1988. Molecular regulation of B lymphocyte response. Annu. Rev. Immunol. **6:** 485–512.
34. BROUDY, V. C., K. KAUSHANSKY, J. M. HARLAN & J. W. ADAMSON. 1987. Interleukin 1 stimulates human endothelial cells to produce granulocyte-macrophage colony-stimulating factor and granulocyte colony-stimulating factor. J. Immunol. **139:** 464–468.
35. HIRANO, T., K. YASUKAWA, H. HARADA, T. TAGA, Y. WATANABE, T. MATSUDA, S. KASHIWAMURA, K. NAKAJIMA, K. KOYAMA, A. IWAMATSU, S. TSUNASAWA, F. SAKIYAMA, H. MATSUI, Y. TAKAHARA, T. TANIGUCHI & T. KISHIMOTO. 1986. Complementary DNA for a novel human interleukin (BSF-2) that induces B lymphocytes to produce immunoglobulin. Nature **324:** 73–77.

Regulation of Human Megakaryocytopoiesis

MICHAEL W. LONG

University of Michigan
Department of Pediatrics
Division of Hematology/Oncology
Room M7510 MSRB I, Box 0684
Ann Arbor, Michigan 48109

INTRODUCTION

A number of *in vitro* systems are available for the study of both human and murine megakaryocyte development. With these assays, it is possible to cultivate megakaryocyte progenitor cells,[1-12] immature megakaryocytes,[13-15] and mature megakaryocytes.[16,17] However, certain developmental characteristics of this lineage handicap the acquisition of insights into the regulation of megakaryocyte differentiation. Most notably, megakaryocyte progenitor cells are restricted in their proliferative potential generating colonies averaging 15–25 cells in size.[2,10] This reduction in proliferative capacity may be a function of the megakaryocyte's ability to endoreduplicate; *i.e.*, progressively double its DNA content without undergoing cytokinesis.[10] Until recently, the lack of knowledge concerning factors regulating megakaryocyte development has resulted in colonies of relatively immature cells, further handicapping the usefulness of clonigenic assays in the analysis of megakaryocyte growth and development. Finally, megakaryocyte progenitor cells and early differentiated cells of this lineage (immature megakaryocytes), are heterogeneous cell populations[2,13,18] additionally complicating the interpretation of the regulatory mechanisms operative in this lineage.

Information concerning the *in vitro* requirements for proliferation and differentiation of megakaryocytes suggests that two factors (activities) are required to regulate this process (FIG. 1). Megakaryocyte colony stimulating activity (Mk-CSA) is an obligate requirement for colony formation and primarily affects the proliferative phases of megakaryocyte development, driving the formation of small colonies of immature cells.[13,19] For optimal colony development (*i.e.*, colonies of large size containing mature megakaryocytes), a second auxiliary factor is required which has little or no role in proliferative events, but stimulates latter phases of differentiation. This distally acting factor (referred to in the murine system as megakaryocyte potentiator activity) lacks the solitary ability to stimulate colony formation, but augments CSA-driven megakaryocyte development by increasing the number of cells per colony, and the size, DNA content, cytoplasmic content, and nuclear complexity of developing megakaryocytes.[2,9-12,20,21] The recent purification of murine interleukin-3 (IL-3) by Ihle and co-workers[22-24] represents the first pure molecule to show megakaryocytic colony-stimulating activity,[25] as well as to support granulocyte and erythrocyte colony development.[24,26] While this purified hematopoietic growth factor is capable of supporting megakaryocyte colony formation, the colonies formed are typically small in size (containing 10–20 cells), the cells are relatively immature, and the cloning efficiency is low.[27] Optimal colony formation, in addition to purified IL-3, still requires the addition of an auxiliary activity to enhance colony development.[27]

Recent work from this laboratory has shown that tumor promoting phorbol diesters (PDE) substitute for biological sources of this megakaryocyte synergistic activity.[2,20] In other cell systems, such as the leukemic cell line HL60, or the pheochromocytoma cell line PC12, 4-β-phorbol 12-myristate 13-acetate (PMA) can induce cell differentiation.[28,29] Among hematopoietic cells, PMA stimulates granulocyte colony development.[30-33] In the erythroid lineage these agents stimulate early, burst-forming progenitor cells,[34] have little effect on later (day 3) colony-forming cells[34] but inhibit terminal phases of erythroid differentiation.[34,35] Data from our laboratory indicates that tumor-producing phorbol diesters stimulate both megakaryocyte progenitor cells and the differentiation of immature megakaryocytes.[2,20]

The development of *in vitro* assays for human megakaryocyte development lags behind the murine studies, undoubtedly due to the relative scarceness of this resource (human bone marrow) and further technical difficulties in growing megakaryocytes. Hu-

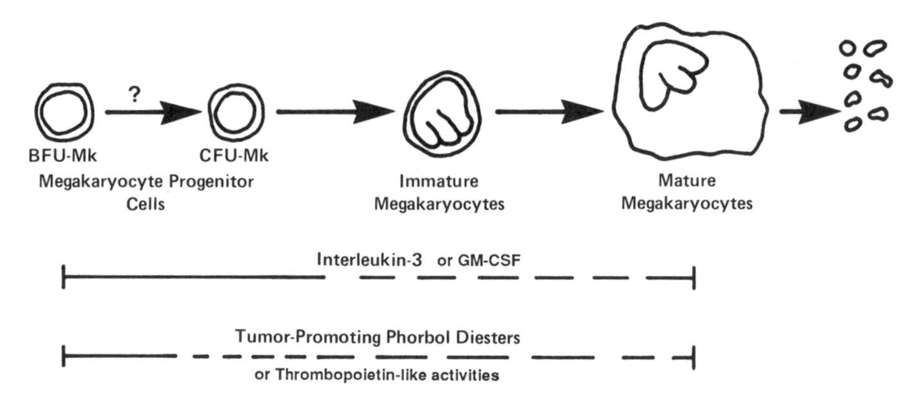

FIGURE 1. Regulation of megakaryocytopoiesis. The cellular elements of this lineage are: the burst-forming progenitor cell (BFU-Mk), an early progenitor which antedates the more mature colony forming cell (CFU-Mk);[2,27] the immature megakaryocyte, a lymphoid-size cell expressing megakaryocyte specific antigens, or enzymes and having a high nucleus:cytoplasm ratio[4,13,18,20] and the easily recognizable mature megakaryocyte. Factors regulating the development of these cells are IL-3 and GM-CSF as well as thrombopoietin-like synergistic co-regulators (see text). *Solid lines* indicate obligate requirement for growth factor, *broken lines* dual regulation.[2,25,27]

man megakaryocytes, for example, require large concentrations of plasma for growth,[36,37] and fetal calf sera is often inhibitory.[36,38] Nonetheless, human megakaryocyte progenitor cells can now be grown in a variety of semisolid support systems: plasma clot,[36,37] methylcellulose,[36,37] and agar.[21] Such studies report various positive and negative regulatory influences *in vitro*. However, large differences in stimuli, kinetics of development, and cloning efficiency are reported and systematic comparison of culture conditions and stimuli are lacking.

In regards to the factors regulating human megakaryocytopoiesis, a number of studies have shown circulating Mk-CSA in disorders such as aplastic anemia[3,37] or amegakaryocytic thrombocytopenia.[1] Other investigations have indicated that disorders such as reactive thrombocytosis, or primary thrombocythemia, lack detectable Mk-CSA and suggest that these disorders may involve an auxiliary factor.[1,39] Thus, these clinical data provide a powerful rationale for examination of these plasma activities (and similar activities from CM) in human megakaryocyte development.

We report that the optimal *in vitro* development of human megakaryocytes requires two synergistic regulatory activities; megakaryocyte colony-stimulating activity is an obligate requirement for megakaryocyte proliferation and stimulates the formation of small colonies of immature megakaryocytes; optimal megakaryocyte development requires an auxiliary activity which influences megakaryocyte size, cytoplasmic content, antigenic expression, and maturational stage.

MATERIALS AND METHODS

Preparation of Bone Marrow Cells

Following informed consent, bone marrow aspirates are obtained from the posterior iliac crest of normal volunteers. Volunteers were reportedly free of pharmaceuticals for three weeks prior to the aspiration.

Bone marrow cells are aspirated into syringes containing 100 U of preservative-free sodium heparin (University of Michigan) in normal saline. In order to prevent peripheral blood dilution, aspirate volumes were kept at 0.5–0.8 ml. Aspirates were immediately mixed with 100 μL of McCoy's 5A medium (see below) containing 100–125 U DNase I (Sigma, St. Louis, MO). The addition of DNase to the samples improves the recovery of cells following density separation. The aspirate is diluted with McCoy's 5A medium (Gibco, Grand Island, NY), layered over ficoll/diatrizoate (density = 1.077 gcm^{-3}, Histopaque, Sigma Chemical Co., St. Louis MO) and spun for 20 min at 400 g. The interface cells are collected, washed × 2, and subjected to two rounds of plastic adherence. For adherence, low density marrow cells are incubated (3×10^5/ml) in McCoy's 5A medium containing 5% FCS for one hour at 37°C, after which the suspension cells are removed and subjected to a second adherence step. The resultant nonadherent, low-adherent cells (NALD; ≤0.1% macrophages) are used as target cells in all experiments.

Hematopoietic Progenitor Cell Assays

Progenitor cells are cultured in a modification of a previously described agar system.[2,20] These cultures are performed as described for murine cells,[2,20] except that 30% human plasma replaces the FCS and PGE_2 is omitted. Briefly, NALD cells are cultured in supplemented McCoy's 5A medium containing 30% heparinized human plasma and 0.25% bactoagar (Difco, Detroit, MI). Bone marrow cells are co-cultivated with conditioned media from a human carcinoma cell line, human lung cell lines or PHA-LCM (*vide infra*). Incubation conditions are identical to those for methylcellulose cultures. Unless otherwise stated, all cultures are plated at a limiting cell density of 5×10^4 NALD cells/ml (the CFU-Mk assay, in agar, is linear over a range of $25–100 \times 10^3$ cells/ml, r = 0.95, $r^2 = 0.90$, y intercept = −0.3).

Megakaryocyte colonies are identified by their morphological characteristics as described elsewhere.[36–38] Morphological identification of megakaryocyte colonies (as presented in Results) was validated by immunoperoxidase labeling using platelet-megakaryocyte specific antibodies (see below). Individual marrow sample replicate cultures were air dried[7] and stained by immunoperoxidase labeling. These parallel cultures gave equivalent results to those based on morphological identification for each of the regulatory activities examined (data not shown).

The sample number reported (n) represents the number of individual marrows examined. For each marrow, all experimental conditions are cultured as 3–5 replicate cultures/condition. Values are expressed as mean ± SD of the mean replicate culture values averaged over the indicated number (n) of individual marrow samples.

Conditioned Media

Various conditioned media are used as sources of hematopoietic regulatory activities. These are: PHA-LCM, a standard source of such regulatory activities,[36,37] and media conditioned by 3 human cell lines: a bladder carcinoma cell line, 5637 (a generous gift of the late Dr. Jorgen Flogh, Sloan Kettering Institute, Rye, NY), and 2 lung cell lines (A549 and Hel 299 from American Type Tissue Culture Collection, Rockville, MD).

Human Phytohemagglutinin Stimulated Leukocyte Conditioned Medium (PHA-LCM)

Human leukocyte-rich plasma is obtained by admixing peripheral blood (buffy coat) with 2% methylcellulose (1:2) and the leukocyte-rich plasma collected 30–60 min later. The cells are washed 3 × with IMDM, and cultured (1 × 10^6/ml) in IMDM containing 10% FCS and 1% PHA (Welcome Diagnostics, Greenville, NC). After 7 days, conditioned medium is harvested by centrifugation, filter sterilized (0.22μ) and stored at −80°C until used.

Human Lung Conditioned Medium (HLuCM)

Human lung cells are grown in RPMI 1640 and containing 10% FCS (Hyclone, Ogden, UT) supplemented with 1.0% essential amino (50×), 0.75% nonessential amino acids (100×) and 0.75% vitamins (100×), v/v (all from Gibco), 7.5% sodium bicarbonate, and 0.25% glucose. These cells are grown to confluence, and the medium conditioned for 5 days. Supernatant medium is decanted, centrifuged, filter sterilized and stored at −80°C.

Human Bladder Carcinoma (5637) Conditioned Medium (BlCM)

Adherent human bladder carcinoma cells are grown in RPMI 1640 containing 0.75% (w/v) glutamine (100×, Gibco) and 2% FCS. Cells are cultured in 75-cm^2 flasks on Cytodex-3 beads (Pharmacia, Piscataway, NJ). Cells are grown to confluence (on beads), and the medium conditioned for 3–4 days. Following centrifugation, the conditioned medium is concentrated 5-fold by ultrafiltration (Amicon YM-10 membrane, Amicon Corp., Lexington, MA), filter sterilized, and stored at −80°C.

Optimal concentrations of these conditioned media are determined using reciprocal titrations as reported elsewhere.[14] These CMs are examined over a range of 0.1–30%. Each CM is titrated alone, and in the presence of the putative, synergistic regulator. The addition of known synergistic activities (*e.g.*, PMA) serves to unmask relatively weak CSAs. These concentration:responsiveness studies thus detect differential thresholds (either by examining broad ranges of CM alone, or by co-culture) or inhibitors (by dilution). However, such studies cannot exclude the presence of an inhibitory activity which has an identical response curve in both types of titration experiment. While improbable, such an event may reduce or mask lineage specific or nonspecific activities. The concentrations routinely used are: BlCM, 7.5%, HLuCM, 10%, PHA-LCM, 10% (final concentrations).

Tumor-Promoting Phorbol Diesters

Phorbol diesters (Sigma) are prepared as stock solutions of 10^{-5} M in dimethylsulfoxide and kept in the dark at −20°C until used. Fresh stock solutions are prepared every

3 weeks. Immediately before use, a working solution is made using culture medium as the diluent.

Human Plasma Samples

Plasma is obtained from normal volunteers following informed consent. Plasma is anticoagulated with preservative-free heparin and immediately spun (1500 g $\times$ 20 min) to produce platelet poor plasma. This plasma is filter sterilized (0.22 μ) and used immediately. Normal plasma, alone, does not stimulate colony development, but is screened for the optimal concentration supporting colony growth (30–40%, final concentration).

Liquid Suspension Cultures

Bone marrow NALD cells are prepared as described above. Following plastic-adherence, the cells are subjected to nylon wool adherence to remove residual adherent cells and B lymphocytes.[20] These human marrow NALD-B cells are cultivated (1–2 $\times$ 10^5/ml) in the presence of regulatory activities for 5 days. Culture conditions are as described above.

Human Recombinant Hematopoietic Growth Factors

Recombinant human IL-3, GM-CSF and IL-6 were the generous gift of Dr. Steven Clark, Genetics Institute, Cambridge, MA. The rh-IL-3 (lot 1039-86; sp. act 0.5–1.0 $\times$ 10^7 U/mg) is *E. coli* expressed and 99% pure. Human, rh-GM-CSF (lot CD19D2-701; sp. act. 1 $\times$ 10^7 U/mg) is *E. coli* expressed and 99% pure.

Solid-Phase Radioimmunoassay (SPRIA) for Hematopoietic Cell Antigens

In order to rapidly and precisely define factor-related changes in megakaryocyte development, we developed a SPRIA to detect megakaryocyte antigens.[40] Briefly, the SPRIA uses a filtration membrane as a solid support (Durapore, Millipore Corp., Bedford, MA) and antibody-bound ^{125}I-staphylococcal protein A (SPA, ICN, Irvine, CA) as a detection system. Intact, cultivated bone marrow cells (NALD-B) are loaded into microtiter wells, and all washes and incubations performed *in situ*. Antibody binding is thus performed in a solid-phase, *i.e.*, antigens remain membrane associated on intact cells which are affixed to the polyvinylidene membrane. Additionally, this procedure prevents cell loss due to repetitive centrifugation, and wash steps.

Antibody Specificity

Three platelet/megakaryocyte specific antibodies are used in these studies.[40] These were: Anti-Factor VIIIag (Dako, Santa Barbara, CA), a monospecific heterologous antibody adsorbed 3 $\times$ with human WBCs, RBCs, and mouse liver powder; and two monoclonal antibodies: antiplatelet glycoprotein IIb/IIIa (a generous gift of Dr. Robb Todd, University of Michigan), and antiplatelet glycoprotein Ib (Dako). The final concentration of each antibody was: anti-Factor VIIIag 1:1000, anti-GP IIb/IIIa 1:30, and anti-Ib 1:20. In order to improve detection, a cocktail of these antibodies (each at optimal concentration) was used. All results reported are based on the signal detected using this antibody cocktail.

Megakaryocyte Morphology

Megakaryocytes are identified using avidin-biotin immunoperoxidase conjugates[41] to label megakaryocyte colonies *in situ* or cytocentrifuge preparations of cells grown in suspension. Analysis of bone marrow cellular reactivity using these antibodies by immunocytochemistry shows only mature and immature megakaryocyte positive for the peroxidase reaction.[40] As well, antibody specificity in the RIA system was examined by using purified human PMNs, HL60 cells, and input NALD cells. None of these latter controls shows specific binding differing from zero.

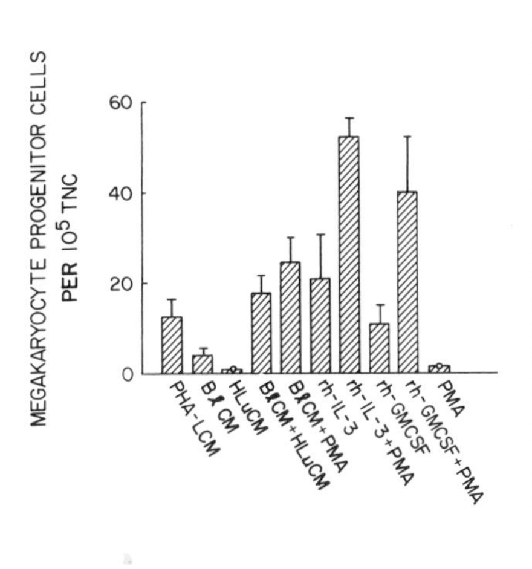

FIGURE 2. Responsiveness of human megakaryocyte progenitor cells to recombinant growth factors and various conditioned media. Human progenitor cells cultured in agar as described in Methods. Values are mean ± SD of 10 individual marrow samples cultured at 3–5 replicate cultures per condition. BlCM, human bladder cell line 5637 CM (7.5% final concentration); PMA, phorbol myristate acetate; PHA-LCM, phytohemagglutinin stimulated human leukocyte conditioned medium (7.5–10% v/v, final concentration); HLuCM, human lung conditioned medium; rh-IL-3, recombinant human interleukin-3 tested at 1 ng/ml; rh-GM-CSF, recombinant granulocyte/macrophage CSF tested at 300 ng/ml. Target cells are NALD-B cells. High concentrations (20% v/v of HLuCM stimulate 7–10 granulocyte clusters/10^5 TNC; contrasts of BlCM vs BlCM plus HLuCM or PMA or rh-IL-3 versus rh-IL-3 plus PMA are significant ($p \leq 0.05$, Student *t* test). Bars marked with a zero show no colony development.

RESULTS

Human Megakaryocyte Colony Development In Vitro

In vitro human megakaryocyte colony development requires at least one regulatory activity: an obligate megakaryocyte colony-stimulatory activity necessary for progenitor cell maintenance and proliferation. Conditioned medium both from the human bladder carcinoma line 5637 (BlCM) and from PHA-LCM are required for colony formation but support limited proliferation of bone marrow megakaryocyte progenitor cells (CFC-Mk) (FIG. 2). These conditioned media were active in final concentrations of 2.5–20% (optimal = 7.5%) and support 5–12 CFC-Mk/10^5 NALD-B cells. Colonies stimulated by optimal concentrations of these CMs contained 3–20 megakaryocytes of relatively small size.

Conditioned medium from the human lung cell line A549 (HLuCM) lacks the solitary ability to promote CFU-Mk growth. However, co-culture of HLuCM with conditioned medium from 5637 bladder carcinoma cells significantly augments megakaryocyte colony formation resulting in 17.6 ± 7/10^5 TNC megakaryocyte progenitor cells contrasted to 3.9 ± 3 colonies formed in the presence of BlCM alone (n = 10, $p \leq 0.05$, Student *t* test)

(FIG. 2). Megakaryocyte colonies grown in the presence of these two conditioned media contain 3–200 megakaryocytes which were larger than megakaryocytes stimulated with bladder carcinoma conditioned medium alone (not shown).

In order to identify recombinant hematopoietic factors capable of regulating megakaryocyte development, we next examined recombinant human IL-3 and GM-CSF for solitary ability to stimulate megakaryocyte colony development or for their actions in augmenting CSF-driven colony formation. Interleukin-3 alone supports equivalent colony formation to that seen with PHA-LCM or BlCM (FIG. 2). Nonetheless, addition of PMA to IL-3 containing cultures significantly augments megakaryocyte colony formations yielding 52 ± 4 colonies/10^5 TNC. This augmentation megakaryocyte colony formation occurs over an IL-3 concentration range of 100–1000 pg/ml (data not shown). Similarly, granulocyte/macrophage CSF stimulated equivalent megakaryocyte colony formation but requires an approximate 100-fold increase in concentration (FIG. 2). Synergistic interactions also occur between GM-CSF and phorbol diesters. At 300 ng/ml GM-CSF, approximately 4 megakaryocyte colonies are developed/10^5 TNC, whereas 10^{-8} M PMA yields 48 ± 23 colonies (mean ± SD; n = 4). Additionally, recombinant erythropoietin, IL-1, or IL-6, do not show solitary colony-stimulating activity or the ability to synergize with sources of Mk-CSA such as interleukin-3 (data not shown).

The requirement for two synergistic activities is confirmed in studies on the effects of phorbol diesters on human megakaryocyte colony development. Phorbol myristate acetate can substitute for HLuCM (but not for (BlCM), a source of Mk-CSA) and increase megakaryocyte colony numbers. In the presence of a source of Mk-CSA, PMA potentiates megakaryocyte colony development over a range of 10^{-9}–10^{-7} M (FIG. 3A). Addition of 10^{-8} PMA to optimal amounts of Mk-CSA causes a 9-fold increase in megakaryocyte colony formation. Examination of the structure activity relationship of phorbol diesters indicates that the tumor-promoting phorbol diesters PMA and phorbol dibutyrate (PdBU) are capable of substituting for the biological source of the synergistic activity whereas the inactive parent compound phorbol, or the vehicle DMSO are not (FIG. 3B).

In order to precisely identify the biological effects of the synergistic activity observed in human lung conditioned medium, human low-density nonadherent cells were cultivated in suspension culture in the presence of sources of Mk-CSA alone or co-cultivated with this activity and either PMA or a source of the biological stimulatory activity (human lung conditioned medium).

Morphological analysis of megakaryocytes grown in the suspension phase cultures indicates a substantial increase in the percentage of mature megakaryocytes stimulated by the addition of secondary synergistic activity (FIG. 4). When cultivated in the presence of Mk-CSA alone, only immature megakaryocytes are seen. These cells have a high nucleus: cytoplasm ratio, are small in size and contain low amounts of megakaryocyte antigen (as visualized by immunolocalization). The addition of either source of synergistic co-regulator results in the reversal of the nucleus:cytoplasm ratio, an increase in size, increased nuclear lobulation, and a visual increase in the content of megakaryocyte specific antigens per cell.

The observation of increased megakaryocyte antigen content suggests that the synergistic co-regulator is capable of regulating megakaryocyte antigenic expression. Two alternative approaches indicate that one of the biological roles of the synergistic co-regulator is the regulation of megakaryocyte antigens. Suspension phase cultures of non-adherent low-density (NALD-B) cells, incubated in the presence of Mk-CSA alone or co-incubated with Mk-CSA and human lung conditioned medium increase the number of antigen positive cells. Megakaryocyte-CSA alone is capable of increasing the number of antigen positive cells roughly four-fold over input levels (FIG. 5) The addition of optimal concentrations (2.5%) of human lung conditioned medium doubles of the frequency of antigen positive megakaryocytes contrasted with cells cultivated in Mk-CSA alone.

In order to more precisely quantitate the regulation of megakaryocyte antigenic content, we developed a solid phase radioimmunoassay (SPRIA) capable of detecting as few as 300 megakaryocytes.[40] This SPRIA indicates that Mk-CSA alone stimulates low amounts of antigen expression, whereas co-cultivation of Mk-CSA and 10^{-8} M PMA results in a 14-fold increase in specific binding compared to unstimulated cells (TABLE 1).

DISCUSSION

Optimal human megakaryocyte development *in vitro* requires the presence of at least two regulators which promote proliferation and differentiation of these cells. Conditioned

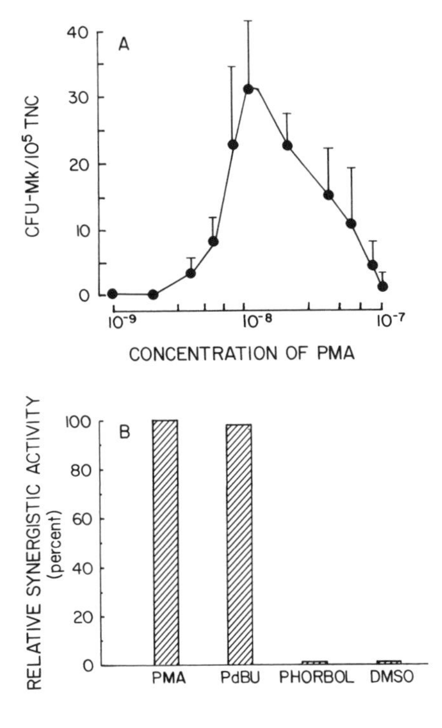

FIGURE 3. **(A)** Responsiveness of human megakaryocyte progenitor cells to phorbol myristate acetate. Progenitor cell assays performed as in text. All assays contain optimal concentrations (7.5% v/v) BICM. Values are mean ± SD (n = 4). **(B)** Structure-activity relationship between various phorbol esters and megakaryocyte-potentiating ability. For comparison, the maximum synergistic activity of 10^{-8} M PMA, when co-cultured with optimal amounts of BICM (7.5% final concentration) is expressed as 100. All phorbol esters tested at 10^{-8} M. Values are means of 4 individual marrow samples each cultured at 3–5 replicate culture/experiment/condition. Optimal value of BICM + PMA represents 24.6 ± 11.3 CFU-Mk/10^5 TNC. PMA, phorbol 12-myristate 13-acetate, PdBU, phorbol 12, 13-dibutyrate; PHORBOL, the parent alcohol of PMA; DMSO, dimethylsulfoxide.

medium from bladder carcinoma cells (line 5637), peripheral blood leukocytes (PHA-LCM), or the recombinant growth factors, IL-3 or GM-CSF, supports limited megakaryocyte development generating colonies of immature megakaryocytes. Colony development is greatly enhanced by the addition of conditioned medium from a neoplastic lung cell line (A549), or tumor promoting phorbol diesters suggesting the existence of dual or multiple regulators in this process. The synergistic co-regulator lacks the solitary ability to stimulate progenitor cell proliferation, but appears to amplify megakaryocyte maturation by increasing the number and size of developing megakaryocytes, the expression of antigenic determinants, and cytoplasmic and nuclear maturation.

We examined the actions of recombinant human hematopoietic growth factors to precisely establish which factors in the complex condition media may control megakaryocyte proliferation. Both rh-IL-3 and GM-CSF support the proliferation of megakaryocyte progenitor cells but have limited effects on maturation. Such observations extend and confirm recent reports by Quesenberry and co-workers showing that murine rh-IL-3 and GM-CSF support megakaryocyte colony formation.[25] While human megakaryocyte colony numbers are higher in the presence of recombinant growth factors than conditioned media, colony development is suboptimal unless another regulatory factor is present. Such an activity is found in HLuCM and its substitute PMA. The addition of phorbol diesters to recombinant growth factors results in similar increases in colony number and size as was observed with the biological activity present in human lung

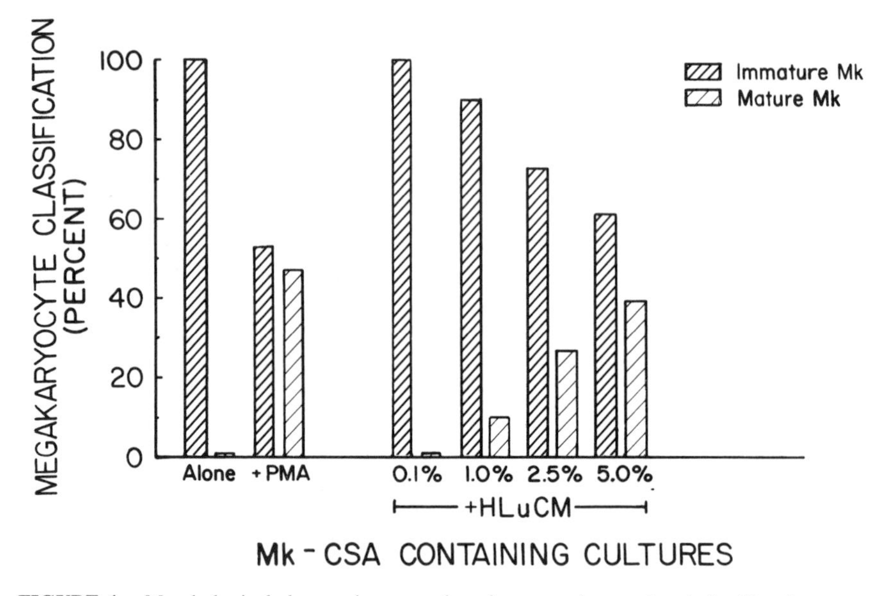

FIGURE 4. Morphological changes in suspension phase megakaryocytopoiesis. Megakaryocytes grown in suspension phase cultures as in Methods. Quantitative values are mean ± SD (n = 3) except for BlCM + HLuCM titration which are 5 separate cultures from an individual marrow donor. Abbreviations as in FIGURE 2. Megakaryocytes are identified by immunoperoxidase and classified as: immature cells ≤18 diameter, high nucleus:cytoplasm content[18] and low amounts of antigen (trace - 1+), and mature cells being ≥20 diameter, low nucleus:cytoplasm ratio and large amounts of antigen (+ + to + + + +); for classification approximately 100 megakaryocytes (mature and immature) were counted per condition/donor. Percentages are amount (v/v) of HLuCM added to a constant concentration of BlCM (7.5%, v/v).

conditioned medium. Thus the use of recombinant growth factors plus PMA as a substitute for the biological co-regulator allows unambiguous confirmation of the need for two factors in megakaryocyte development.

The observed increases in megakaryocyte antigenic expression also substantiates our observations of synergistic regulation of megakaryocyte maturation. In suspension cultures, both HLuCM and PMA elevate total antigenic expression over that seen with Mk-CSA alone. Notably, the number of total antigen positive cells (*i.e.*, immature plus mature megakaryocytes) increases greatly. The addition of Mk-CSA alone results in a 4-fold increase in megakaryocyte numbers over input levels. Co-culture of Mk-CSA with

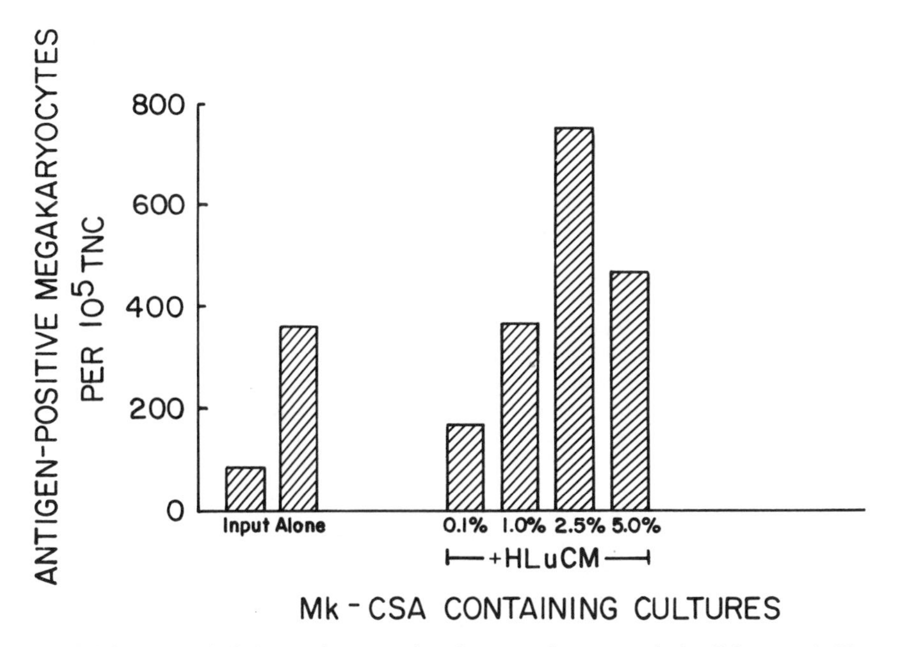

FIGURE 5. Numerical changes in suspension phase magakaryocytopoiesis. Cultures and abbreviations as in FIGURE 4.

HLuCM yields a 9-fold increase in megakaryocyte numbers. Finally, substitution of PMA for HLuCM causes a 37-fold increase (Long *et al.*, unpublished observations). The increases observed in both colony numbers and the total number of antigenic positive megakaryocytes suggest that synergistic regulator(s) either stimulate proliferation, without supporting colony formation, or, alternatively, regulate the expression of megakaryocyte antigens making previously unrecognizable cells antigen positive. This latter hypothesis was confirmed in SPRIA studies showing that antigen positivity per cell increases in the presence of Mk-CSA and the synergistic co-regulator. However, these studies do not allow distinction between actions on the progenitor cell, which is capable of proliferation, or antigen negative immature megakaryocytes, which then become antigen positive; both target cell populations exist in these cultures.[16]

TABLE 1.
Regulation of the Antigenic Content of Developing Megakaryocytes[a]

Stimuli	Specific Binding ^{125}I-SPA (pg/10^5 TNC)
Input	2.0
After incubation[b]	20.7 ± 5
Mk-CSA[c]	45.7 ± 24
Mk-CSA + PMA[d]	350.0 ± 16

[a]Solid-phase radioimmunoassay (SPRIA) performed as in Methods. Stimulation of antigenic content after 5 days of culture. Values are mean ± SD (n = 11, except for input, which is a single sample). Values are means of triplicate SPRIA determinations for each donor.

[b]Incubation of unstimulated NALD-B cells in the presence of bovine serum albumin (BSA).

[c]Mk = CSA, megakaryocyte colony-stimulating activity.

[d]PMA, phorbol myristate acetate.

Two-tiered regulation of human megakaryocyte development was first described by Harker and Finch who proposed that separate activities regulate the expansion versus the size of developing megakaryocytes.[42] In the murine system, early after thrombocytopenia, megakaryocytes increase in size, cytoplasmic content, and maturity.[43–46] With prolonged stimulation changes are seen in megakaryocyte number.[47,48] These observations suggest that two phases of *in vivo* regulation occur, an early thrombopoietin response affecting megakaryocyte/platelet development, and a later Mk-CSA responsive phase affecting megakaryocyte number (proliferation).

In vitro, Hoffman and co-workers observe that purified Mk-CSA acts at the level of the megakaryocyte progenitor cell stimulating the formation of colonies of small immature megakaryocytes.[19] They further suggest that activities such as thrombopoietin regulate more distal events. Consistent with this, Gewirtz *et al.* failed to detect Mk-CSA in patients with reactive thrombocytosis and hypothesized the presence of a second regulator in these patients.[39] We show that a synergistic co-regulator functions with Mk-CSA to regulate megakaryocyte size, cytoplasmic content, and antigen maturity. Thus, the *in vitro* actions of this synergistic regulator mimic some of the known *in vivo* effects of thrombopoietin. Discovering whether these regulators are different molecules or members of the same family of regulators must await purification of both factors.

The eventual purification of an activity regulating the latter phases of megakaryocyte development will allow functional comparison between this and other factors influencing differentiation. Rosenberg and co-workers recently purified a protein, megakaryocyte stimulatory factor (MSF), having the ability to stimulate platelet factor 4 synthesis, a functional characteristic associated with megakaryocyte maturation.[49,50] It is premature to establish identity between this molecule and the activity in HLuCM. While the latter is not purified, the similarities are that both influence megakaryocyte maturation. However, the activity in HLuCM works at the progenitor level to augment CSA-driven colony formation, stimulates increases in size, cytoplasmic content, nuclear complexity, and antigenic content in developing megakaryocytes. These functions were not examined for MSF (except that it is not a CSA). Further work at the biochemical level (for our activity) and the cellular level (for MSF) is required to identify similarities or co-factor type interactions between these factors.

We conclude that optimal human megakaryocyte development requires the presence of two synergistic regulatory activities. Regulation of megakaryocyte proliferation is an activity of IL-3 or GM-CSF, either of which is obligate for colony formation and stimulates proliferation and limited differentiation of megakaryocytes. Full megakaryocyte development requires a synergistic co-regulator which, lacking CSA, influences characteristics of megakaryocyte maturation such as: number, size, cytoplasmic content, antigenic content, and maturity. However, some of the studies here utilize conditioned medium rather than purified proteins, and accessory cell-depleted bone marrow rather than purified cells, making direct assessment of the role of either factor and their target cells difficult. Thus it cannot be concluded that accessory cells are not involved in these studies or that more than one synergistic regulator is present. Given these limitations, further characterization of the role of co-regulator and its targets must await purification of this synergistic regulatory factor and the use of purified target cell populations.

REFERENCES

1. Hoffman, R., E. Mazur, E. Bruno & V. Floyd. 1981. Assay of an activity in the serum of patients with disorders of thrombopoiesis that stimulates formation of megakaryocytic colonies. N. Engl. J. Med. **305:** 533–538.

2. LONG, M.W., L. L. GRAGOWSKI, C. H. HEFFNER & L. A. BOXER. 1985. Phorbol diesters stimulate the development of an early murine progenitor cell. J. Clin. Invest. **76:** 431–438.
3. MAZUR, E. M., R. HOFFMAN & E. BRUNO. 1981. Regulation of human megakaryocytopoiesis. J. Clin. Invest. 68**: 733–741.**
4. MAZUR, E. M., R. HOFFMAN, J. CHASIS, S. MEROLES & E. BRUNO. 1981. Immunofluorescent identification of human megakaryocytic colonies using an antiplatelet glycoprotein antiserum. Blood **57:** 277–286.
5. VAINCHENKER, N., W. J. GUZCHARD & J. BRETON-GORIUS. 1968. Growth of human megakaryocyte colonies in culture from fetal, neonatal and adult peripheral blood cells: An ultrastructural analysis. Blood Cells **5:** 25–31.
6. WILLIAMS, N. & H. JACKSON. 1968. Regulation of proliferation of murine megakaryocyte progenitor cells by cell cycle. Blood **52:** 163–168.
7. WILLIAMS, N., H. M. JACKSON, R. R. EGER & M. W. LONG. 1981. The separate roles of factors in murine megakaryocyte colony formation. *In* Megakaryocyte Biology and Precursors: Cloning and Cellular Properties. B. L. EVATT, R. F. LEVINE & N. WILLIAMS, Eds.: 69–75. North-Holland. Amsterdam.
8. WILLIAMS, N. & R. LEVINE. 1982. The origin, development and regulation of megakaryocytes. Br. J. Haematol. **52:** 173–180.
9. WILLIAMS, N., H. JACKSON, A. P. C. SHERIDA, M. J. MURPHY, JR., A. ELSTE & M. A. S. MOORE. 1978. Regulation of megakaryocytopoiesis in long-term murine bone marrow cultures. Blood Cells **5:** 43–48.
10. WILLIAMS, N. & H. JACKSON. 1982. Kinetic analysis of megakaryocyte numbers and ploidy levels in developing colonies from mouse bone marrow cells. Cell Tissue Kinet. **15:** 483–494.
11. WILLIAMS, N., R. R. EGER, H. M. JACKSON & D. J. NELSON. 1982. Two-factor requirement for murine megakaryocyte colony formation. J. Cell. Physiol. **110:** 101–104.
12. WILLIAMS, N., H. JACKSON, P. RALPH & I. NAKOINZ. 1981. Cell interactions influencing murine marrow megakaryocytes: Nature of the potentiator cell in bone marrow. Blood **57:** 157–160.
13. LONG, M.W. & N. WILLIAMS. 1982. Immature megakaryocytes in the mouse: Physical characteristics, cell cycle status and *in vitro* responsiveness to thrombopoietic stimulatory factor. Blood **59:** 569–757.
14. LONG, M.W. & N. WILLIAMS. 1982. Differences in the regulation of megakaryocytopoiesis in the murine bone marrow and spleen. Leukemia Res. **6:** 721–728.
15. LONG, M. W., N. WILLIAMS & T. P. MCDONALD. 1982. Immature megakaryocytes in the mouse: *In vitro* relationship to megakaryocyte progenitor cells and mature megakaryocytes. J. Cell. Physiol. **112:** 339–344.
16. RABELLINO, E. M., R. B. LEVENE, L. L. K. LEUNG & R. L. NACHMAN. 1981. Human megakaryocytes. II. Expression of platelet proteins in early marrow megakaryocytes. J. Exp. Med. **154:** 88–100.
17. RABELLINO, E. M., R. L. NACHMAN, N. WILLIAMS, R. J. WINCHESTER & G. D. ROSS. 1974. Human megakaryocytes. I. Characterization of membrane and cytoplasmic components of isolated marrow megakaryocytes. J. Exp. Med. **149:** 1273–1287.
18. LONG, M. W. & N. WILLIAMS. 1981. Immature megakaryocytes in the mouse: Morphology and quantification by acetylcholinesterase staining. Blood **58:** 1032–1039.
19. HOFFMAN, R., H. H. YANG, E. BRUNO & J. E. STRANEVA. 1985. Purification and partial characterization of a megakaryocyte colony-stimulating factor from human plasma. J. Clin. Invest. **75:** 1174–1182.
20. LONG, M. W., J. E. SMOLEN, P. SZCZEPANSKI & L. A. BOXER. 1984. Role of phorbol diesters in *in vitro* murine megakaryocyte colony formation. J. Clin. Invest. **74:** 1686–1692.
21. LONG, M. W., R. HUTCHINSON, C. HEFFNER & L. GRAGOWSKI. 1985. Human megakaryocyte colony development *in vitro* requires two synergistic activities. Clin. Res. **33:** 347a.
22. IHLE, J. H., J. KELLER, L. HENDERSON, F. KLEIN & E. PALASZYNSKI. 1982. Procedures for the purification of interleukin-3 to homogeneity. J. Immunol. **129:** 2431–2436.
23. LEE, J. C., A. J. HAPEL & J. N. IHLE. 1982. Constitutive production of a unique lymphokine (IL-3) by the Wehi-3 cell. J. Immunol. **128:** 2393–2398.
24. PRYSTOWSKY, M. B., G. OTTEN, M. NAUJOKAS, J. VARDIMAN, J. N. IHLE, E. GOLDWASSER

& F. W. FITCH. 1984. Multiple hemopoietic lineages are found after stimulation of mouse bone marrow precursor cells with interleukin-3. Am. J. Pathol. **117:** 171–179.
25. GOODMAN, J. W., E. A. HALL, K. L. MILLER & S. G. SHINPOCK. 1985. Interleukin-3 promotes erythroid burst formation in "serum-free" cultures without detectable erythropoietin. Proc. Natl. Acad. Sci. USA **82:** 3291–3295.
26. QUESENBERRY, P. J., J. N. IHLE & E. MCGRATH. 1985. The effect of interleukin-3 and GM-CSA-2 on megakaryocyte and myeloid clonal colony formation. Blood **65:** 214–217.
27. LONG, M. W., C. H. HEFFNER & L. L. GRAGOWSKI. 1986. *In vitro* differences in responsiveness of early (BFU-Mk) and late (CFU-Mk) murine megakaryocyte progenitor cells. *In* Megakaryocyte Development and Function: 179–186. Alan R. Liss. New York, NY.
28. ROVERA, J. T. J. O'BRIEN & L. DIAMOND. 1979. Induction of differentiation in human promyelocytic leukemia cells by tumor promoters. Science **204:** 868–870.
29. END, D., N. TOLSON, M. Y. YU & G. GUROF. 1982. Effects of 12-O-tetradecanoyl phorbol-13-acetate on rat pheochromocytoma (PC 12) cells: Interactions with epidermal growth factor and nerve growth factor. J. Cell. Physiol. **111:** 140–148.
30. STUART, R. K. & J. A. HAMILTON. 1980. Tumor-promoting phorbol esters stimulate hematopoietic colony formation *in vitro*. Science **208:** 402–404.
31. STUART, R. K., L. L. SENSENBRENNER, R. K. SHADDUCK, A. WAHEED & C. CARAMATTI. 1983. Phorbol ester-stimulated murine myelopoiesis. Role of colony-stimulating factors. J. Cell. Physiol. **117:** 30–38.
32. STUART, R. K., J. A. HAMILTON, L. L. SENSENBRENNER & M. A. S. MOORE. 1981. Regulation of myelopoiesis *in vitro:* Partial replacement of colony-stimulating factors by tumor-promoting phorbol esters. Blood **57:** 1032–1042.
33. GRIFFIN, J. D., P. LARCOM & D. W. KUFE. 1985. TPA induces differentiation of purified human myeloblasts in the absence of proliferation. Exp. Hematol. **13:** 1025–1032.
34. FIBACH, E., P. A. MARKS & R. A. RIFKIND. 1980. Tumor promoters enhance myeloid and erythroid colony formation by normal mouse hematopoietic cells. Proc. Natl. Acad. Sci. USA **77:** 4152–4155.
35. FIBACH, E., R. GAMBARI, P. A. SHAW, G. MANAITIS, R. C. REUBEN, S. SHIGERU, R. A. RIFKIND & P. MARKS. 1979. Tumor promoter-mediated inhibition of cell differentiation: Suppression of the expression of erythroid functions in murine erythroleukemia cells. Proc. Natl. Acad. Sci. USA **70:** 1906–1910.
36. MESSNER, H. A., N. JAMAL & C. IZAGUIRRE. 1982. The growth of large megakaryocyte colonies from human bone marrow. J. Cell. Physiol. Suppl. **1:** 45–51.
37. SOLBERG, L. A., JR., N. JAMAL & H. A. MESSNER. 1985. Characterization of human megakaryocytic colony formation in human plasma. J. Cell. Physiol. **124:** 67–74.
38. KIMURA, H., S. A. BURSTEIN, D. THORNING, J. S. POWELL, L. A. HARKER, P. J. FIALKOW & J. W. ADAMSON. 1984. Human megakaryocytic progenitors (CFU-M) assayed in methylcellulose: Physical characteristics and requirements for growth. J. Cell. Physiol. **118:** 87–96.
39. GEWIRTZ, A. M., E. BRUNO, J. ELWELL & R. HOFFMAN. 1983. *In vitro* studies of megakaryocytopoiesis in thrombocytotic disorders of man. Blood **61:** 384–389.
40. LONG, M. W. & C. H. HEFFNER. 1988. Detection of human megakaryocyte antigens by solid-phase radioimmunoassay. Exp. Hematol. **16:** 62–70.
41. HSU, S. M., L. RAINE & H. FANGER. 1981. The use of avidin-biotin-peroxide complex (ABC) in immunoperoxidase technique—a comparison between ABC and unlabeled antibody (PAP) procedures. J. Histochem. Cytochem. **29:** 577–580.
42. HARKER, L.A. & C. N. FINCH. 1969. Thrombokinetics in man. J. Clin. Invest. **48:** 963–974.
43. ODELL, T. T. & C. SHELTON. 1979. Increasing stimulation of megakaryocytopoiesis with decreasing platelet count. Proc. Soc. Exp. Biol. Med. **161:** 531–533.
44. ODELL, T. T., J. R. MURPHY & C. W. JACKSON. 1976. Stimulation of megakaryocytopoiesis by acute thrombocytopenia in rats. Blood **48:** 765–775.
45. ODELL, T. T., C. W. JACKSON, T. J. FRIDAY & D. E. CHARSHA. 1969. Effects of thrombocytopenia on megakaryocytopoiesis. J. Haematol. **17:** 91–101.
46. EBBE, S. 1976. Biology of Megakaryocytes. *In* Progress in Hemostasis and Thrombosis, Vol. III. T. H. SPAET, Ed.: 211–229. Grune & Stratton. New York, NY.
47. HARKER, L. A. 1968. Kinetics of thrombopoiesis. J. Clin. Invest. **47:** 458–465.

48. ODELL, T. T., C. W. JACKSON, T. J. FRIDAY & D. E. CHARSHA. 1969. Effects of thrombocytopenia on megakaryocytopoiesis. Br. J. Maematol. **17:** 91–101.
49. TAYRIEN, G. & R. D. ROSENBERG. 1987. Purification and properties of a megakaryocyte stimulatory factor present both in the serum-free conditioned medium of human embryonic kidney cells and in thrombocytopenic plasma. J. Biol. Chem. **262:** 3262–3268.
50. GREENBERG, S. M., D. J. KUTER & R. D. ROSENBERG. 1987. *In vitro* stimulation of megakaryocyte maturation by megakaryocyte stimulatory factor. J. Biol. Chem. **262:** 3269–3277.

Regulation of Human Fetal Hemoglobin Gene Expression

JAMES KAYSEN, MARYANN DONOVAN-PELUSO,
SANTINA ACUTO, DAVID O'NEILL, AND ARTHUR BANK[a]

Columbia University
College of Physicians and Surgeons
Department of Medicine
and
Department of Genetics and Development
Hammer Health Sciences Center
701 West 168th Street
New York, New York 10032

INTRODUCTION

The expression of the β-like globin genes undergoes two major "switches" during development. FIGURE 1 depicts the general organization of the human β globin cluster found on chromosome 11. During the first 12 weeks of gestation, the ε gene is expressed and embryonic hemoglobin is produced. Then the ε gene is turned off and the γ genes are activated and fetal hemoglobin becomes the major hemoglobin in the fetus. The γ genes remain on until birth, when the switch to the adult globins β and δ occurs. This switch is complete by the end of the first year and the composition of hemoglobin becomes stable.[1,2]

The transcription of the β-like globin genes, as with other genes, is regulated by cis-acting DNA elements which interact with trans-acting protein factors.[2,3] Several cis-acting DNA elements have been identified in the flanking regions of the globin genes by mutation and gene transfer studies. A CAAT region, a GC-rich region, and a TATA box have been identified in the promoters of the β-like genes.[1-4] Enhancer elements 3′ to the γ and β genes have also recently been reported.[5,6] Protein factors which bind to these types of cis-acting DNA elements have been identified for many genes and cell types. Some have been partially or completely purified.[7-10] Cis-acting DNA elements and trans-acting factors may participate in the formation of activating structures with RNA polymerase II. These trans-acting proteins may work locally or at a distance of several hundred nucleotides away due to long acidic side chains, which interact with RNA polymerase II and aid in the efficient and accurate initiation of transcription.[11-13] There have been preliminary reports of protein factors binding to the promoters and enhancers of the β-like globin genes. Gel mobility shift studies suggest that proteins bind to the regions around the −175 Aγ and −202 Gγ mutations associated with hereditary persistence of fetal hemoglobin (HPFH).[14,15] These findings suggest that these mutations may lead to increased binding of a positive factor, or decreased binding of negative factors, leading to increased γ gene expression.

[a]To whom correspondence should be addressed.

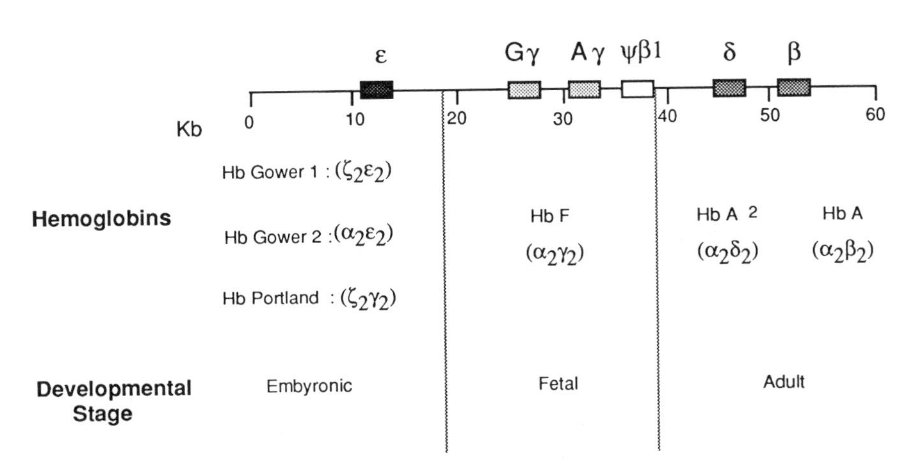

FIGURE 1. Organization of the human β globin locus on chromosome 11 and the hemoglobins produced at each stage of development.

Expression of Hybrid Globin Genes in K562 Cells

In order to determine the cis-acting sequences required for γ globin gene expression and responsible for preventing β expression in K562 cells, hybrid genes containing elements of the γ and β were constructed and transfected into K562 cells by calcium phosphate precipitation or protoplast fusion.[16,17] The transcription of these hybrid genes was studied by RNase protection after stable integration.[18] The β gene shown in FIGURE 2, like all other globin genes, is composed of an untranslated 5′ region, U; three exons, E1, E2, and E3; two introns I1, I2; and an untranslated 3′ region, the expression and induction of these genes are indicated in the E and I columns on the right side of FIGURE 2. After transfection into K562 cells, the β gene is not expressed (E-). After the addition of hemin, a compound that increases endogenous globin gene expression and promotes erythroid differentiation in K562 cells, the β gene also does not induce (I-). A γβ fusion gene that contains 5′ γ sequences (hatched regions and double lines) is both expressed and induced. This indicates that the 5′ γ sequences on this construct are sufficient for expression of this gene although the potential additional role of 3′ β sequences exists (see below). A γ gene with a β IVS2 is expressed but not inducible, suggesting a role for IVS in response to hemin induction. A construct containing a β gene with a γIVS2 is both expressed and induced when the transcripts are analyzed with a 3′ probe. However, this gene is not initiated at the proper 5′ cap site, indicating that γ5′ sequences are required for accurate initiation of transcripts in K562 cells. A hybrid gene containing a γ promoter and β structural sequences ($\gamma^P\beta$) is both expressed and induced. This result suggests that the γ 5′ sequences are most important for the γ expression in the fetal erythroid environment. However, the 5′ γ sequences alone may not be sufficient for γ induction, since the $\gamma^P\beta$ fusion gene contains 3′ β enhancer sequences that may also contribute to its expression and induction.[19,20]

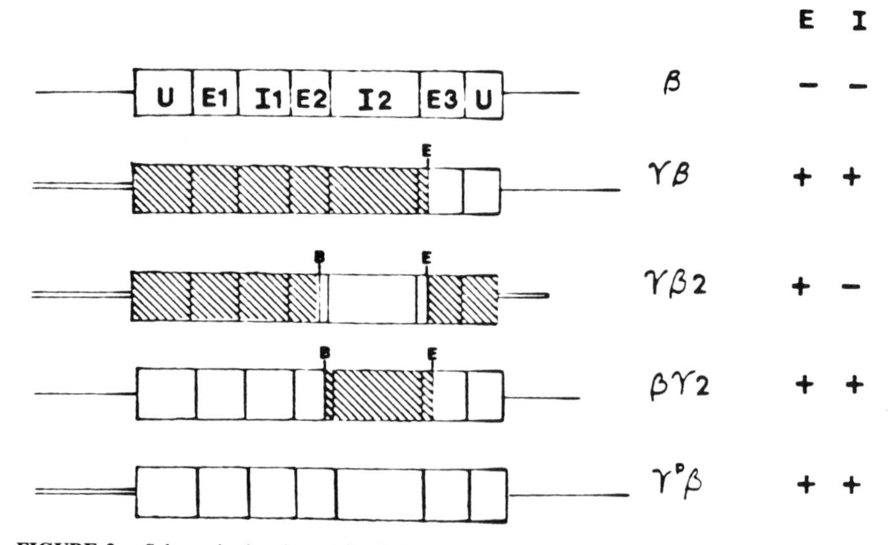

FIGURE 2. Schematic drawing of the fusion globin genes transfected into K562 cells. The prototype gene β is shown on the *top line*. U = untranslated regions; E1 = exon 1; I1 = intron 1; E2 = exon 2; I2 = intron 2; E3 = exon 3. Fusion genes were constructed containing portions of the γ gene (*hatched regions*) with those of the β gene (*open regions*). Four fusion genes were analyzed, γβ, γβ2, βγ2, and $\gamma^P\beta$. Their expression (E) and induction (I) are shown in the columns at the right.

The Role of Globin Gene Promoters in the Expression of Hybrid Genes in Erythroid and Nonerythroid Cells

To study the effects of γ and β globin promoters in erythroid and nonerythroid cells, hybrid genes containing these promoters linked to the neomycin resistances (neo^R) were transfected into K562 or Hela cells. The transfected cells were then selected with G418, a neomycin analog. The number of resultant clones were taken as a measure of gene expression.[21] A hybrid gene containing the γ promoter (FIG. 3, hatched region) linked to the neo^R (γ neo^R) is expressed in K562 cells but not in Hela cells. This result indicates erythroid specific expression of genes containing the γ promoter. The β promoter (FIG. 3, open region) linked to neo^R gene (β neo^R) was much less active in K562 cells than γ neo^R. This result indicates that K562 cells either lack the positive factors needed for β expression or contain negative factors inhibiting β expression. These studies demonstrate the importance of 5′ γ promoter sequences on γ expression. While the γ neo^R gene is expressed in K562 cells it is not inducible with hemin. When a 3′ β enhancer is added to

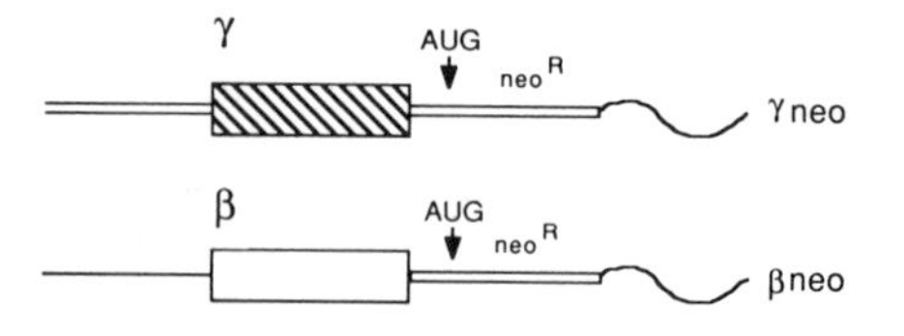

FIGURE 3. Schematic drawing of two neo^R fusion. 5′ γ (*hatched regions*) and 5′ β (*open regions*) were ligated on to the neo^R gene then transfected into K562 or Hela cells. Expression was measured by the number of G418 (a neomycin analog) resistant clones.

TABLE 1. DNA Mobility Gel Shift Assay

Label DNA fragment with ^{32}P.
Incubate labeled fragment with nuclear extract.
Run reaction mixture on acrylamide gel.
Dry gel and analyze by autoradiography.
Results:
Free DNA runs fast.
DNA protein complexes are retarded and run more slowly.

γ neo^R, induction occurs indicating an interaction between 5′ γ and 3′ β globin elements.[20,23]

Protein Factors Binding 5′ to the Aγ Globin Gene

Protein binding to 5′AγDNA sequences was studied using the gel mobility shift assay as shown in TABLE 1.[23-25] FIGURE 4 shows a map of the 5′ region of the Aγ globin gene. The items indicated are the −196 and −198 mutation associated with HPFH (asterisks), the octamer motif ATTGCAT (OCTA) found in immunoglobin, histone, and small nuclear RNA promoters,[3,10] and the restriction sites used to generate DNA fragments for the gel shift assay. FIGURE 5 shows an experiment using a fragment spanning −300 to −161 5′ to the cap site (γ139). The first lane P shows the migration of the γ139 probe alone. The other lanes show the results when extracts from HL-60, uninduced K562 (K−), induced K562 (K+), uninduced mouse erythroleukemia MEL (M−), and induced MEL (M+) cells were used. A non-cell type specific band (FIG. 5, arrow A) was seen with all extracts. Erythroid specific bands common to both K562 and MEL (FIG. 5, arrow B), and only in K562 cells (FIG. 5, arrow C) were also present. The intensity of the K562 specific band decreases with hemin induction (FIG. 5, arrow C). These results are consistent with the binding of a protein that modulates γ gene expression and whose release from DNA permits the increased expression of the γ gene accompanying hemin induction. The experiment in FIGURE 6 shows that the binding of the protein factors is DNA sequence specific. Lane 1 shows the migration of γ139 probe alone. Lane 2 shows γ139 and K562 extract. In Lanes 3–5, increasing amounts of unlabeled γ139 have been added as competitor. All of the bands are competed out by unlabeled γ139 DNA, including band C, which decreases with induction. By contrast, Lanes 6–8 show the results when nonspecific DNA sequences (sp65 plasmid) are used, no decrease in the intensity of the bands is seen. FIGURE 7 shows a gel shift assay with a 42 base pair subfragment of γ139 from −204 to −161 (γ42) 5′ of the Aγ gene. FIGURES 7A and B show the results of binding reactions when Hela (H) and uninduced and induced K562 cells (K− and K+) are used. The results are similar to those seen with γ139; one non-cell type specific band (A) and two erythroid specific bands (B and C), one of which decreases with induction are

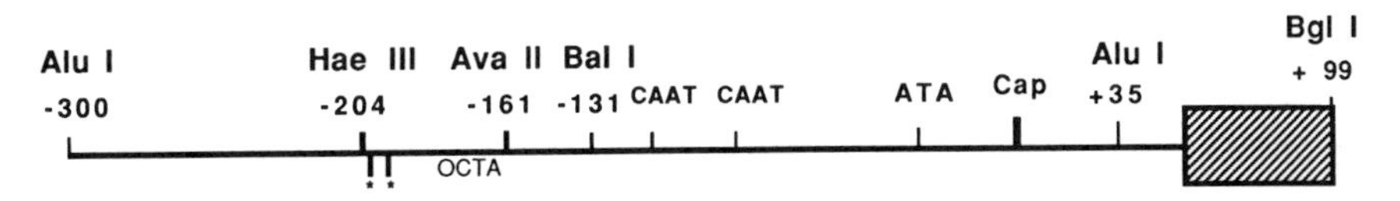

FIGURE 4. Map of the 5′ human Aγ globin gene. Indicated are; restriction enzymes used to generate binding probes, mutations associated with HPFH (−196, −198) (*asterisks*), and the octamer ATTTGCAT(OCTA).

observed. These results further localize the binding of the erythroid and nonerythroid specific factors, and are contrary to recent reports.[15]

Another small DNA fragment immediately 3′ to γ42 (FIG. 8) was assayed for DNA binding activity. The 30 base pair fragment, spanning −161 to −131 (γ30) 5′ to the Aγ cap, contains the CACC box common to both γ and β genes.[15] The results in FIGURE 8 show two bands which are not cell type specific and do not change with induction. The region near the first exon of the Aγ gene from +35 to +99 was also studied using the gel shift assay (FIG. 9). This region showed two binding activities, one non-cell type specific (FIG. 9, arrow I) and one found only in Hela cells (FIG. 9, arrow II).

These results, as well as those from other laboratories, suggest the presence of several

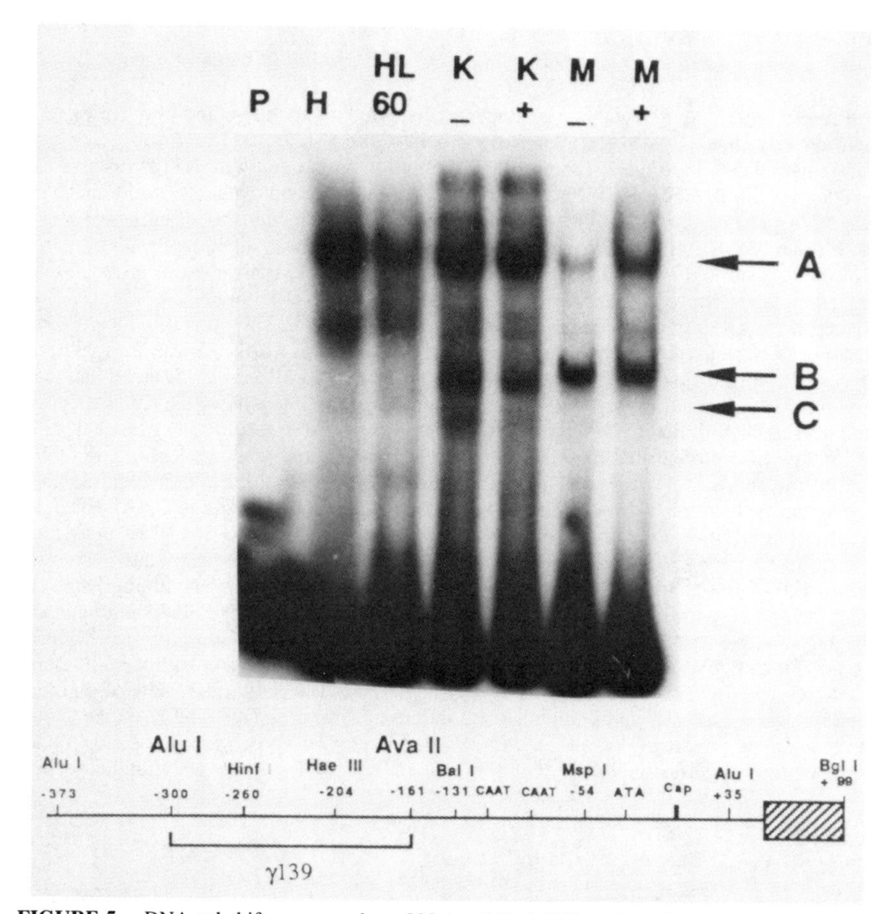

FIGURE 5. DNA gel shift assay on the −300 to −161 (γ139) region of the Aγ gene. *Lanes:* P) probe alone; H) Hela cell extract; HL-60) HL-60 cell extract; K−) uninduced K562 cell extract; K+) induced K562 cell extract; M−) uninduced MEL cell extract; M+) induced MEL cell extract. *Arrow* A: band present in all extracts. *Arrow* B: band present only in K562 and MEL cell extracts. *Arrow* C: band present only in K562 cell extracts that decreases with induction.

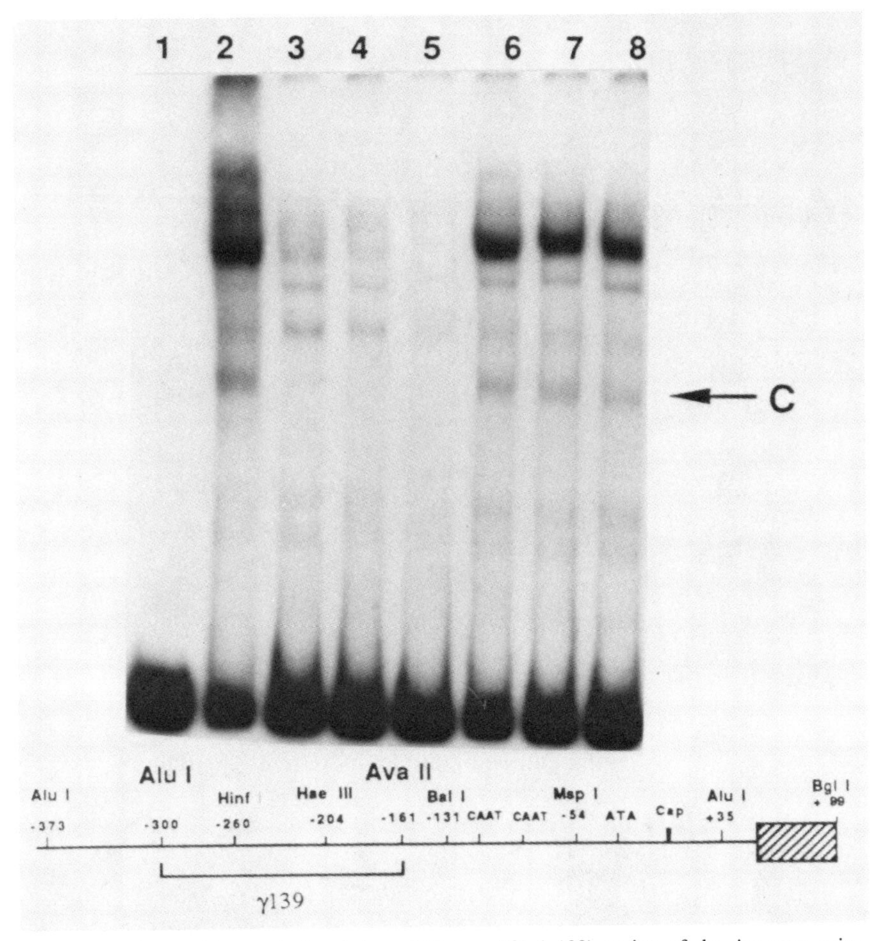

FIGURE 6. Competition analysis of the −300 to −161 (γ139) region of the Aγ gene using uninduced K562 cell extracts. *Lanes:* 1) probe alone; 2) K562 extract with no competitor; 3–5) competition of binding with 150, 300, and 600 ng of unlabeled γ139 fragment; 6–8) competition of binding with 150, 300, and 600 ng of unlabeled plasmid sp65 plasmid DNA. All bands were specificially competed by unlabeled γ139 DNA including band C seen in FIGURE 2 but not by nonspecific sp65 DNA.

DNA binding proteins in extracts from erythroid cells associated with the 5′ region of the Aγ gene.[14,15]

CONCLUSION

In our research on the expression of the γ and β genes, we have found that: (1) sequences 5′ to the γ gene are required for the expression of the γ gene in K562 cells; (2) 5′ β sequences do not allow this tissue specific expression; (3) that sequences in γIVS2

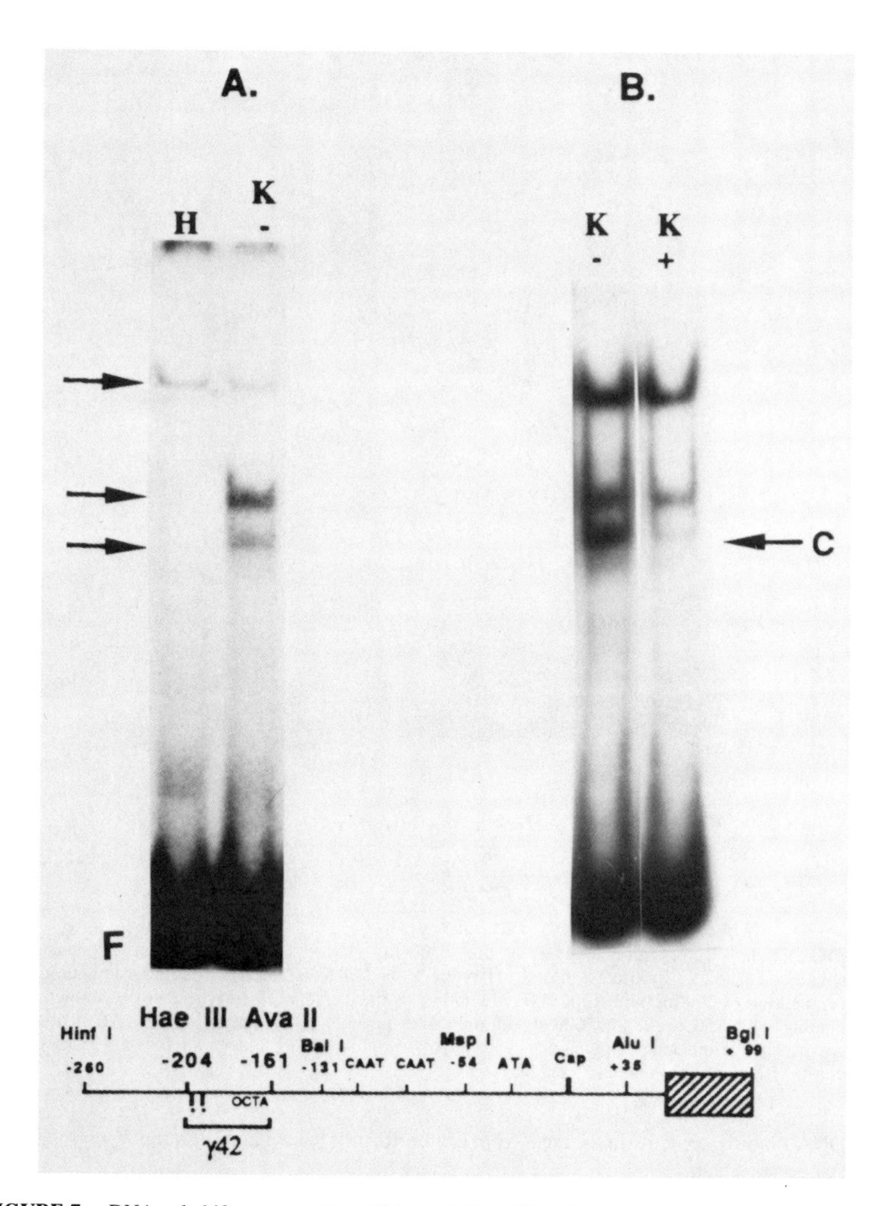

FIGURE 7. DNA gel shift assay on the −204 to −161 (γ42) region of the Aγ gene. *Lanes:* **(A)** F = free or unbound DNA; H) Hela cell extract; K−) uninduced K562 cell extract. **(B)** K−) uninduced K562 cell extract; K+) induced K562 cell extract. *Arrow* A: band present in Hela and K562 extracts. *Arrow* B: band present only in K562 extracts. *Arrow* C: band present only in K562 extracts that decreases with induction.

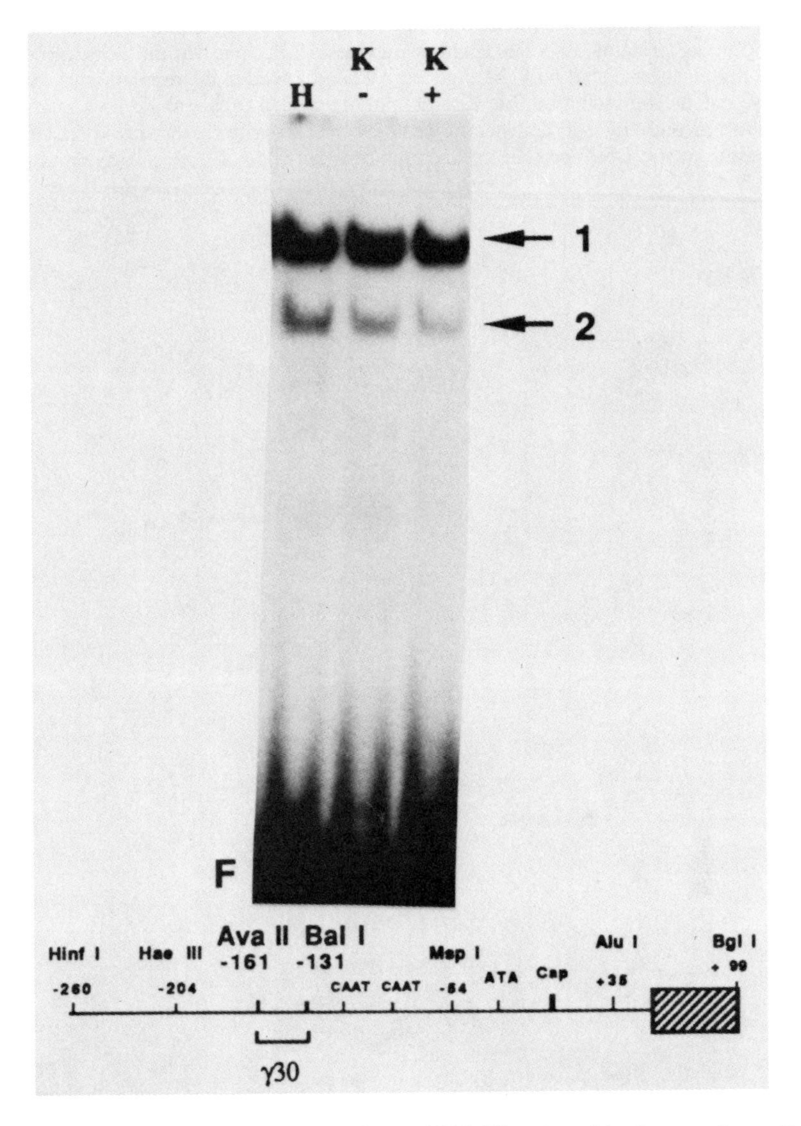

FIGURE 8. DNA gel shift assay on the −161 to −131 (γ30) region of the Aγ gene. *Lanes:* F = free DNA; H) Hela cell extract; K−) uninduced K562 cell extract; K+) induced K562 cell extract. *Arrows* 1 and 2: bands present in both Hela and K562 extracts.

are also important for the expression and induction of the γ gene; and (4) sequences 5′ to the γ gene are the most important factor in their expression.

Studies with sequences 5′ to the Aγ gene using the DNA gel shift assay have shown: (1) the binding of several cell type specific and non-cell type specific factors; (2) the binding of at least one protein present in K562 cells that decreases with induction; and (3) a binding activity found in Hela cells that is absent in K562 cells.

Our findings, as well as those from other laboratories, suggest the presence of several DNA binding proteins associated with sequences 5′, 3′, and within the γ globin gene. These proteins associated with the γ gene may be involved in the regulation of expression of this gene during induction, but further study is needed to determine how many factors are involved and how the nonspecific proteins interact with the erythroid specific ones. The nature of the DNA protein interactions and the roles of both positive and negative

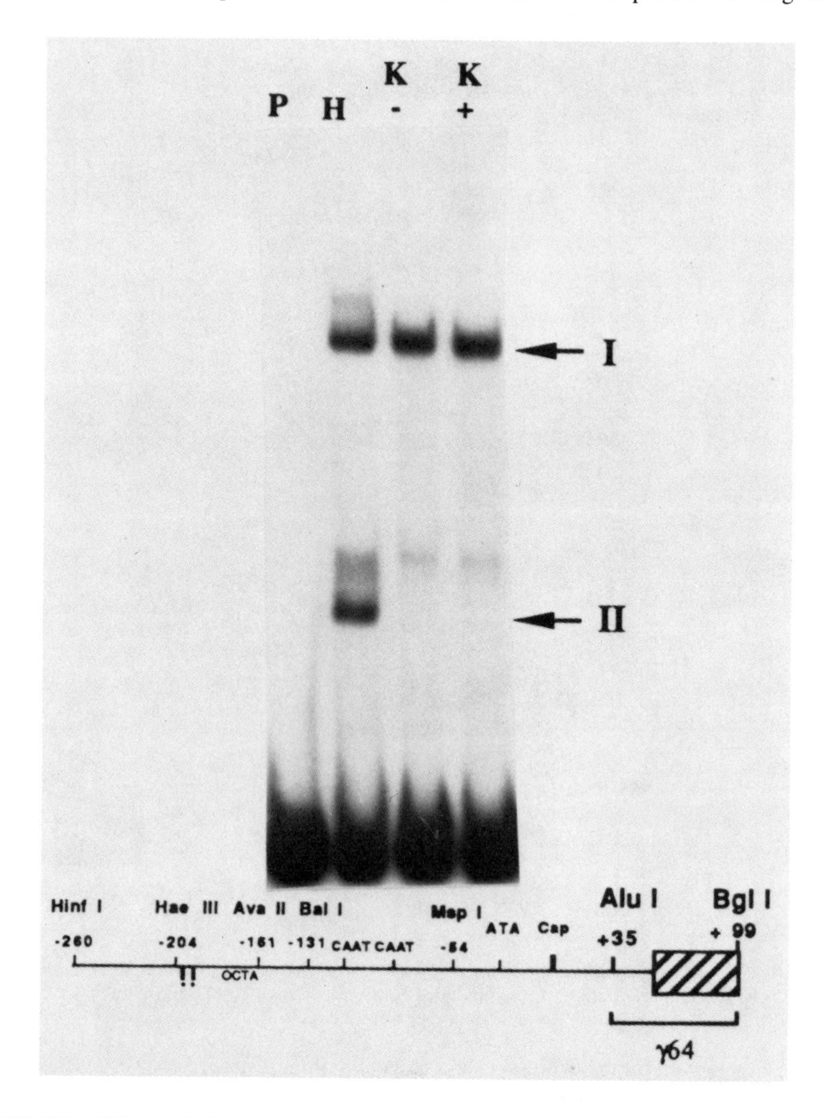

FIGURE 9. DNA gel shift assay on the +35 to +99 (γ64) region of the Aγ gene. *Lanes:* P) probe alone; H) Hela cell extract; K−) uninduced K562 cell extract; K+) induced K562 cell extract. *Arrow* I: band present in all extracts. *Arrow* II: band present only in Hela cell extracts.

factors on the regulation of γ globin genes also remain to be elucidated. With a better understanding of γ gene regulation, it may be possible to identify new methods for increasing γ globin gene expression as a means of treating β thalassemia and sickle cell disease.

SUMMARY

An understanding of the mechanism involved in the regulated expression of the human γ and β globin genes requires the detailed definition of the cis-acting DNA sequences and trans-acting protein factors responsible for their developmental stage specific expression. To determine the critical cis-acting elements, hybrid genes containing elements of the γ and β globin genes were transfected into K562 cells, a human erythroleukemia line. The regulated expression of the γ and β genes was also studied by transferring hybrid genes containing the γ or β promoters linked to the neomycin resistance gene (neo^R) into erythroid (K562) cells and nonerythroid (Hela) cells. DNA sequences found to be important to the expression of the γ gene were assayed for the presence of transacting factors by studying the binding of protein factors using the gel mobility shift assay. The results suggest that there are multiple cis-acting elements 5′ and 3′ to the γ and β genes, and perhaps within these genes contributing to their regulation. In addition, there are multiple trans-acting protein factors interacting with these regions which may determine their transcriptional regulation in erythroid cells.

REFERENCES

1. Collins, F.S. & S. M. Weissman, 1984. Prog. Nucleic Acids Res. Mol. Biol. **31:** 316–461.
2. Stamatoyannopoulos, G. & A. W. Nienhuis. 1987. Hemoglobin switching. *In* The Molecular Basis of Blood Diseases. G. Stamatoyannopoulos, A. W. Nienhuis, P. Leder & P. Majerus, Eds.: 65–105, W. B. Saunders. Philadelphia, PA.
3. Jones, N.C., P. W. Rigby & E. B. Ziff. 1988. Genes Dev. **2:** 267–281.
4. Myers, R.M., K. Tilly & T. Maniatis. 1986. Science **232:**615–618.
5. Trudel, M., J. Magram, L. Brukner & F. Constantini. 1987. Mol. Cell. Biol. **7:** 4024–4029.
6. Bodine, D.M. & T. J. Ley. 1985. Blood (Suppl.) **66:** 68A.
7. Dynan, W.S. & R. Tijan. 1985. Nature **316:** 774–778.
8. McKnight, S. & R. Tijan. 1986. Cell **46:** 795–805.
9. Treisman, R. 1986. Cell **46:** 567–574.
10. Sen, R. & D. Baltimore. 1986. Cell **46:** 705–716.
11. Sigler, P.B. 1988. Nature **333:** 210–212.
12. Allison, L.A., M. Meyle, M. Shales & C. J. Ingles, 1985. Cell **49:** 599–610.
13. Zehring, W.A., J.M. Lee, J. R. Weeks, R.S. Jokerst & A. L. Greenleaf. 1988. Proc. Natl. Acad. Sci. USA **85:** 3698–3702.
14. Stykes, K. & R. E. Kaufman. 1987. Blood (Suppl.) **70:** 80A, #176.
15. Mantovani, R., N. Malgarelli, B. Gilioni, S. Cappellini, & S. Ottolenghi. 1987. Nucleic Acids Res. **15:** 9349–9364.
16. Graham, F.L., P.J. Abrahams, C. Mulder, H.L. Heijneker, S.O. Warnaor, F.A. deVries, W. Fiers & A. J. Vandere. 1975. Cold Spring Harbor Symp. Quant. Biol. **39:** 637–650.
17. Parker, B.A. & G. R. Stark. 1979. J. Virol. **31:** 360–369.
18. Donovan-Peluso, M., S. Acuto, M. Swanson, C. Dobkin & A. Bank. 1987. J. Biol. Chem. **262:** 17051–17057.
19. Donovan-Peluso, M. 1988. Preliminary results.

20. Donovan-Peluso, M., S. Acuto, D. O'Neill & A. Bank. 1987. Blood (Suppl.) **70:** 73A.
21. Acuto, S., M. Donovan-Peluso & A. Bank. 1987. Biochem. Biophys. Res. Commun. **143:** 1099–1106.
22. Donovan-Peluso, M., S. Acuto & A. Bank. 1988. Regulation of fetal globin gene expression in human erythroleukemia (K562) cells. *In* Molecular Biology of Hemopoiesis. M. Tavassleoli, N. G. Abraham, J. L. Ascensao, E. D. Zanjani & A. Levine, Eds. Plenum. New York, NY.
23. Maxim, A.M. & W. Gilbert. 1980. Methods Enzymol. **65:** 499–560.
24. Strauss, F. & A. Varshavsky. 1984. Cell **37:** 889–901.
25. Sinh, H., R. Sen, D. Baltimore & P. Sharp. 1986. Nature **319:** 154–156.

Cell Membrane and Volume Changes during Red Cell Development and Aging[a]

NARLA MOHANDAS[b] AND WARREN GRONER[c]

[b]*Department of Laboratory Medicine and Cancer Research Institute*
Moffitt 1282/Box 0128
University of California, San Francisco
San Francisco, California 94143-0128
and
[c]*Technicon Instruments Corporation*
511 Benedict Avenue
Tarrytown, New York 10591-5097

The human red cell, after entering the circulation from the bone marrow as a reticulocyte, spends its lifetime of approximately 120 days performing its function of oxygen delivery. During this time, the red cell is constantly tested for its ability to undergo marked cellular deformation. This ability to deform is essential for optimal cell function, since the resting diameter of the human red cell far exceeds that of the capillaries and splenic endothelial slits through which it must pass. The remarkable ability of the red blood cell to undergo the necessary deformations to negotiate capillary and splenic channels while withstanding continuous circulatory stresses is programmed into its geometry, the viscosity of the intracellular hemoglobin milieu, and the material properties of the membrane.

The mature circulating human red cell is in fact optimally designed with respect to various cellular characteristics that regulate deformability. By possessing excess surface area in relation to cell volume, the discoid red cell is able to undergo marked deformations while maintaining a constant surface area. Through careful regulation of cell water content and hence cell hemoglobin concentration, the red cell minimizes the contribution of cytoplasmic viscosity to cell deformation. A highly deformable yet remarkably stable membrane in turn allows the red cell to undergo extensive passive deformation while resisting membrane fragmentation. A great deal of progress has been made in our understanding of membrane, shape and volume changes in various red cell disorders and their potential contribution to altered cell function and reduced cell survival. However, much less is known regarding membrane and volume changes during red cell development from its precursor cell and during the normal cell aging process. In this paper, we will briefly review cell membrane and volume changes during red cell development and aging.

Reticulocyte Maturation and the Evolution of the Red Cell

The reticulocyte—the immediate precursor of the mature red cell—is formed from orthochromatic normoblast by the process of nuclear extrusion (FIG. 1a).[1] Following its genesis, two distinct morphologic stages can be recognized during the transformation of the reticulocyte into a mature red cell.[2,3] The young reticulocyte is motile and multilob-

[a]This work was supported in part by National Institutes of Health Grant DK 26263.

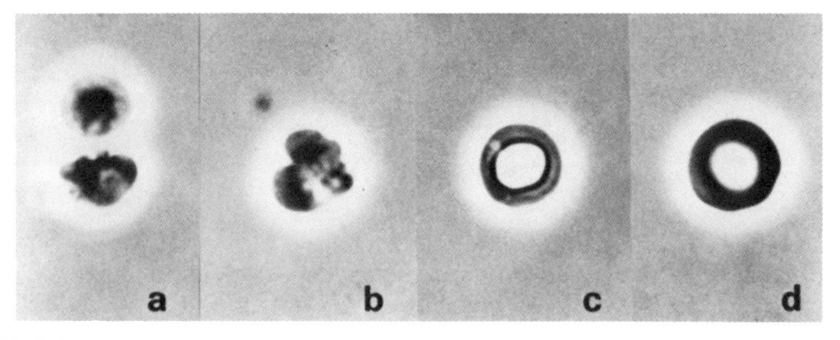

FIGURE 1. Phase-contrast micrographs of cells during reticulocyte maturation. **(a)** An orthochromatic erythroblast expelling its nuclei; **(b)** motile and multilobular immature reticulocyte; **(c)** cup-shaped, nonmotile mature reticulocyte; and **(d)** mature discoid red cell.

ular (FIG. 1b). The membrane of this immature reticulocyte is extremely convoluted and the cytoplasm contains multiple organelles. The older, mature form of reticulocyte is nonmotile and cup-shaped (FIG. 1c). This dramatic change in cell shape during the transformation of the reticulocyte into mature discoid red cells occurs over a span of 72 hours, first in the bone marrow and then in the circulation. During this time, organelles in the cytoplasm including mitochondria and ribosomes are lost, and marked changes in membrane organization involving both cytoskeletal and skeletal proteins take place.

The role of cytoskeletal components during the genesis and development of reticulocytes has recently been defined.[4] It has been found that the process of nuclear extrusion, which is essential for the generation of the young, anucleate reticulocyte from its nucleated precursor cell, is blocked, both *in vivo* and *in vitro*, by colchicine. This data strongly implies that microtubules play an important role in the nuclear extrusion process. In contrast, colchicine had no effect on the motility of reticulocytes. Young reticulocytes, as well as the cytoplasmic portion of the erythroblast arrested at the stage of nuclear extrusion, were motile in the presence of colchicine. However, cell motility was markedly inhibited by cytochalasin B, a microfilament-disrupting agent, implying a role for these structures in regulating reticulocyte motility. It is interesting to note that these cytoskeletal structures are not functional in the mature red cell. Thus it appears that cytoskeletal structures including microtubules and microfilaments play an important role in early stages of reticulocyte development, but are lost during its maturation to the mature red cell stage.[4]

Changes in membrane skeletal assembly during reticulocyte maturation were assessed by quantitating changes in membrane deformability and mechanical stability, two properties regulated by skeletal protein interactions.[4,5] Membrane mechanical stability was determined by monitoring the rate of fragmentation of resealed membranes at a defined value of applied shear stress in an ektacytometer. As shown in the left-hand panel of FIGURE 2, membranes prepared from populations of cells enriched for immature reticulocytes showed marked decreases in mechanical stability, as reflected by a more rapid decrease in deformability index with time. This result implies that immature reticulocyte membranes are mechanically unstable compared to those of mature red cells.[4] To document that the membranes acquire mechanical stability during maturation, this membrane property was measured for populations of cells enriched for reticulocytes of increasing maturity. As shown in the right-hand panel of FIGURE 2, the membranes become progressively more stable during reticulocyte maturation. Changes in membrane deformability—a distinctly different property of the membrane—were also evaluated during reticulocyte maturation. Membranes prepared from immature reticulocytes were found to be approximately ten times more rigid than those prepared from mature red cells.[4,6] However, membranes prepared from mature forms of reticulocytes were found to have deformability characteristics similar to those of red cells. The progressive acquisition of membrane mechanical stability and improved deformability imply that major structural remodeling of the membrane occurs during reticulocyte maturation.[4] Since skeletal protein organization regulates these two membrane properties, these findings further imply that substantial skeletal protein remodeling must take place during the evolution of the red cell from its precursor.

Accompanying these changes in shape, skeletal and cytoskeletal components are a variety of other modifications.[7–11] Reticulocytes lose 90 percent of their insulin receptors, all of their transferrin receptors, and 60 percent of their sodium-potassium pumps during maturation into discoid red cells.[7,9] There is also a loss of membrane phospholipids and cell water but no loss of hemoglobin.[11] An additional important feature of the remodeling of maturing reticulocytes is the loss of membrane surface area. Two mechanisms have been proposed for this loss: 1) external loss of membrane, or exocytosis, and 2) internalization of segments of membrane in the form of sealed endocytic vacuoles.[12,13] However, the relative contributions of these two processes to membrane loss is yet to be defined. Finally, it is generally assumed that reticulocytes are less dense than mature red cells as a consequence of a reduced cell hemoglobin concentration. While it is true that immature reticulocytes usually have a larger volume and are less dense than mature red cells, these cells exhibit a broad distribution of cell volume and density values. Moreover, it is becoming increasingly clear that red cells of varying cell densities are generated during maturation of reticulocytes (our unpublished observations). Thus, contrary to widely held beliefs, it is likely that the heterogeneity in cell volumes and cell densities seen for red cells in circulation may in large part be the result of changes in cell hydration state that accompany maturation of reticulocytes.

The brief review of cellular and membrane changes that accompany reticulocyte maturation outlined here raises several interesting issues for the study of the biology of reticulocytes as well as red cells. To date, most studies of reticulocytes have considered

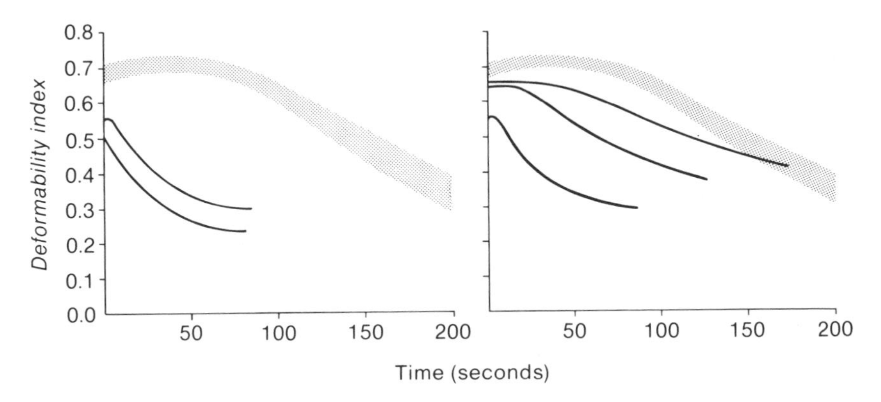

FIGURE 2. Reticulocyte membrane mechanical stability. *Left-hand panel:* the deformability index of membranes from red cell populations enriched for immature reticulocytes (>85%) from two individuals (*solid lines*) decreased more rapidly with time when compared with mature red cell membranes (*shaded area*). *Right-hand panel:* populations of red cells enriched with increasing numbers of mature reticulocytes were prepared from one of these samples. As the percent of mature reticulocytes increased and the percent of immature reticulocytes decreased (>85%, *bottom solid line;* 35%, *middle solid line;* 18% *top solid line*), the membrane mechanical stability increased.

them as a homogeneous population of cells. The marked changes in various cellular characteristics that occur over the short period of three days during reticulocyte maturation imply that a great deal of attention needs to be paid to the age distribution of reticulocytes in the interpretation of any biochemical and biophysical characterization of these cells. In contrast to extensive studies on biochemical and structural characterization of mature red cells, very little information is available regarding reticulocytes. It appears that characterization of the molecular basis for changes in cellular properties during reticulocyte maturation will improve our understanding of red cell biology.

Red Cell Aging

The normal human red cell has a life span of about 120 days, after which it is removed from the circulation. Although several mechanisms have been proposed to explain the recognition and removal of senescent red blood cells from the circulation, there is still no consensus regarding a cardinal cellular modification that serves as a signal for clearance of these cells. The two cellular changes that have received the most attention as possible determinants of red cell life span are: 1) deformability loss accompanying red cell aging in circulation leading to mechanical entrapment of the cells in the reticuloendothelial system, and 2) cell surface changes leading to binding of immunoglobulins to the senescent cell surface and consequent immunologic clearance of the cells by the macrophages of the reticuloendothelial system.

The concept that red cell density increases with increasing cell age in the circulation, although controversial, has found wide acceptance. Density separation of red cells has been used in most of the studies on cell aging.[14–18] During *in vivo* aging of human red blood cells, it is thought that progressive loss of total cation and water content decreases cell volume and increases cell density.[19] Representative density distribution of red cells obtained from a normal individual on analytical Stractan density gradients is shown in FIGURE 3. A majority of normal red cells have cell densities in the range of 1.0829–1.0964 g/ml. However, populations of cells with cell density values of less than 1.0829 g/ml and greater than 1.0964 g/ml are also seen. The volume and hemoglobin concentration values for red cells isolated from the various density interfaces are shown in TABLE 1. It can be seen that increase in cell density is the result of a decrease in cell volume and a consequent increase in cell hemoglobin concentration. Rheological measurements on red cells isolated following density fractionation has shown a progressive decrease in cellular deformability with increasing cell density (FIG. 4).[20–23] As red cell density increases, there is loss of membrane surface area and decrease in volume, causing the cells to become more spherical.[21] This combination of surface area loss and increased cytoplasmic viscosity resulting from cellular dehydration significantly compromise the ability of dense red cells to deform. Several investigators have suggested that the extent of deformability loss observed for the dense normal red cells is of sufficient magnitude to allow for their removal by a filtration mechanism in the reticuloendothelial system.[20,22,24] Although it seems reasonable to suggest that the rheologically compromised dense senescent red cells are sequestered from the circulation, the validity of this hypothesis rests on the crucial assumption that increasing cell density parallels increasing cell age.

Two lines of evidence are often used to justify a direct relationship between cell age and cell density. First, the activity of several glycolytic enzymes progressively decreases with increasing cell density.[25] Because the red cells do not have the ability to synthesize these enzymes *de novo*, decreased enzyme activity with increasing cell density is thought to reflect age-dependent loss of enzyme activity. Second, measurement of the distribution of iron-59 and glycine 2-^{14}C cohort-labeled red cells on density gradients show that, over time, labeled cells are found in layers of progressively increasing cell density.[16–18] Al-

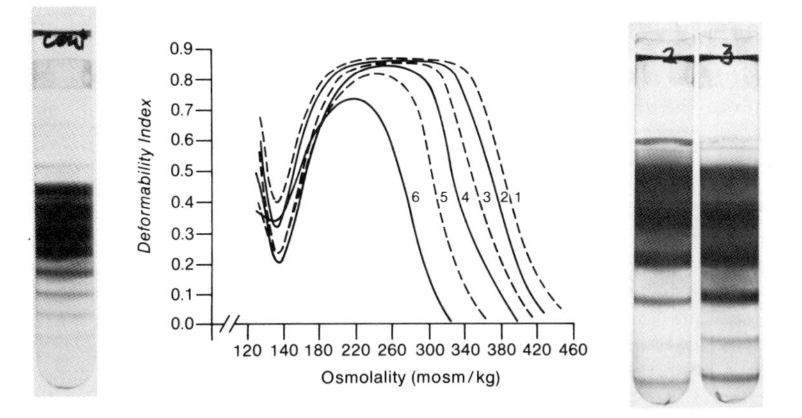

FIGURE 3 (*left*). Distribution of normal human red cells on a discontinuous Stractan density gradient. The gradient consisted of 8 layers ranging from 1.0785 to 1.1056 g/ml density in approximately 0.045 g/ml increments.
FIGURE 4 (*center*). Osmotic deformability profiles generated by ektacytometry for subpopulations of normal red cells isolated on discontinuous Stractan density gradients. Curves 1–6 represent samples of progressively increasing cell density and mean cell hemoglobin concentration (MCHC). Progressive reduction in the maximally attained deformability index implies loss of surface area and the shift of the curves to lower osmolality values in the hypertonic region implies increased cytoplasmic viscosity.
FIGURE 5 (*right*). Distribution of red cells from a control mouse (*left*) and from a mouse which was hypertransfused for 56 days (*right*) on a discontinuous Stractan density gradient.

though both types of evidence suggest that cell age and cell density are related, lingering doubts persist regarding the interpretation of these experimental data. For example, because the glycolytic enzyme levels of the reticulocyte are severalfold higher than in mature red cells, the decrease in enzyme levels with increasing cell density may result from the differential distribution of the reticulocytes in the density gradient, rather than from variations in enzyme activity with post-reticulocyte red cell age.[26] The cohort radioisotope labeling experiments are accompanied by a broad distribution of labeled cells in the density gradient persisting throughout the course of the experiment. There is substantial overlap in the density distribution of the labeled cells at the beginning and at the end of the experiment. Thus it appears that the technique of density separation may not

TABLE 1. Values of Cell Volume and Cell Hemoglobin Concentration for Density-fractionated Normal Red Cells (as Measured by Technicon H*1 System)

Stractan Density Interface	Cell Volume (fL)	Cell Hemoglobin (g/dL)
<1.0785	106.5	27.6
1.0785–1.0829	98.6	29.8
1.0829–1.088	94.7	31.0
1.088–1.0919	91.4	32.1
1.0919–1.0964	84.3	34.6
1.0964–1.1009	77.0	37.0
1.1009–1.1056	70.6	40.1
>1.1056	69.9	41.1

yield a population of cells with a defined cell age, but only populations with increasing mean cell age. Absolute separation on the basis of age may not be possible using any physical characteristics of the cells,[27] as reticulocytes emerging from the bone marrow may already be heterogeneous in terms of their cellular properties including cell density.

An important contribution to our understanding of the relationship between red cell age and cell density was recently provided by an elegant study by Morrison and his colleagues.[28,29] These investigators examined changes in mouse red cell density distribution as a function of *in vivo* red cell aging. They began with 64 consanguineous mice and exsanguinated half of them in order to obtain blood to hypertransfuse the other half, thus suppressing erythropoiesis. Before the hematocrit of these hypertransfused mice fell to levels at which endogenous erythropoiesis might begin, half of this group of mice were sacrificed and used to hypertransfuse the other half. This maneuver was repeated until only two mice of the original group of 64 remained. All of the circulating red cells in these two mice were senescent, in that they were removed from circulation in the following two to three days. As shown in FIGURE 5, the mean red cell density of these senescent red cells was only slightly increased compared to the cell density at the beginning of hypertransfusion, with substantial overlap in the density profiles. This data strongly suggests that density heterogeneity is not a direct reflection of heterogeneity of cell age but more a reflection of the heterogeneous density distribution of young red cells leaving the marrow to begin their life in the circulation.

The concept that red blood cell senescence involves exposure of hidden antigenic sites capable of binding naturally occurring autologous IgG antibodies, leading to the recognition and phagocytosis of these cells has also received increasing attention in recent years. Several laboratories have shown increased *in vivo* binding of IgG to subpopulations of red cells with increased cell densities.[30–34] The major weakness of these studies, as with deformability studies, is that all of the data on cell senescence is based on the unproved concept that cell density and cell age are directly related. Despite this criticism, it seems reasonable to conclude that loss of cell surface area, increased cytoplasmic viscosity and exposure of hidden antigenic sites may all play a role in the removal of senescent red cells from the circulation. The defining of the relative contributions of each of these factors to account for the finite life span of the red cell remains a challenge.

SUMMARY

This paper provides a summary of our understanding of cell membrane and volume changes during red cell development and aging. Cytoskeletal structures which include microtubules and microfilaments appear to play key roles in the genesis of the anucleate reticulocyte from its nucleated precursor cell, as well as in the early stages of reticulocyte development. The maturation of reticulocyte into red cell is accompanied by marked changes in cell shape and extensive remodeling of the membrane skeleton, resulting in the mature red cell acquiring a highly deformable yet remarkably stable membrane. The volume and cell density heterogeneity seen for circulating red cells also appears to be the result of the membrane changes that occur during reticulocyte maturation. Following its genesis from reticulocyte, the mature red cell undergoes further membrane and volume changes during its life span of 120 days. While it is clear that surface area loss, decrease in cell volume and cell surface modifications leading to binding of immunoglobulins accompany red cell aging, the cardinal cellular modification responsible for the removal of senescent red cells from the circulation is yet to be defined.

ACKNOWLEDGMENTS

We would like to thank James Harris for his expert assistance in preparing this manuscript. We are most grateful to Drs. T. Mueller and C. Jackson for allowing us to perform cell density analysis of senescent mouse red cell samples.

REFERENCES

1. Bessis, M. & M. Bricka. 1952. Aspect dynamique des cellules du sang. Son étude par microcinematographie en contraste de phase. Rev. Hematol. **7:** 407–420.
2. Mel, H. C., M. Prenant & N. Mohandas. 1977. Reticulocyte motility and form: studies on maturation and classification. Blood **49:** 1001–1009.
3. Coulombel, L., G. Tchernia & N. Mohandas. 1979. Human reticulocyte maturation and its relevance to erythropoietic stress. J. Lab. Clin. Med. **94:** 467–474.
4. Chasis, J. A., M. Prenant & N. Mohandas. 1988. Skeletal and cytoskeletal remodeling during reticulocyte maturation. Blood **72** (Suppl. 1): 24A (abstract).
5. Chasis, J. A. & N. Mohandas. 1986. Erythrocyte membrane deformability and stability: two distinct membrane properties that are independently regulated by skeletal protein associations. J. Cell Biol. **103:** 343–350.
6. Leblond, P. F., P. L. LaCelle & R. I. Weed. 1971. Cellular deformability: a possible determinant of the normal release of maturing erythrocytes from the bone marrow. Blood **37:** 40–46.
7. Frazier, J. L., J. H. Caskey, M. Yaffe & P. A. Seligman. 1982. Studies on the transferrin receptor on both human reticulocytes and nucleated human cells in culture. J. Clin. Invest. **69:** 853–860.
8. Eng, J., L. Lee & R. S. Yalow. 1980. The influence of age of erythrocytes on their insulin receptors. Diabetes **29:** 164–166.
9. Wiley, J. S. & C. C. Shaller. 1977. Selective loss of calcium permeability on maturation of reticulocytes. J. Clin. Invest. **59:** 1113–1119.
10. Patel, V. P., A. Ciechanover, O. Platt & H. F. Lodish. 1985. Mammalian reticulocytes lose adhesion to fibronectin during maturation to erythrocytes. Proc. Natl. Acad. Sci. USA **82:** 440–444.
11. Come, S. E., S. B. Shohet & S. H. Robinson. 1972. Surface remodelling of reticulocytes produced in response to erythroid stress. Nature New Biol. **236:** 157–159.
12. Holyrode, C. P. & F. H. Gardner. 1970. Acquisition of autophagic vacuoles by human erythrocytes: physiologic role of the spleen. Blood **36:** 556–562.
13. Johnstone, R. M. & B. T. Pan. 1983. Fate of the transferrin receptor during maturation of sheep reticulocytes in vitro: selective externalization of the receptor. Cell **33:** 967–975.
14. Piomelli, S., G. Lurinsky & L. R. Wasserman. 1967. The mechanism of red cell aging. I. Relationship between cell age and specific gravity evaluated by ultracentrifugation in a discontinuous density gradient. J. Lab. Clin. Med. **69:** 659–674.
15. Corash, L. M., S. Piomelli, H. C. Chen, C. Seaman & E. Gross. 1974. Separation of erythrocytes according to age on a simplified density gradient. J. Lab. Clin. Med. **84:** 147–151.
16. Luthra, M. G., J. M. Friedman & D. A. Sears. 1979. Studies of density fractions of normal human erythrocytes labeled with iron-59 in vivo. J. Lab. Clin. Med. **94:** 879–896.
17. Bennett, G. D. & M. M. B. Kay. 1981. Homeostatic removal of senescent murine erythrocytes by splenic macrophages. Exp. Hematol. **9:** 297–307.
18. Clark, M. R. 1988. Senescence of red blood cells: progress and problems. Physiol. Rev. **68:** 503–554.
19. Brugnara, C. & D. C. Tosteson. 1987. Cell volume, K transport and cell density in human erythrocytes. Am. J. Physiol. **252** (Cell Physiol 21): C269–C276.
20. Nash, G. B. & S. J. Wyard. 1981. Changes in surface area and volume measured by micropipette aspiration for erythrocyte ageing in vivo. Biorheology **17:** 479–484.

22. CLARK. M. R., N. MOHANDAS & S. B. SHOHET. 1983. Osmotic gradient ektacytometry: comprehensive characterization of red cell volume and surface maintenance. Blood **61:** 899–910.
23. LINDERKAMP, O. & H. J. MEISELMAN. 1982. Geometric, osmotic and membrane mechanical properties of density-separated human red cells. Blood **59:** 1121–1127.
24. EVANS, E., N. MOHANDAS & A. LEUNG. 1984. Static and dynamic rigidities of normal and sickle erythrocytes: major influence of cell hemoglobin concentration. J. Clin. Invest. **73:** 477–488.
25. WILLIAMS, A. R. & D. R. MORRIS. 1980. The internal viscosity of the human erythrocyte may determine its lifespan in vivo. Scand. J. Hematol. **24:** 57–62.
26. SEAMAN, C., S. WYSS & S. PIOMELLI. 1980. The decline of energetic metabolism with aging of the erythrocyte and its relationship to cell death. Am. J. Hematol. **8:** 31–42.
27. BEUTLER, E. 1985. How do red cell enzymes age? A new perspective. Annotation. Br. J. Haematol. **61:** 377–384.
28. VAN DER VEGT, S. G. L., A. M. T. RUBEN, J. M. WERRE, D. M. A. PALSMA, C. W. VERHOEF, J. DE GIER & G. E. J. STAAL. 1985. Counterflow centrifugation of red cell populations: a cell-age related separation technique. Br. J. Haematol. **61:** 393–403.
29. MORRISON, M., C. W. JACKSON, T. J. MUELLER *et al.* 1983. Does red cell density correlate with red cell age? Biomed. Biochem. Acta **42:** S107–S111.
30. MUELLER, T. J., C. JACKSON, M. E. DOCKTER & M. MORRISON. 1987. Membrane skeletal alterations during in vivo mouse red cell aging: increase in the band 4.1a:4.1b ratio. J. Clin. Invest. **79:** 492–499.
31. ALDERMAN, E. M., H. H. FUDENBERG & R. E. LOVINS. 1981. Isolation and characterization of an age-related antigen present on senescent human red blood cells. Blood **58:** 341–349.
32. KAY, M. M. B., S. R. GOODMAN, K. SORENSEN *et al.* 1983. Senescent cell antigen is immunologically related to band 3. Proc. Natl. Acad. Sci. USA **80:** 1631–1635.
33. KHANSARI, N. & H. H. FUDENBERG. 1983. Phagocytosis of senescent erythrocytes by autologous monocytes: requirement of membrane-specific autologous IgG for immune elimination of aging red blood cells. Cell. Immunol. **78:** 114–121.
34. LOW, P. S., S. M. WAUGH, K. ZINKE & D. DRENCKHAHN. 1985. The role of hemoglobin denaturation and band 3 clustering in red blood cell aging. Science **227:** 531–533.
35. GALILI, U., I. FLECHNER, A. KNYSZYNSKI, D. DANON & E. RACHMILEWITZ. 1986. The natural anti-α-galactosyl IgG on human normal senescent red blood cells. Br. J. Haematol. **62:** 317–324.

The Use of Recombinant Erythropoietin in the Treatment of the Anemia of Chronic Renal Failure[a]

JOSEPH W. ESCHBACH, N. REBECCA HALEY,
AND JOHN W. ADAMSON

Division of Hematology
School of Medicine
University of Washington
Seattle, Washington 98195

Anemia was first recognized as a complication of chronic renal failure by Richard Bright in 1836. Alan Erslev was the first to show convincingly that there was an erythropoietic stimulating factor in anemic blood,[1] and in 1957 Leon Jacobson and colleagues concluded that the kidney was the source of this factor,[2] now known as erythropoietin (Epo). The purification of human urinary Epo by Miyaki and colleagues in 1977[3] made it possible for Lin and colleagues in 1985,[4] using genetic engineering techniques, to isolate and clone the gene for human Epo. As a result of these advances, we now better understand the pathophysiology of the anemia of chronic renal failure,[5,6] as well as have a means by which to treat it.[7–10] This paper briefly reviews the results of therapy with recombinant human Epo (rHuEpo) and, more specifically, the effects of rHuEpo on marrow function as quantitated by ferrokinetic and other measures.

Clinical trials using rHuEpo, produced by AMGen (Thousand Oaks, CA), involve anemic hemodialysis patients in the United States, Canada and Europe. In Seattle we have treated 65 patients over the past 2 years and, in all but one case, rHuEpo successfully eliminated the anemia and the need for red cell transfusions and resulted in marked clinical improvement. This experience has been confirmed recently by the results of a multicenter trial in the United States.[11]

At a dose of 150 μ/kg, given 3×/wk, the hematocrit typically increases at a rate of 1.8 volume % per week. Therefore, most anemic hemodialysis patients will reach a hematocrit of 35 within 8–12 weeks. Once the hematocrit goal is attained, the dose of rHuEpo is decreased to 75 μ/kg, 3×/wk, and readjusted at 2–4-week intervals, as necessary, to maintain a hematocrit between 33 and 38.

Erythropoiesis in patients with chronic renal failure is stimulated by rHuEpo in a dose-dependent fashion. In our patients, erythropoiesis was quantitated by ferrokinetics (the erythron transferrin uptake; ETU[12]) (FIG. 1), the reticulocyte response (FIG. 2), and by changes in the hematocrit (FIG. 3).

Erythropoiesis was evaluated again in our patients after the goal hematocrit was

[a]Portions of this work were supported by Research Grants DK 33488 and DK 19410 and Clinical Research Center Grant FR0037 from the National Institutes of Health, Department of Health and Human Services, and by patient care funds from AMGen Corporation, Thousand Oaks, CA, and Ortho Pharmaceutical Corporation, Raritan, NJ.

attained because the dose of rHuEpo necessary to maintain a normal or near normal hematocrit varied. Approximately 80% of patients require no more than 100 μ/kg, 3×/wk, to maintain a stable hematocrit. A minority of patients, however, have required up to 300 μ/kg, 3×/wk. The reason why these patients require larger amounts of rHuEpo is not clear but, based on ferrokinetic and reticulocyte studies, the erythroid marrow appears to respond appropriately to the higher doses of rHuEpo. Red cell survival, as measured by ^{51}chromium labeling of red blood cells, did not correlate inversely with the maintenance dose of rHuEpo. Aluminum[10,13] and/or parathyroid hormone excess,[14] the latter manifested as osteitis fibrosa, may contribute to the relative refractoriness of some patients to rHuEpo. FIGURE 4 (upper panel) reveals the relatively poor response to rHuEpo in a hemodialysis patient with osteitis fibrosa diagnosed by thoracic vertebral changes on X-ray and a serum alkaline phosphatase value of >800 U/L. Iron deficiency, infection

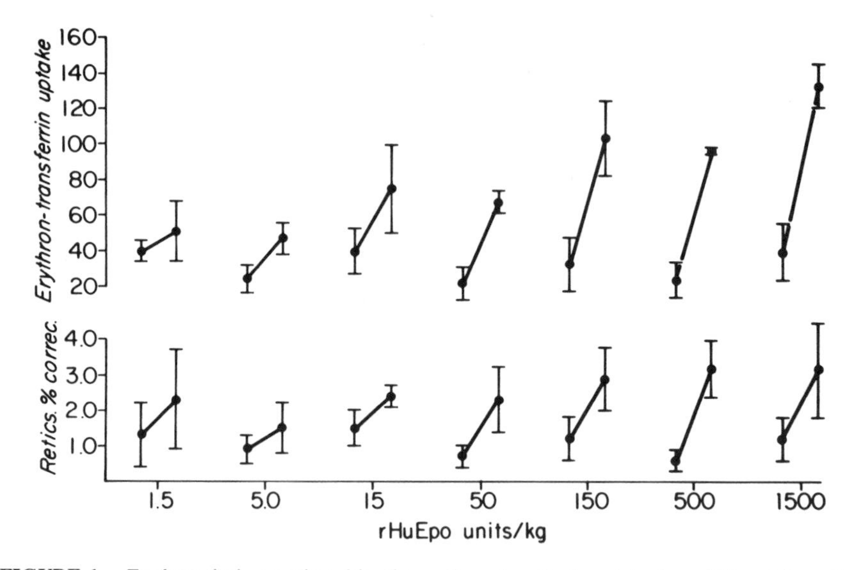

FIGURE 1. Erythropoiesis quantitated by the erythron transferrin uptake (μmol/L) and the reticulocyte count, corrected for the anemia, in hemodialysis patients treated with various amounts of recombinant human erythropoietin (rHuEpo). Each connected pair of points represents the mean (±SD) values observed before treatment and after 4 doses of rHuEpo. Three to 5 patients were studied at each dose.

and the inflammatory effect of surgery are also conditions that will blunt the effectiveness of rHuEpo.[15] FIGURE 4 (lower panel) depicts the marked fall in hematocrit after artificial hip replacement despite the continuation of regular doses of intravenous rHuEpo. This is probably an extreme example of the inflammatory effect on erythropoiesis in patients undergoing major surgery.

Erythroid function, as quantitated by ferrokinetics, is subnormal in patients with renal failure not yet requiring dialysis therapy.[16] The ETU, normally 60 ± 12 μmol/L,[12] was 36 μmol/L in 16 predialysis patients with a mean hematocrit of 25. In comparison, subjects with normal renal function who have a hemolytic anemia with a similar hematocrit have an ETU of 400 ± 130 μmol/L.[12] After hemodialysis is initiated, patients either stabilize and slowly increase their hematocrit with time or they begin to require repeated

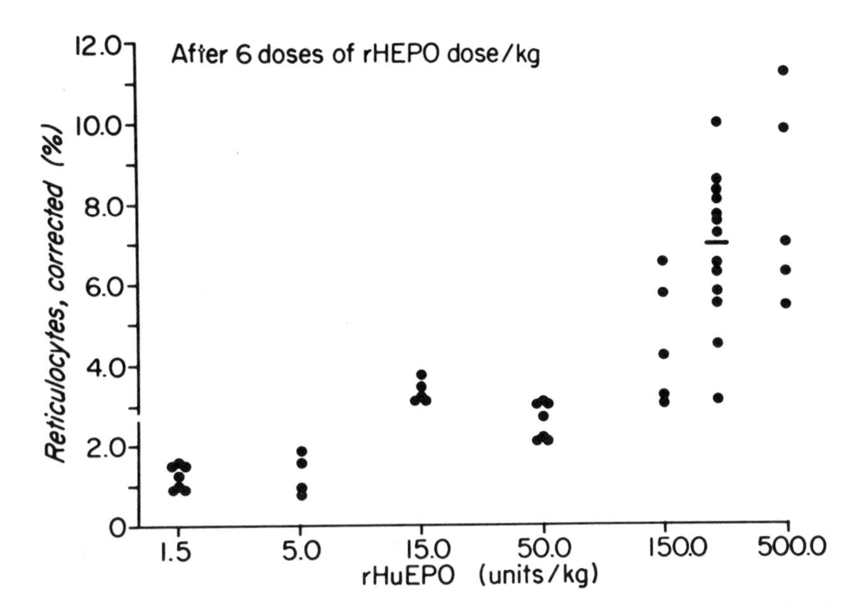

FIGURE 2. The reticulocyte count, corrected for the anemia, after 6 doses of rHuEpo. Maximum reticulocytosis usually occurred after the sixth dose. Each dot represents an individual patient receiving the indicated dose of rHuEpo. The unmarked column indicates a dose of 300 U/Kg. The doses were given intravenously 3×/week.

red cell transfusions. Ferrokinetic measurements in these two groups indicate that erythropoiesis increases in the former to a mean ETU of 73 ± 21 μmol/L, whereas the transfusion-dependent patients have a more suppressed erythroid function as indicated by an ETU of 35 ± 11 μmol/L.[12] Regardless of the level of erythroid function, and whether the patients are transfusion-dependent, erythropoiesis immediately increases once rHuEPO therapy is initiated. The degree of response depends on the dose of rHuEpo. rHuEpo also stimulates the anemic, predialysis patient in a similar manner.[17]

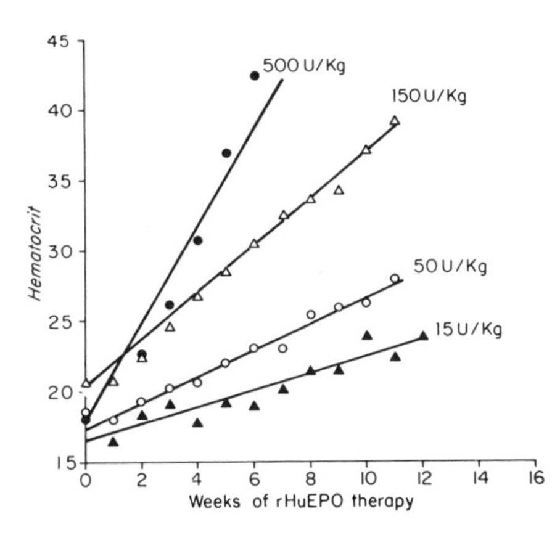

FIGURE 3. The rate of rise in hematocrit of hemodialysis patients treated with different doses of rHuEpo. (From Eschbach *et al.*[8] Reprinted by permission from the *New England Journal of Medicine*.)

Studies with rHuEpo also have shown that the erythroid marrow in uremia is responsive to changes in tissue oxygenation or red cell mass, as is true in the normal state. Previously we reported that red cell transfusions in most dialysis patients further suppressed erythropoiesis as quantitated by ferrokinetics.[16] However, others have suggested that erythroid function is autonomous in renal failure.[18] We performed ferrokinetic studies in two patients with advanced renal failure not yet on dialysis before and after the fourth dose of rHuEpo. As seen in FIGURE 5, the ETU increased from a mean of 36 to 105 μmol/L. After the hematocrit rose from 23 to 37, rHuEpo was discontinued, and 9 days later the ETU had fallen to 15 μmol/L, indicating that a near normal red cell mass will suppress erythroid function in the patient with chronic renal failure unless exogenous Epo is administered continually.

There are other direct marrow and hematologic effects of rHuEpo. Because one of the actions of Epo is to prematurely shift marrow reticulocytes into circulation, and since these younger cells are larger than mature red cells, the mean corpuscular volume (MCV) increases acutely with the administration of rHuEpo.[7] This effect is most marked in the patient with iron overload. Such patients have red cells with a high MCV even before stimulation with rHuEpo.[19]

Transferrin receptor protein has recently been detected in circulation and the levels appear to correlate directly with the level of erythropoiesis. Preliminary studies in hemodialysis patients responding to rHuEpo indicate that transferrin receptor protein increases in proportion to the increase in the ETU.[20]

Studies in a small number of patients with chronic renal failure treated with rHuEpo suggest that red cell survival, which typically is 1/2 to 2/3 of normal in this group, improves significantly, but not to normal. The mechanism for the mild hemolysis observed in renal failure is thought to be due to the retention of some extracorporeal

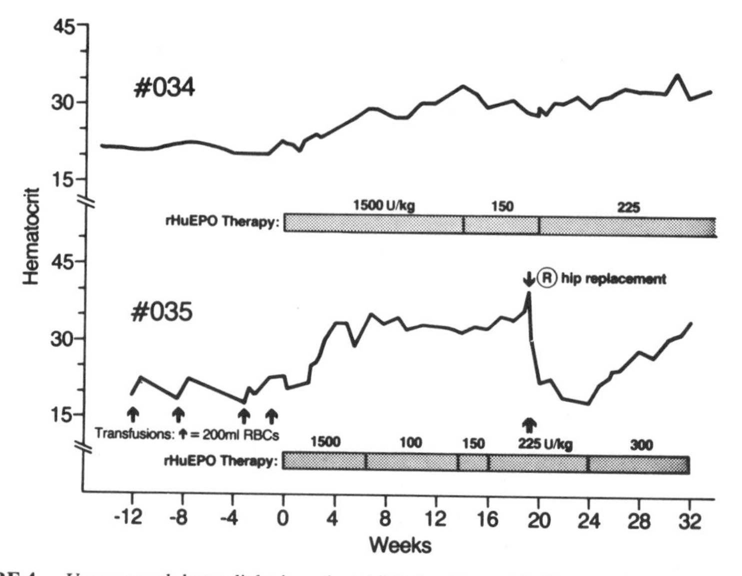

FIGURE 4. *Upper panel:* hemodialysis patient (#034) with osteitis fibrosa who responded poorly to rHuEpo at 1500 U/Kg given 3×/wk. *Lower panel:* hemodialysis patient (#035) who was receiving a constant dose of rHuEPO (225 U/kg, 3×/wk) but who failed to respond during the 5 weeks immediately after hip replacement surgery.

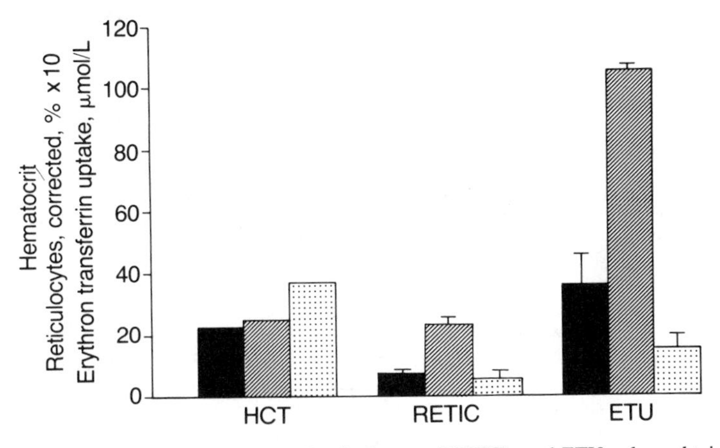

FIGURE 5. Hematocrit (HCT), corrected reticulocytes (RETIC), and ETU values obtained prior to treatment (baseline) (*filled columns*), after 4 doses of rHuEpo (150 U/Kg, 3×/wk) (*hatched columns*), and once the target hematocrit had been reached, 9 days after stopping rHuEpo therapy (*dotted columns*). The studies were performed on two patients and the bars represent the range of values obtained. As is seen, the corrected reticulocyte count and ETU increased 2.5–3.0-fold with rHuEpo therapy. After the drug was stopped, the ETU was 1/2 of the baseline value, indicating that the erythroid marrow was still responsive to changes in O_2 carrying capacity and that the baseline ETU value was very likely set by the degree of anemia present, presumably mediated through endogenous Epo production.

substance.[21] Since it is not clear why correcting the anemia should improve red cell survival, more extensive studies are required.

The platelet count has increased in some studies[9] and was shown to be significantly increased after 4 months of rHuEpo therapy in the multicenter trial in the United States.[11] However, such an increase (224×10^9/L to 261×10^9/L) is still within the normal range of platelet counts and its biological significance is obscure.

Platelet function, as measured by the bleeding time, appears to improve in the patient whose hematocrit increases to >30 with rHuEpo therapy.[22] Whether this is the reason why some hemodialysis patients treated with rHuEpo develop increased clotting within the extracorporeal dialysis circuit when their hematocrit increases remains to be determined. However, this is not a clinically significant problem since a slight increase in heparin prevents the clotting.[8]

In summary, rHuEpo has resulted in significant clinical improvement in the anemic patient with chronic renal failure. This application of recombinant DNA technology has allowed us to better understand the quantitative and qualitative responses of the erythroid marrow to Epo stimulation.

ACKNOWLEDGMENTS

The authors are most appreciative for the technical assistance of Glenda Schneider.

REFERENCES

1. Erslev, A. J. 1953. Humoral regulation of red cell production. Blood **8:** 349–387.
2. Jacobson, L.O., E. Goldwasser, W. Fried & L. Plazk. 1957. Role of the kidney in erythropoiesis. Nature **179:** 633–634.

3. Miyaki, T., C. K. H. Kung & E. Goldwasser. 1977. Purification of human erythropoietin. J. Biol. Chem. **252:** 5558–5564.
4. Lin, F.-K., S. Suggs, C. H. Lin, J. K. Browne, *et al.* 1985. Cloning and expression of the human erythyropoietin gene. Proc. Natl. Acad. Sci. USA **82:** 7580–7584.
5. Eschbach, J. W. & J. W. Adamson. 1986. Anemia of end stage renal disease (ESRD). Kidney Int. **28:** 1–5.
6. Eschbach, J.W. & J. W. Adamson. 1988. Modern aspects of the pathophysiology of renal anemia. Contrib. Nephrol. **66:** 63–70.
7. Winearls, C. G., M. J. Pippard, M. Downing, D. O. Oliver, C. Reid & P. M. Cotes. 1986. Effect of human erythropoietin derived from recombinant DNA on the anemia of patients maintained by chronic haemodialysis. Lancet **2:** 1175–1178.
8. Eschbach, J. W., J. C. Egrie, M. R. Downing, J. K. Browne & J. W. Adamson. 1987. Correction of the anemia of end-stage renal disease with recombinant human erythropoietin: Results of a combined phase I and II clinical trial. N. Engl. J. Med. **316:** 73–78.
9. Bommer, J. C., C. Alexiou, U. Muller-Buhl, J. E. Fert & E. Ritz. 1987. Recombinant human erythropoietin therapy in haemodialysis patients—dose determination and clinical experience. Nephrol. Dial. Transplant. **2:** 238–242.
10. Casati, S., P. Passerini, M. R. Campise, G. Graziani, B. Cesana, *et al.* 1987. Benefits and risks of protracted treatment with human recombinant erythropoietin in patients having haemodialysis. Br. Med. J. **295:** 1017–1020.
11. Eschbach, J. W. & J. W. Adamson. 1988. Correction of the anemia of hemodialysis patients with recombinant human erythropoietin: Results of a multicenter study. Kidney Int. **33:** 189 (A).
12. Cazzola, M., P. Pootrakul, H. A. Hubers, M. Eng, J. Eschbach & C. A. Finch. 1987. Erythroid marrow function in anemic man. Blood **69:** 296–301.
13. Van Wyck, D. B., J. Stivelman, J. Ruiz, M. A. Katz & D. A. Odgen. 1988. Aluminum excess poses modest resistance to recombinant human erythropoietin (rHuEPO) for dialysis anemia. Kidney Int. **33:** 240 (A).
14. Hampl, H., E. Riedel, U. Stabell, P. Scigalla & G. Wendel. 1988. Influence of parathyroid hormone on exogenous erythropoietin stimulated erythropoiesis in hemodialysis patients. Kidney Int. **33:** 224 (A).
15. Eschbach, J. W. & J. W. Adamson. 1988. Recombinant human erythropoietin: Implications for nephrology. Am. J. Kidney Dis. **11:** 203–209.
16. Eschbach, J. W., J. W. Adamson & J. D. Cook. 1970. Disorders of red blood cell production in uremia. Arch. Int. Med. **126:** 812–815.
17. Eschbach, J. W., M. R. Kelly, N. R. Haley, R. Abels & J. W. Adamson. 1989. Correction of the anemia of progressive renal failure with recombinant human erythropoietin. In press.
18. Erslev, A. J., P. J. McKenna, J. P. Capelli, *et al.* 1968. Erythropoiesis in nephrectomized patients. Arch. Intern. Med. **122:** 230–235.
19. Gokal, R., D. J. Weatherall & C. Bunch. 1979. Iron induced increase in red cell size in hemodialysis patients. Q. J. Med. **48:** 393–401.
20. Beguin, Y., H. Huebers, P. Pootrakul, R. Haley, J. W. Eschbach, J. W. Adamson & C. A. Finch. 1987. Plasma transferrin receptor levels as a monitor of erythropoiesis in man: Correlation with ferrokinetics and stimulation by recombinant human erythropoietin. Blood **70:** Suppl 1. Abstr. 51a.
21. Joske, R. A., J. M. McAllister & T. A. J. Prankerd. 1956. Isotope investigation of red cell production and destruction in chronic renal disease. Clin. Sci. **15:** 511–522.
22. Casati, S., P. Passerini, M. Moia, G. Graziani, P. M. Manucci & C. Ponticelli. 1988. Correction of coagulation disturbances in chronic hemodialysis patients treated with human recombinant erythropoietin. Kidney Int. **33:** 622.

In Vivo Biologic Activities of Recombinant Human Granulocyte-Macrophage Colony-Stimulating Factor

SAROJ VADHAN-RAJ,[a] WALTER N. HITTELMAN,[a]
HAL E. BROXMEYER,[b] MICHAEL KEATING,[a] DAVID URDAL,[c]
AND JORDAN U. GUTTERMAN[a]

[a]*Departments of Clinical Immunology and Biological Therapy*
Box 41
The University of Texas M. D. Anderson Cancer Center
1515 Holcombe Boulevard
Houston, Texas 77030
[b]*Department of Medicine*
Indiana University School of Medicine
541 Clinical Drive
Long Clinical Building
Room 379
Indianapolis, Indiana 46223
[c]*Immunex Corporation*
51 University Street
Seattle, Washington 98101

INTRODUCTION

The colony-stimulating factors are a family of glycoproteins that play an important role in the proliferation and differentiation of hemopoietic progenitor cells and functional activation of mature cells *in vitro*.[1–4] The recent cloning and expression of human hemopoietic growth factors as recombinant proteins has prompted examination of their role in regulation of hematopoiesis *in vivo*.[5–10]

Granulocyte-macrophage colony-stimulating factor (GM-CSF) is a multipotential hematopoietin that stimulates the formation of granulocyte, macrophage, eosinophil, erythroid and multipotential colonies in human bone marrow cultures.[8,11] In addition, it enhances multiple functions of neutrophils, including oxidative metabolism, phagocytosis, lysozyme secretion and antibody-dependent cellular cytotoxicity.[12–16] Studies in nonhuman primates have shown that recombinant human GM-CSF is rapidly cleared from circulation when given by intravenous bolus injection and induces leukocytosis when administered by continuous intravenous infusion.[17] In this report, we review the *in vivo* biologic effects of GM-CSF in patients with bone marrow failure and in patients with malignancy.[18,19]

MATERIALS AND METHODS

Patient Characteristics

Of the 25 patients entered on the study, 14 were males and 11 were females. The median age was 58 years (range, 20 to 79) and median performance status was 70%

(range, 50 to 90). Diagnoses included solid tumors in six patients and hematological diseases in 19 patients. Five of the six patients with solid tumors had normal blood counts, while one patient with solid tumor and 19 patients with hematological diseases had one or more cytopenias. Of these 20 patients with cytopenias, 15 had neutropenia (neutrophils $<1000/mm^3$), 17 had thrombocytopenia (platelets $\leq 100{,}000/mm^3$) and 20 had anemia requiring red cell transfusions at <1- to 5-week intervals.

All patients had histologically confirmed advanced malignancy and/or bone marrow failure. Informed consent was obtained from all patients. For inclusion in the study, patients were required to have Karnofsky performance status of ≥50%, life expectancy of ≥12 weeks, serum creatinine ≤2 mg/100 ml, proteinuria ≤2+, serum bilirubin ≤1.5 mg/100 ml, and prothrombin time ≤1.3 times control.

Granulocyte-Macrophage Colony-Stimulating Factor

The recombinant human GM-CSF used in this study was prepared and provided by Immunex Corporation (Seattle, WA). The gene for human GM-CSF was cloned from a cDNA library constructed by using mRNA from the human T-Cell line HUT-102, and the clone was expressed in yeast.[5] The purified recombinant protein had a specific activity of 5×10^7 colony-forming units/mg of protein.

Study Design

Patients were treated with various fixed doses of GM-CSF given by continuous i.v. infusion daily for two weeks. After a two-week rest period, the i.v. treatment cycles were repeated. Dose levels were 15, 30, 60, 120, 250 or 500 $\mu g/m^2$/day. If favorable responses were observed during the two treatment cycles, patients became eligible for maintenance treatment at the same schedule, and at a GM-CSF dose adjusted to maintain white count in the normal range. Treatment was discontinued in any patient who had severe toxicity and was resumed at a reduced dose upon full recovery.

Laboratory Tests

Patients were evaluated before treatment with a complete history and physical examination, complete blood cell count, WBC differential count, serum biochemical profile, urinalysis, and prothrombin time. All patients had a chest X-ray and an EKG before and after two treatment cycles. In addition, bone marrow aspirate and biopsy specimens were obtained before and after each treatment cycle.

Analysis of Hemopoietic Progenitor Cells In Vitro

Bone marrow cells were separated into a low density fraction and plated at 1×10^5 cells/ml in agar culture medium with 100 units of recombinant human GM-CSF as described previously.[20] Granulocyte-macrophage colonies (CFU-GM) were scored after 7 and 14 days of incubation. The assays for erythroid (BFU-E) and multipotential (CFU-GEMM) progenitor cells were performed as described previously.[20]

Assays for Antibodies to Recombinant Human GM-CSF

Serum samples from patients before and after two cycles of GM-CSF treatment were analyzed for the presence of antibodies to recombinant human GM-CSF by using an ELISA assay. The serum samples which showed reactivity to GM-CSF by an ELISA assay were then tested for the capacity to neutralize GM-CSF activity in the human bone marrow proliferation assay.

Premature Chromosome Condensation Technique

Karyotype of mature cells from peripheral blood of patients with myeloid diseases was determined by analysis of premature chromosome condensation (PCC), as described previously.[21] Briefly, peripheral blood cells were separated into mononuclear cells and polymorphonuclear cells by two-step ficoll-hypaque gradient. Mitotic Chinese hamster ovary cells (CHO cells) were used to induce PCC in peripheral blood cells after cell fusion. The CHO cells were prelabeled with BUDR to distinguish their chromosomes from the prematurely condensed chromosomes. The number of chromosomes in fused cells was determined and the karyotypic abnormalities were detected by G-banding.

RESULTS

Patients with Solid Tumors

Treatment with GM-CSF was associated with marked increases in total leukocyte count in a dose-dependent manner,[19] with increases up to 3-fold at a lower dose-level (30 μg/m^2/day) and up to 7-fold at a higher dose-level (250 μg/m^2/day) (TABLE 1). Although mature neutrophils and bands accounted for most of the increase in white blood cells, significant increases were also seen in eosinophils, monocytes, and lymphocytes. Thekinetics of the rise in white blood cells and differentials are shown in FIGURE 1. The rise in white count was rapid and sustained for the duration of treatment. The counts returned to baseline level upon discontinuation of infusion. No significant changes were observed in hemoglobin, hematocrit levels, and in platelet counts (TABLE 2).

Patients with Hematologic Diseases

Treatment resulted in an elevation in white blood cell count that reached a level 2- to 70-fold higher than the baseline level (TABLE 3). The elevation in white blood cell count appeared to be dose dependent, although a wide range of responses were seen in different patients treated at the same dose. Responses in white blood cells were seen in patients with bone marrow failure of diverse etiologies, including aplastic anemia, myelofibrosis, myelodysplastic syndromes, and smouldering myeloid leukemia. In eight patients with myelodysplastic syndromes (MDS), dose-dependent increases were seen in white blood cells, up to 30,000/mm^3 at lower doses ($\leq$120 μg/m^2/day) and up to 100,000/mm^3 at higher doses ($\geq$250 μg/m^2/day).[18] Although the predominant effects were seen in mature cell types at all dose levels (FIG. 2), some transient increases were seen in immature cells

TABLE 1. Effects of GM-CSF Treatment on White Blood Cell Counts in Patients with Solid Tumors

Patient Number	Dose of GM-CSF μg/m²	White Blood Cell Count[a] (× 10³/mm³)	
		Before	After[b]
1	30	3.7	6.6
2	60	6.4	13.6
3	120	7.1	22.9
4	120	11.0	37.7
5	250	6.4	45.0

[a]Blood counts from 5 of 6 patients with solid tumors are shown here. The sixth patient with small cell lung cancer received only 8 days of treatment (500 μg/m²/ day).

[b]Maximal response in counts during first cycle of GM-CSF treatment.

at the highest doses (250 to 500 μg/m²/day). Of the seven patients with smouldering myeloid leukemia, significant increases in mature cells were seen in four patients, while in three patients the predominant effects were in immature cells.

Significant increases in white blood cells and granulocytes were also observed in two patients (one patient with chronic myelogenous leukemia and another with multiple myeloma) who had failed to recover after autologous bone marrow transplantation. An example is shown in FIGURE 3.

Multilineage Response

In addition to responses in white blood cells, six patients with cytopenias experienced a multilineage response after repeated cycles of treatment. The response was characterized by significant increases in platelet counts (>2-fold increase + >100,000/mm³) and a reduced requirement for transfusions. Three of these six patients (two with MDS and one with AML) did not require any red cell or platelet transfusions for 17 to >40 weeks.

Effects on Bone Marrows

Responses in peripheral blood counts were accompanied by increases in bone marrow cellularity and myeloid:erythroid cell ratios. The growth patterns of progenitor cells from

TABLE 2. Peripheral Blood Counts in Patients with Solid Tumors Treated with GM-CSF

Parameter	Mean ± SD	
	Baseline	After[a] Treatment
Hemoglobin (g/dl)	12.96 ± 1.42	11.94 ± 1.69
Hematocrit (%)	38.80 ± 4.60	37.02 ± 5.95
Reticulocytes (%)	1.20 ± 0.05	2.32 ± 1.24
Platelets (× 10³/mm³)	237.00 ± 90.01	257.00 ± 138.80

[a]Blood counts after two cycles of GM-CSF treatment.

bone marrows and their response to stimulation by GM-CSF *in vitro* did not change significantly after two cycles of treatment with GM-CSF. However, the frequency of the colony-forming cells (CFU-GM) in S-phase (DNA synthesis) as determined by the tritiated thymidine suicide technique *in vitro* was significantly increased from baseline mean level of 8 ± 3% to 46 ± 4% after treatment.[20]

Karyotypic Analysis of Mature Cells

The PCC analysis of the granulocytes of patients with MDS showed that mature cells were derived from both "normal" (diploid) and "abnormal" (with karyotypic abnormalities) clones. Neutrophils from six of the seven patients studied showed the same karyotypic abnormalities as was previously documented in the bone marrow, while the seventh patient with trisomy eight who had been previously rendered hypoplastic and diploid with chemotherapy, contained only diploid elements in the mature cells.[22]

TABLE 3. Effects of GM-CSF on White Blood Cell Counts in Patients with Hematological Diseases

		White Blood Cells Count[a] ($\times 10^3/mm^3$)	
Number of Patients	Dose of GM-CSF $\mu g/m^2$	Before	After[b]
5	15–60	1.7 (1.2–3.6)	5.3 (2.2–9.5)
4	120	1.9 (1.7–2.4)	10.0 (9.3–26.1)
5	250	1.9 (1.1–4.7)	13.0 (4.5–96.5)
5	500	1.5 (0.2–2.6)	16.7 (4.3–106.0)

[a]Median (range) values are shown.
[b]Maximal response in counts during first cycle of GM-CSF treatment.

Antibodies to GM-CSF

Serum samples from 19 patients were analyzed for the presence of antibodies by an ELISA assay. Sera from two patients showed antibody titers after two cycles of GM-CSF treatment. Only one of these two patients' sera showed reactivity by western blot analysis. However, sera from both these patients did not inhibit activity of either *E. coli* or yeast-derived GM-CSF in the human bone marrow proliferation assay.

Side Effects

Treatment was generally well tolerated. Side effects were mild to moderate in intensity and commonly included constitutional symptoms, bone pain, and gastrointestinal disturbances (TABLE 4).

DISCUSSION

The major objective of this study was to evaluate the clinical safety and biologic activities of recombinant human GM-CSF in patients with malignancy and in patients with bone marrow failure.[18,19] Treatment with GM-CSF resulted in a dose-dependent increase in white count consisting predominantly of neutrophils, eosinophils, and monocytes. The rise in white count was rapid, suggesting that the initial response was probably due to demargination of cells and release of mature cells from the bone marrow. However, the elevation in count was sustained for the duration of infusion, suggesting that a second component of response was due to stimulation of proliferation in the myeloid compartment. This notion was derived on the basis of two observations. First, the fraction of colony-forming units (CFU-GM) in S-phase increased significantly after GM-CSF treatment. Second, there was evidence for marked increase in bone marrow cellularity after treatment.

The goal of GM-CSF treatment in patients with primary or secondary bone marrow failure was to restore hematopoiesis and to reduce the risk of infection and hemorrhage. Significant stimulation of myelopoiesis was seen in patients from different disease cate-

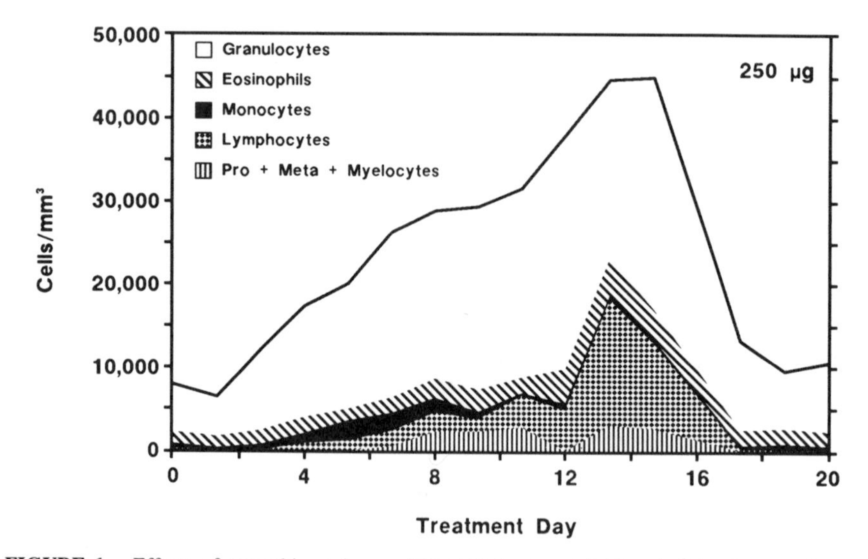

FIGURE 1. Effects of recombinant human GM-CSF treatment (250 $\mu g/m^2$/day i.v. continuous infusion from days 1 to 14) on white blood cell counts and differential cell counts in a patient with solid tumor.

gories, including myelodysplastic syndromes, smouldering myeloid leukemia, myelofibrosis, aplastic anemia, and in patients who had failed to recover after autologous bone marrow transplantation. Furthermore, evidence for multilineage stimulation was observed in six patients characterized by significant improvement in platelet counts and a reduced requirement for red cell transfusions. These results are encouraging particularly in patients with myelodysplastic syndromes since most attempts to stimulate hematopoiesis with steroids, androgens, and hematinics have been unsuccessful, and the treatment with differentiation-inducing agents has been plagued with a high incidence of toxicity and a low response rate.[18]

Since myelodysplastic syndromes are clonal hemopoietic disorders, we were interested to know whether neutrophils appearing after GM-CSF treatment were derived from normal stem cells or from the neoplastic clone. To approach this question, we determined

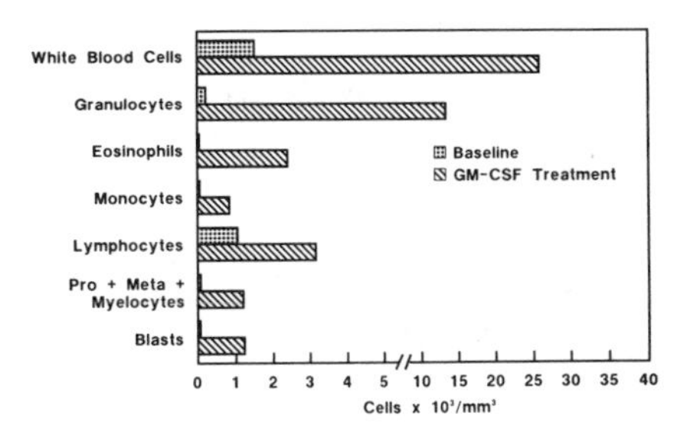

FIGURE 2. White blood cell count and differential cell count (mean values) before (baseline) and after GM-CSF treatment (maximum response) in 8 patients with myelodysplastic syndromes.

the karyotype of mature cells using the technique of premature chromosome condensation, since many of these patients exhibit the karyotypic abnormalities in their abnormal clone. Our findings indicate that both normal and abnormal clones were stimulated to proliferate and mature during GM-CSF treatment. Furthermore, the ratios of normal cells to abnormal cells in proliferating fraction in the bone marrow and in nondividing mature cells in the peripheral blood were similar (Hittelman, W. N. and Vadhan-Raj, S., unpublished observation) suggesting that proliferation was coupled with differentiation.

Although the predominant effects were in mature cell types in patients with myelodysplastic syndrome and in some patients with smouldering myeloid leukemia, significant increases were seen in immature cells in three other patients (two with relapsed AML and one with myelofibrosis evolving into AML). It is possible that this may be related to the inability of leukemic cells in these patients to undergo maturation. Future trials in patients with preleukemic conditions and in patients with smouldering myeloid leukemia with significant leukemic infiltrates might, therefore, focus on combinations of agents that stimulate hematopoiesis (*i.e.*, colony stimulating factors) with therapies that can eradicate (*i.e.*, chemotherapy), or suppress (*i.e.*, Interferons) the neoplastic clone.

Treatment by continuous i.v. infusion was generally well tolerated. Intermediate doses (120 to 250 μg/m²/day) of GM-CSF were associated with minimal side effects and yet produced consistent biologic effects. The drug was rapidly cleared from circulation when given by i.v. bolus injection with a terminal phase half-life of approximately 30 minutes at the highest dose tested (Vadhan-Raj, S., manuscript in preparation), indicating that prolonged infusions, as utilized in our study, might be necessary to achieve the optimal biologic effects. These observations will have to be considered when designing future clinical studies.

The material used in this study was derived from yeast and is glycosylated like the natural protein. One potential concern with the use of this material in humans, therefore,

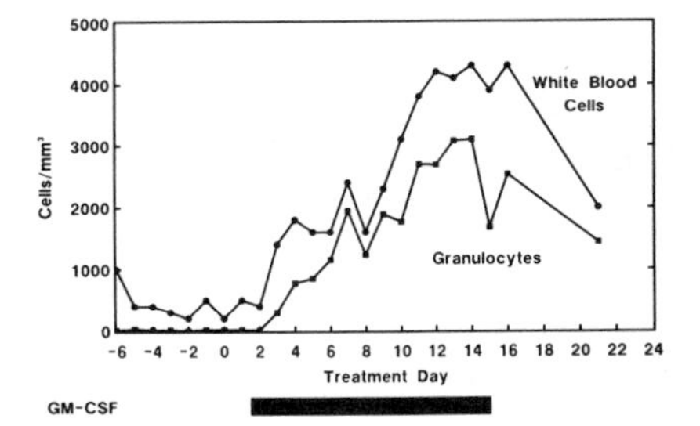

FIGURE 3. Effects of recombinant human GM-CSF treatment on white blood cell and granulocyte counts in a patient with multiple myeloma who failed to recover within two months after autologous bone marrow transplantation. *Solid bar* indicates GM-CSF treatment (500 μg/m²/day) by continuous i.v. infusion.

is the development of antibodies to the carbohydrate moiety. It was, therefore, encouraging that no neutralizing antibodies were found in sera of the patients tested after two cycles of treatment. However, longer follow-up and analysis of samples from a larger number of patients will be necessary to determine whether this molecule is immunogenic.

In summary, the results of our studies[18,19] indicate that GM-CSF is an effective stimulator of hematopoiesis *in vivo*. Although stimulation of myelopoiesis was the major component of response, multilineage stimulation occurred in some patients. Based on these findings, GM-CSF might play a role as a stimulator of hematopoiesis in several clinical arenas, including primary bone marrow failure (*i.e.*, aplastic anemia, myelodysplasias, and myelofibrosis), cytopenias related to chemo/radio therapy, and in the setting of bone marrow transplantation. Further, in combination with other cytokines such as interleukin-1 or interleukin-3, GM-CSF may be more effective in stimulating hematopoiesis along the multilineage pathways. With the increased availability of recombinant human gene products, it will now be possible to evaluate the potential of the above approach for restoring hematopoiesis.

TABLE 4. Side effects of GM-CSF (N = 21)[a]

	Number of Patients	%
Bone pain	15	71
Fatigue	15	71
Fever	9	43
Chills	4	19
Myalgias	9	43
Headache	7	33
Reduced appetite	7	33
Nausea/vomiting	5	24
Indigestion/diarrhea	3	14

[a]Twenty-one patients received at least 1 cycle of GM-CSF treatment and were considered evaluable for toxicity.

SUMMARY

We evaluated the biologic effects of recombinant human granulocyte-macrophage colony-stimulating factor (GM-CSF) in 25 patients with malignancy and/or bone marrow failure of diverse etiologies. The continuous infusion of GM-CSF (15 to 500 $\mu g/m^2$/day) elicited marked leukocytosis (2- to 70-fold increase), consisting primarily of neutrophils, eosinophils, and monocytes. Six patients with cytopenias experienced a multilineage response characterized by significant increases in platelet counts and improvement in erythropoiesis. Response in blood counts was accompanied by significant increases in bone marrow cellularity, myeloid:erythroid cell ratios, and frequency of cycling progenitors, indicating an effect at the stem cell level. By premature chromosome condensation analysis, neutrophils from patients with myeloid diseases were found to be derived from normal as well as abnormal clones. Side effects were generally mild and commonly included constitutional symptoms and bone pain. These results indicate that GM-CSF is a significant stimulus for hematopoiesis *in vivo* and might play an important role in several clinical arenas.

REFERENCES

1. Burgess, A. W. & D. Metcalf. 1980. The nature and action of granulocyte-macrophage colony-stimulating factors. Blood **56:** 947.
2. Nicola, N. A. & M. Vadas. 1984. Hemopoietic colony-stimulating factors. Immunol. Today. **5:** 76.
3. Metcalf, D. 1985. The granulocyte-macrophage colony-stimulating factors. Science **229:** 16.
4. Broxmeyer, H. E. 1986. Biomolecule-cell interactions and the regulation of myelopoiesis. Int. J. Cell Cloning **4:** 378.
5. Cantrell, M. A., D. Anderson, D. P. Cerretti, V. Price, K. McKereghan, R. J. Tushinski, D. Y. Mochizuki, A. Larsen, K. Grabstein, S. Gillis & D. Cosman. 1985. Cloning, sequence, and expression of a human granulocyte-macrophage colony-stimulating factor. Proc. Natl. Acad. Sci. USA **82:** 6250.
6. Wong, G. W., J. S. Witrek, P. A. Temple, K. M. Wilkens, A. C. Leary, D. P. Luxenberg, S. S. Jones, E. L. Brown, R. M. Kay, E. C. Orr, C. Showmaker, D. W. Colde, R. J. Kaufman, R. M. Hewick, E. A. Wang & S. C. Clark. 1985. Human GM-CSF: Molecular cloning of the complementary DNA and purification of the natural and recombinant proteins. Science **228:** 810.
7. Lee, F., T. Yokota, T. Otsuka, L. Gemmell, N. Larson, J. Loh, K. Arai & D. Rennick. 1985. Isolation of cDNA for a human granulocyte-macrophage colony-stimulating factor by functional express in mammalian cells. Proc. Natl. Acad. Sci. USA **82:** 4360.
8. Kaushansky, K., P. J. H. O'Hara, K. Berkner, G. M. Segal, F. S. Hagen & J. W. Adamson. 1986. Genomic cloning, characterization, and multilineage growth-promoting activity of human granulocyte-macrophage colony-stimulating factor. Proc. Natl. Acad. Sci. USA **85:** 3101.
9. Souza, L. M., T. C. Boone, J. Gabrilove, P. H. Lai, K. M. Asebo, D. C. Murdock, V. R. Chazin, J. Bruszewski, H. Lu, K. K. Chen, J. Barendt, E. Platzer, M. A. S. Moore, R. Mertelesmann & K. Welte. 1986. Recombinant human granulocyte colony-stimulating factor: Effect on normal and leukemic myeloid cells. Science **232:** 61.
10. Nagata, S., M. Tsuchiya, S. Asano, Y. Kaziro, T. Yamazaki, O. Yamamoto, Y. Hirata, N. Kubota, M. Oheda, H. Nomura & M. Ono. 1986. Molecular cloning and expression of cDNA for human granulocyte colony-stimulating factor. Nature **319:** 415.
11. Sieff, C. A., S. G. Emerson, R. E. Donahue, D. G. Nathan, E. A. Wang, G. G. Wong & S. C. Clark. 1985. Human recombinant granulocyte-macrophage colony-stimulating factor: A multilineage hematopoietin. Science **230:** 1171.
12. Weisbart, H. H., D. W. Golde, S. C. Clark, G. G. Wong & J. C. Gasson. 1985. Human granulocyte-macrophage colony-stimulating factor is a neutrophil activator. Nature **315:** 361.
13. Fleischmann, J., D. W. Golde, R. H. Weisbart & J. C. Gasson. 1986. Granulocyte-macrophage colony-stimulating factor enhances phagocytosis of bacteria by human neutrophils. Blood **68:** 708.
14. Metcalf, D., C. G. Begley, G. R. Johnson, N. A. Nicola, M. A. Vadas, A. F. Lopez, D. J. Williamson, G. G. Wong, S. C. Clark & E. A. Wang. 1986. Biologic properties *in vitro* of a recombinant human granulocyte-macrophage colony-stimulating factor. Blood **67:** 37.
15. Burgess, A. W., G. Begley, G. R. Johnson, A. F. Lopez, D. J. Williamson, J. J. Mermod, R. J. Simpson, A. Schmitz & J. F. DeLamarter. 1987. Purification and properties of bacterially synthesized human granulocyte-macrophage colony-stimulating factor. Blood **69:** 43.
16. Vadas, M. A., N. A. Nicola & D. Metcalf. 1983. Activation of antibody-dependent cell mediated cytotoxicity of human neutrophils and eosinophils by separate colony-stimulating factors. J. Immunol. **130:** 795.
17. Donahue, R. E., E. A. Wang, D. K. Stone, R. Kamen, G. G. Wong, P. K. Sehgal, D. G. Nathan & S. C. Clark. 1986. Stimulation of hematopoiesis in primates by continuous infusion of recombinant human GM-CSF. Nature **321:** 872.
18. Vadhan-Raj, S., M. Keating, A. LeMaistre, W. N. Hittelman, K. McCredie, J. M. Trujillo, H. E. Broxmeyer, C. Henney & J. U. Gutterman. 1987. Effects of recombinant human granulocyte-macrophage colony-stimulating factor in patients with myelodysplastic syndromes. N. Engl. J. Med. **317:** 1545.

19. VADHAN-RAJ, S., S. BUESCHER, A. LEMAISTRE, M. KEATING, R. WALTERS, H. KANTARJIAN, W. N. HITTELMAN, H. E. BROXMEYER & J. U. GUTTERMAN. 1988. Stimulation of hematopoiesis in patients with bone marrow failure and in patients with malignancy by recombinant human granulocyte-macrophage colony stimulating factor. Blood **71:** 134.
20. BROXMEYER, H. E., S. COOPER, D. E. WILLIAMS, G. HANGOC, J. U. GUTTERMAN & S. VADHAN-RAJ. 1988. Growth characteristics of marrow hematopoietic progenitor/precursor cells from patients on a phase I clinical trial with purified recombinant human granulocyte-macrophage colony-stimulating factor. Exp. Hematol. **16:** 594.
21. HITTELMAN, W. N., L. C. BROUSSARD, G. DOSIK & K. B. MCCREDIE. 1980. Predicting relapse of human leukemia by means of premature chromosome condensation. N. Engl. J. Med. **303:** 479.
22. HITTLEMAN, W. N., I. PETKOVIC, P. AGBOR & S. VADHAN-RAJ. 1988. Maturation of abnormal elements in patients with myelodysplastic syndrome after GM-CSF treatment. Proc. Am. Assoc. Cancer Res. **29:** 38.

Granulomonopoiesis and Production of Granulomonopoietic Regulating Factors in Hodgkin's Disease and Non-Hodgkin's Lymphomas[a]

XAVIER LÒPEZ-KARPOVITCH, MARIA ROCIO CÀRDENAS, EDUARDO LOBATO-MENDIZABAL, AND JOSEFA PIEDRAS

Department of Hematology
Instituto Nacional de la Nutrición Salvador Zubirán
Vasco de Quiroga 15
Delegación Tlalpan, 14000, Mexico, D.F.

INTRODUCTION

The development of *in vitro* semisolid culture techniques has given insight into the interactions underlying the regulation of normal hematopoiesis. Hodgkin's disease (HD) and non-Hodgkin's lymphomas (NHL) are suitable biologic models to study the relationship that exists between lymphoid and myeloid systems, since patients with these lymphoproliferative malignancies often show quantitative and functional cellular abnormalities, not only of T and B lymphocytes but also of monocytes.[1–4]

Cloning of colony-forming units of granulocytes and monocytes/macrophaphes (CFU-GM) allows the evaluation of this segment of the pluripotent committed cell compartment and of its growth under the influence of cytokines such as the colony-stimulating activity (GM-CSA) produced by monocytes[5] and the colony-inhibiting activity (GM-CIA) purportedly produced by autologous rosette-forming T-cells (Tar cells),[6] a postthymic precursor T-cell subpopulation.[7]

In this work we employed this methodology in an attempt to determine whether granulomonopoiesis and production of granulomonopoietic regulating factors, or both, were involved in the pathogenesis of lymphoproliferative malignant diseases. By studying HD and NHL patients devoid of treatment we were able to assess the role of disease on GM-CSA and GM-CIA production but not on CFU-GM growth.

MATERIAL AND METHODS

Subjects

Seven patients with HD and 5 with NHL were studied. Histologic classification of HD and NHL was in accord with the recommendations made at the Rye Conference[8] and the Working Formulation,[9] respectively. Clinical staging was performed following the Ann Arbor classification.[10] The histologic, clinical, and hematologic profiles of the patients are listed in TABLE 1. None of the patients had previously received chemotherapy and/or radiotherapy and all were free of treatment at the time of the study.

[a]This work was supported in part by the Consejo Nacional de Ciencia y Technologìa (CONACyT).

Neutrophilia (>7.25 × 10^9/L) was present in 3 patients, neutropenia (<1.7 × 10^9/L) in 1, monocytosis (>0.95 × 10^9/L) in 5, and monocytopenia (<0.21 × 10^9/L) in 1 patient.

Seven apparently healthy volunteers served as controls. Leukocyte counts were within normal values in all of them. Approximately 60 mL of peripheral blood (PB) were obtained from all controls and patients and collected in sterile plastic syringes containing preservative-free heparin (25 IU/mL).

Cell Separation Procedures

Buffy coat (BC) cells, adherent (Adh) cells, Tar cells, and the nonadherent-T-lymphocyte-depleted (Adh^-T^-) cell fraction were isolated through centrifugation, adherence techniques, rosette formation, and Ficoll-Hypaque gradients as described in detail elsewhere.[6,11] The purity of each mononuclear cell fraction was >96%, as established by rosetting assays and immunophenotyping with Leu-1 (anti-CD5) and Leu M3 (anti-CDw14) monoclonal antibodies which recognize T-cells and monocytes, respectively. Cell viability, assessed with the trypan blue dye exclusion test, was >98%.

TABLE 1. Clinical and Hematologic Findings of Patients with Hodgkin's Disease and Non-Hodgkin's Lymphomas

Patient No.	Sex/Age	Histologic Diagnosis[a]	Clinical Staging	Marrow Infiltration	Neutrophils (10^9/L)	Monocytes (10^9/L)
1	M/28	HD; lymphocyte depleted	IV B	Yes	4.70	0.67
2	F/39	HD; lymphocyte depleted	III B	No	14.94	1.54
4	F/16	HD; mixed cellularity	III B	No	11.78	1.37
5	M/42	HD; mixed cellularity	III B	No	6.67	1.50
9	M/20	HD; nodular sclerosis	IV B	Yes	8.95	0.60
10	F/23	HD; mixed cellularity	III B	No	6.71	0.34
12	M/44	HD; mixed cellularity	IV B	Yes	5.03	0.11
3	M/43	NHL; diffuse mixed, cleaved and noncleaved (IG)	IV B	No	6.27	1.23
6	M/36	NHL; small noncleaved cell, diffuse (HG)	I A	No	4.39	0.31
7	M/56	NHL; lymphoblastic (HG)	IV B	Yes	1.49	2.98
8	M/82	NHL; plasmacytoid lymphocytic (LG)	IV B	No	4.72	0.76
11	M/37	NHL; diffuse large noncleaved (IG)	I B	No	5.25	0.66

[a]HD = Hodgkin's disease; NHL = non-Hodgkin's lymphoma. (Morphologic categories of NHL: LG = low-grade; IG = intermediate-grade; HG = high-grade lymphomas.)

Preparation of Conditioned Media

Phytohemagglutinin-leukocyte-conditioned medium (PHA-LCM), a standard source of GM-CSA, was prepared with BC cells from various healthy donors, and Tar-cell-conditioned media (Tar-CM) from controls and patients were produced as previously described.[6,12] Adherent-cell-conditioned media (Adh-CM) from controls and patients were produced by incubating Adh cells at a concentration of 10^6/mL in RPMI medium containing 10% heat-inactivated fetal calf serum (FCS) for 96 h at 37°C in a humidified atmosphere of 7.5% CO_2. Tar-CM were absorbed with human erythrocytes for 18 h at 4°C to remove PHA and incubated for 1 h at 56°C to inactivate interferon.[13] All conditioned media were sterilized through 0.45-μ filters and stored at −20°C before assaying for GM-CSA and GM-CIA.

CFU-GM Assay

To evaluate CFU-GM growth in PB samples of controls and patients, Adh^-T^-cells at a final concentration of 2×10^5 cells/mL were plated in duplicates into 35-mm plastic Petri dishes in a 1-mL mixture containing alpha-medium, 0.8% methylcellulose (MTC), 0.22% FCS and 20% PHA-LCM. The plates were incubated for 7 days at 37°C in a humidified atmosphere of 7.5% CO_2.

GM-CSA and GM-CIA Assays

Release of GM-CSA by Adh cells and of GM-CIA by Tar cells of controls and patients were tested incubating 2×10^5 Adh^-T^-cells, obtained from a single normal donor, suspended in 1 mL of 0.8% MTC containing alpha-medium and 20% FCS. Each conditioned medium, Adh-CM or Tar-CM, was incorporated into cultures at a 20% concentration. Culture conditions were the same as those employed for CFU-GM assay. GM-CSA and GM-CIA were expressed in percentage of the number of cell aggregates obtained with a standard source of GM-CSA (PHA-LCM), the same batch being used for the whole study. Cultures without conditioned medium served as blanks.

Enumeration of Cell Aggregates

The culture dishes were scored for the number of clones as determined by the sum of clusters (aggregates of 3–39 cells) and colonies (aggregates of >40 cells) under an inverted microscope equipped with phase contrast. The granulocyte or monocyte/macrophage nature of clones was assessed *in situ* by peroxidase and alpha-naphtyl-acetate stains, respectively.[14,15]

Statistical Analysis

Individual values were analyzed with the U test of Mann & Whitney and rank correlation coefficient of Spearman, when appropriate.

RESULTS

Since Tar-CM contain PHA and lectin-activated T lymphocytes release several substances that inhibit clone formation, such as interferon,[16] we thought it important to study the effect of previous absorption with human erythocytes (18 h at 4°C to remove PHA) and heat inactivation (1 h at 56°C to inactivate interferon) upon the GM-CIA of crude

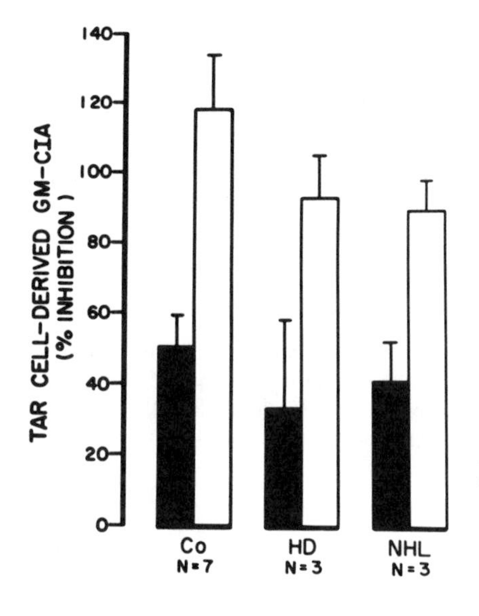

FIGURE 1. Inhibition of clone formation by Tar-CM of controls (Co), patients with Hodgkin's disease (HD), and patients with non-Hodgkin's lymphomas (NHL) before (*filled bars*) and after erythrocyte-absorption and heat inactivation (*open bars*). CM-target cells were peripheral blood non-adherent-T-lymphocyte depleted cells (2×10^5) obtained from a single normal donor. Inhibition was calculated by the following formula: (No. of clones with 20% (v/v) PHA-LCM—No. of clones without CM)—(No. of clones with PHA-LCM plus Tar-CM 20% (v/v) mixture—No. of clones without CM) $\times$ 100/(No. of clones with PHA-LCM—No. of clones without CM). The data are expressed as the mean $\pm$ SD.

Tar-CM (FIG. 1). We were able to do this in 13 subjects, 7 controls and 6 patients. The GM-CIA in Tar-CM increased after both, absorption and heat inactivation in all samples studied. These data suggest that PHA-treated Tar cells release a colony-inhibiting factor different from interferon.

TABLE 2 displays CFU-GM growth, Adh-cell-derived GM-CSA and Tar cell-derived GM-CIA in patients and controls. Two of 5 patients with HD and 3 of 5 with NHL had CFU-GM values below those recorded in the control group, whereas 2 patients, one of each group, showed a high number of clones. Although the mean values of CFU-GM in HD were higher and in NHL lower than in the control group, the differences did not reach statistical significance when compared with controls. The difference between HD and NHL groups was not significant.

Ten out of 11 patients, 6 with HD and 5 with NHL, had GM-CSA values below those observed in the control group. The mean percentage of GM-CSA was significantly decreased in both HD ($p = 0.002$) and NHL ($p = 0.003$) patients as compared to controls. The difference between HD and NHL patients was not significant.

Four of 7 patients, two from each group, showed GM-CIA values below those found in the control group. The mean percentage of GM-CIA was significantly decreased in both HD ($p = 0.012$) and NHL ($p = 0.017$) patients as compared to controls. The difference between HD and NHL groups was not significant.

There was no association between *in vitro* assays with histologic diagnosis, clinical staging and bone marrow infiltration. Correlation coefficients between CFU-GM with GM-CSA and GM-CIA were not significant.

DISCUSSION

Employing a PB cell population highly enriched for hemopoietic pluripotent cells, such as Adh^-T^-cells, we were unable to detect a significant change in the number of granulocyte/macrophage progenitor cells (CFU-GM) in HD and NHL patients as compared with controls. This finding suggests that CFU-GM is not involved in the pathogenesis of HD and NHL.

In order to investigate monocyte function, which has been shown by others to be impaired in HD and NHL,[3,4] we measured the GM-CSA in CM obtained from Adh cells of patients with lymphomas. Herein, the decreased production of GM-CSA by unstimulated monocytes in HD and NHL may indicate that they are defective of producing cytokines despite the normal number of their precursors. An alternative explanation for this finding could be the presence of prostaglandins of the E series, which are known to inhibit CFU-GM growth,[17] in the Adh-CM of our patients. Indeed, Bockman[18] has shown that monocytes of HD patients synthetize and release large amounts of prostaglandin E. GM-CSA regulates monocyte-prostaglandin E synthesis.[17] Thus, normal or increased GM-CSA would be expected in our patients' Adh-CM, however, no such phenomenon occurs suggesting that GM-CSA-driven prostaglandin synthesis is altered in HD and NHL.

Several authors have demonstrated the physiologic role of T cells and T cell-derived cytokines in the regulation of CFU-GM.[19,20] It has recently been shown that normal Tar

TABLE 2. CFU-GM Growth in Patients and Controls and Effect of Conditioned Media Prepared from Adherent Cells and Tar Cells of Patients and Controls on Normal Peripheral Blood CFU-GM Growth

Patient No.	CFU-GM (No. Clones)[a]	Adh Cell GM-CSA (% Stimulation)[b]	Tar Cell GM-CIA (% Inhibition)[c]
Hodgkin's disease			
1	370.5	37.5	ND[d]
2	106.5	13.2	ND[d]
4	52.5	39.6	107.0
5	181.0	45.1	86.9
9	ND[d]	138.2	85.9
10	84.5	95.8	94.6
12	ND[d]	60.1	ND[d]
Mean	159.0	61.3[e]	93.6[e]
Non-Hodgkin's lymphomas			
3	197.5	28.0	98.6
6	107.5	29.9	ND[d]
7	80.0	ND[d]	86.4
8	38.5	15.3	80.8
11	80.0	51.4	ND[d]
Mean	100.7	31.1[e]	88.6[e]
Controls (N = 7)			
Mean	135.7	180.2	118.1
Range	101.0–182.0	112.5–395.8	88.1–140.0

[a]Sum of granulomonocytic clusters and colonies per 2 × 10^5 nonadherent-T-lymphocyte depleted (Adh^-T^-) peripheral blood cells using 20% (v/v) PHA-LCM as a standard source of GM-CSA.

[b]Number of granulomonocytic clones per 2 × 10^5 Adh^-T^- peripheral blood cells, isolated from a single normal donor, using 20% (v/v) adherent cell-CM. The results are expressed as the percent of cell aggregates obtained with 20% (v/v) PHA-LCM.

[c]Inhibition of granulomonocytic clone formation by absorbed and heat-inactivated Tar-CM. CM-target cells were peripheral blood Adh^-T^- cells (2 × 10^5) obtained from a single normal donor. Inhibition was calculated as in FIGURE 1.

[d]ND = not done.

[e]Significant difference (Mann & Whitney's U test) compared with controls ($p < 0.05$).

cells, a postthymic precursor subpopulation, inhibit CFU-GM via a soluble factor different of interferon.[6] Because Tar cells were decreased in 9 of 11 patients with HD and NHL (Lòpez-Karpovitch et al., unpublished data), we studied the production of Tar-derived GM-CIA by lymphoma patients and found it significantly decreased. This finding is in line with previous observations showing that antigen- or mitogen-induced production of lymphokines, such as leukocyte inhibitory factor and interferon, are significantly depressed in HD.[21,22]

In sum, the data presented herein suggest that cytokine-dependent mechanisms regulating normal CFU-GM proliferation are impaired in HD and NHL.

SUMMARY

Quantitative and functional abnormalities of T and B lymphocytes and monocytes have been described in Hodgkin's disease (HD) and non-Hodgkin's lymphomas (NHL), thus making both diseases suitable models to study the interactions between lymphoid and myeloid systems. We evaluated the growth of colony-forming units of granulocytes and monocytes/macrophages (CFU-GM), as well as the colony-stimulating activity (GM-CSA) produced by monocytes and the colony-inhibiting activity (GM-CIA) released by autologous rosette-forming T-cells (Tar cells), a postthymic precursor subpopulation, in peripheral blood samples from 7 patients with HD and 5 with NHL.

CFU-GM growth in HD and NHL patients was similar to that observed in controls. However, GM-CSA and GM-CIA were significantly decreased in both HD ($p = 0.002$ and $p = 0.012$, respectively) and NHL ($p = 0.003$ and $p = 0.017$, respectively) patients as compared to controls. These data suggest that cytokine-dependent mechanisms regulating normal CFU-GM proliferation are impaired in HD and NHL.

REFERENCES

1. Noorloos, A. B., T. Splinter, A. Van Beck, F. Décary, P. Van Heerde, A. V. D. Borne & K. Melief. 1977. Immunological studies on non-Hodgkin's lymphomas. Br. J. Haematol. **35:** 676.
2. Ben-Bassat, H., S. Penchas, A. Polliack, S. Mitrani-Rosenbaum, E. Naparstek, Y. Matzner, A. Kedar, D. Shouval, A. Eldor, M. Prokocimer & N. Goldblum. 1980. Changes in the Con-A-induced redistribution pattern of lymphocytes: A possible aid in the differential diagnosis between malignant lymphoma and other diseases. Blood **55:** 205–210.
3. Estevez, M., L. Sen, A. E. Bachman, A. Pavlovsky. 1980. Defective function of peripheral blood monocytes in patients with Hodgkin's disease and non-Hodgkin's lymphomas. Cancer **46:** 299–302.
4. Romagnani, S., P. L. R. Ferrini & M. Ricci. 1985. The immune derangement in Hodgkin's disease. Semin. Hematol. **22:** 41–55.
5. Golde, D. W. & M. J. Cline. 1972. Identification of the colony-stimulating cell in human peripheral blood. J. Clin. Invest. **51:** 2981–2983.
6. López-Karpovitch, X., M. R. Padrós-Semorile, R. Rojas & L. Martinez-Sánchez. 1985. Release of granulocyte-macrophage colony-inhibiting activity by normal human postthymic precursor cells. Am. J. Hematol. **20:** 247–256.
7. Palacios, R., D. Alarcón-Segovia, L. Llorente, A. Ruiz-Arguelles & E. Diaz-Jouanen. 1981. Human postthymic precursor cells in health and disease. I. Characterization of the autologous rosette-forming T cells as postthymic precursors. Immunology **42:** 127–135.
8. Lukes, R. J., L. F. Craver, T. C. Hall, H. Rappaport & P. Rubin. 1966. I. Report of the nomenclature committee. Cancer Res. **26:** 1311.
9. National Cancer Institute. 1982. NCI-sponsored study of classifications of non-Hodgkin's

lymphomas. Summary and description of a working formulation for clinical usage. The Non-Hodgkin's Lymphoma Pathologic Classification Project. Cancer **49:** 2112–2135.
10. CARBONE, P. P., H. S. KAPLAN, K. MUSSHOFF, D. W. SMITHERS & M. TUBIANA. 1971. Report of the Committee on Hodgkin's disease staging classification. Cancer Res. **31:** 1860–1861.
11. LÓPEZ-KARPOVITCH, X., M. R. PADRÓS-SEMORILE & X. ALVAREZ-HERNÁNDEZ. 1984. Ferritin in normal human peripheral blood T-lymphocyte subpopulations. Cell. Immunol. **86:** 255–260.
12. LÓPEZ-KARPOVITCH, X., M. R. PADROŚ-SEMORILE, R. HURTADO-MONROY, M. R. CÁRDENAS, R. NIESVIZKY, L. MARTINEZ-SÁNCHEZ & R. ROJAS. 1985. Comparison of colony-stimulating activities for evaluating in vitro granulomonopoiesis in normal subjects and in patients with myeloid leukemias. Rev. Invest. Clin. (Méx.) **37:** 323–327.
13. VALLE, M. J., G. W. JORDAN, S. HAAHR & T. C. MERIGAN. 1975. Characteristics of immune interferon produced by human lymphocyte cultures compared to other interferons. J. Immunol. **115:** 230–233.
14. ZUCKER-FRANKLIN, D. & G. GRUSKY. 1976. The identification of eosinophil colonies in soft agar cultures by differential staining for peroxidase. J. Histochem. Cytochem. **24:** 1270–1272.
15. YAM, L. T., C. Y. LI & W. H. CROSBY. 1971. Cytochemical characterization of monocytes and granulocytes. Am. J. Clin. Pathol. **55:** 283–290.
16. GREENBERG, P. L. & S. MOSNY. 1977. Cytotoxic effects of interferon in vitro on granulocytic progenitor cells. Cancer Res. **37:** 1794–1799.
17. PELUS, L. M., H. E. BROXMEYER, J. I. KURLAND & M. A. S. MOORE. 1979. Regulation of macrophage and granulocyte proliferation. Specificities of prostaglandin E and lactoferrin. J. Exp. Med. **150:** 277–292.
18. BOCKMAN, R.S. 1980. Stage-dependent reduction in T colony formation in Hodgkin's disease. Coincidence with monocyte synthesis of prostaglandins. J. Clin. Invest. **66:** 523–531.
19. VERMA, D. S., D. A. JOHNSTON & K. B. MCCREDIE. 1983. Evidence for the separate human T-lymphocyte subpopulations that collaborate with autologous monocyte/macrophages in the elaboration of colony-stimulating activity and those that suppress this collaboration. Blood **62:** 1088–1099.
20. VERMA, D. S., D. A. JOHNSTON & K. B. MCCREDIE. 1984. Identification of T lymphocyte subpopulations that regulate elaboration of granulocyte-macrophage colony-stimulating factor. Br. J. Haematol. **57:** 505–520.
21. GOLDING, B., H. GOLDING, R. LOMNITZER, R. JACOBSON, H. J. KOORNHOF & A. R. RABSON. 1977. Production of leukocyte inhibitory factor (LIF) in Hodgkin's disease. Clin. Immunol. Immunopathol. **7:** 114–122.
22. RASSIGA-PIDOT, A. L. & O. R. MCINTYRE. 1974. In vitro leukocyte interferon production in patients with Hodgkin's disease. Cancer Res. **34:** 2995–3002.

Index of Contributors